Biotechnology for Fruit, Vegetable and Spice Crops

Biotechnology has revolutionized horticulture by enhancing the productivity, resilience and nutritional quality of fruit, vegetable and spice crops. This comprehensive volume provides an in-depth exploration of cutting-edge biotechnological advancements that are reshaping horticultural science. From genomics-driven crop improvement to the development of functional foods, this book presents a meticulously curated compilation of research and methodologies addressing key challenges and opportunities in modern horticulture. By integrating molecular techniques, plant–microbe interactions and bioprocess innovations, this book provides a unique perspective on sustainable and precision-driven horticultural practices.

Key Features

- Insights into genomic approaches for understanding abiotic stress tolerance and developing climate-resilient varieties.
- Advances in tissue culture, marker-assisted selection and genome editing for apple, grapevine and potato breeding.
- Molecular and biocontrol strategies for tackling major threats, such as root rot disease in apples.
- Applications of soilless cultivation techniques and plant growth-promoting rhizobacteria (PGPR) to optimize crop yield and quality.
- Biotechnological tools for developing probiotic-enriched fruits and vegetables and the valorization of non-grape fruit wines.

With contributions from leading researchers, this book serves as an essential reference for graduate students, academics and professionals in plant biotechnology, horticulture and food science. It provides a valuable resource for those seeking to harness the power of biotechnology to drive sustainable innovation in horticultural crop production.

Pankaj Kumar currently serves as an Assistant Professor in Biotechnology at the Department of Biotechnology, Dr. Yashwant Singh Parmar University of Horticulture and Forestry in Solan, Himachal Pradesh, India. His academic journey includes post-doctoral studies at the National Institute of Plant Genome Research in New Delhi and the Council of Scientific & Industrial

Research – Institute of Himalayan Bioresource Technology in Palampur, Himachal Pradesh. Dr. Kumar boasts 12 years of expertise in horticulture biotechnology, focusing on plant secondary metabolite enhancement through cell and tissue engineering approaches. His research encompasses the identification of molecular cues associated with enhanced metabolites using a comparative transcriptomics approach. Additionally, he possesses experience in hydro-aero cultivation of medicinal plants for industrial significance. His accolades include the DST INSPIRE JRF/SRF Fellowship, ICAR AICE-SRF (PGS) and ICAR ASRB National Eligibility Test (NET). Dr. Kumar has an extensive publication record with more than 100 research/review papers, book chapters and six books. He has also filed and published four Indian patents and registered four apple mutants with unique, early color and early maturity characteristics in the Indian National Germplasm Repository (INGR 20103, INGR 20104, INGR 20105 and INGR 20106). Recognized for his scientific contributions, Dr. Kumar has received numerous awards, including the Young Scientists Fellowship Award-2021 (ICMR, New Delhi), Scientist's Pool Scheme Award-2020 (CSIR, New Delhi) and the SERB-ACS-NPDF Best Poster Award-2022, etc.

Ajay Kumar Thakur is currently working as a Principal Scientist (Agri. Biotechnology) at ICAR-Central Potato Research Institute, Shimla, H.P. Presently, he is working on biotechnological interventions for potato improvement. He has obtained his graduation, PG and Doctoral degrees from Dr. Y.S. Parmar University of Horticulture and Forestry, Solan, H.P. He got induced into Agricultural Research Services (ARS) in 2008. He has published 72 research/review papers in various journals of international and national repute, authored one book "Agricultural Biotechnology at a Glance", edited three books and contributed eight book chapters and 23 popular articles. He has developed high-efficiency plant regeneration and genetic transformation protocols in a number of crops including *Populus ciliata, P. deltoides, Punica granatum, Capsicum annum* and *Cucumis sativus*. Dr. Thakur is associated with the *Brassica juncea* improvement program using biotechnological interventions from the last 14 years. He has developed a core set of SSR markers for *B. juncea* genomics and has worked on germplasm characterization and association mapping of various agronomically important traits in this oilseed crop. He has also been granted one patent from Indian Patent Office, New Delhi and associated in the development of two high yielding Indian mustard varieties Giriraj and Brajraj, one gobhi sarson variety, AKMS-8141 and five disease-resistant Indian mustard genetic stocks. Dr. Thakur has received many awards from various societies and scientific organizations for his scientific contribution. He is also an elected Member of the Plant Tissue Culture Association of India (PTCAI) and selected Member of the National Academy of Sciences-India (NASI), Allahabad.

Dinesh Kumar Srivastava has retired as Director Extension Education, prior to this he worked as Professor and Head in the Department of Biotechnology, Dr. Yashwant Singh Parmar University of Horticulture and Forestry, Solan, Himachal Pradesh, India. He did his post-doctoral studies in the field of Plant Molecular Biology at the Institute of Molecular Genetics, USSR Academy of Sciences, Moscow, USSR and at Washington University, St. Louis, USA. He has 32 years of research experience and 29 years of teaching experience in the field of Plant Biotechnology and Molecular Biology. He had guided 32 M.Sc. students and nine Ph.D. students for their research work. He has published 110 research papers in journals of national and international repute and 38 research articles, contributed 12 chapters in edited books, and also authored a book on biotechnology. He has delivered 45 invited lectures and attended 47 national and international conferences. His main area of research includes plant tissue culture, genetic transformation and molecular characterization of plants. He is a life member of various national and international academic bodies/societies. He has received many awards for his scientific contributions, including the 'MS Swaminathan Award' from the Society of Plant Research, Meerut, India.

Biotechnology for Fruit, Vegetable and Spice Crops

Edited by

Pankaj Kumar, Ajay Kumar Thakur and Dinesh Kumar Srivastava

CRC Press
Taylor & Francis Group
Boca Raton London New York

CRC Press is an imprint of the
Taylor & Francis Group, an **informa** business

First edition published 2026
by CRC Press
2385 NW Executive Center Drive, Suite 320, Boca Raton FL 33431

and by CRC Press
4 Park Square, Milton Park, Abingdon, Oxon, OX14 4RN

CRC Press is an imprint of Taylor & Francis Group, LLC

ISBN: 978-1-041-03846-7 (hbk)
ISBN: 978-1-041-03848-1 (pbk)
ISBN: 978-1-003-62567-4 (ebk)

DOI: 10.1201/9781003625674

Typeset in Palatino
by Newgen Publishing UK

This book is dedicated to Dr. Atul Mehta, a renowned plant biologist, and mentor to many students and grand students. Aptly nicknamed 'Guru' by his wide family of students, Atul Mehta still lives in their hearts.

Dr. Mehta was born to Jaikirti, a social worker and Ramakant, an accountant in Gaekwad state in Gujarat. Atul was very fond of sports and music since young, largely due to family influence. In his academic career too, he excelled in earning a teaching position at the MS University of Baroda after earning his masters degree. While teaching, he was awarded a scholarship for further study abroad and he chose to work with the renowned botanist Prof. H.E. Street at University of Wales in UK where a new technique of plant tissue culture was emerging. Thus, his background study on basics of botany ironically shifted to embarking upon a deeper insight into study on biological mechanisms. He did his doctoral research on growth patterns in crown gall and normal cells in in vitro *cultures which led to highly acclaimed and well cited classic publications. Upon his return to his hometown in India in 1967, he established a Tissue culture laboratory, the first one in the country, at Botany Department, MS University of Baroda. The initial basic research in plant physiology using organ and tissue culture techniques in Mehta's lab was not generated in a class 10 facility with air-conditioning and laminar work stations, but a simple inoculation chamber in the corner of the main physiology laboratory. It clearly indicated his thirst for excelling even under not so favorable and comforting circumstances available to him.*

A large number of post graduates, doctoral and postdoctoral students learnt their first lesson in plant tissue culture in this nest. Dr. Dinesh Kumar Srivastava, who is one of the Editors of this book, has been mentored by Prof. Mehta for learning basics of plant tissue culture at MS University of Baroda, Vadodara,Gujarat.

To keep pace with advances in applied field of plant tissue culture, he later took a sabbatical leave for postdoctoral work with Prof. Staba at University of Minnesota to work on secondary product metabolism in cultured cells. The school of learning he established at Botany Department, inspired not only his own students and those of his colleagues who all worked as one team, but also was a seat of training courses for students from other Universities in India and abroad. With UGC awarded departmental research support, distinguished scientists from acclaimed Universities frequented interaction, doctoral and postdoctoral research flourished, workshops such as The UNESCO–ICRO sponsored workshop he organized drew researchers from several countries. Leading scientists from around the world were invited as experts in these workshops. Undoubtedly, this inspired young Indian scientists to pursue careers in plant biotechnology. The vibrant environment of the centre buzzed with new ideas and enthusiasm, quickly establishing itself as a centre for plant tissue culture expertise. Funding from the Council of Scientific and Industrial Research, and the University Grants Commission of the Government of India allowed the team to undertake a diverse variety of research projects involving in vitro *techniques of plant tissue culture.*

His visionary endeavours led him to receive the honour of being a recipient of the Jaikrishna Indrajit Award by the Hari Om Ashram in 1978 and Prof. Vishwambhar Puri Medal by the Indian Botanical Society in 1991. The scientific community had the privilege to appreciate him and select him to deliver Memorial lectures, notably among which were Prof. Philip R.White Memorial Lecture at Jadavpur University, Prof. H.E. Street Memorial Lecture at Nagpur University and the Prof. A.R. Chavan Memorial Lecture at the M.S. University of Baroda. He earned the position and remained for many years as a National Correspondent of the International Association of Plant Tissue Culture (IAPTC) for South Asian region.

Besides being an eminent scientist, Prof. Mehta was also an efficient administrator. As a member of the academic senate of the University, he contributed significantly to the academic ambience and in fine-tuning the University administrative procedures and protocols. He also served in the National Committees like DST, DBT, UGC, ICAR and CSIR as an expert member. Besides, he visited several universities and delivered lectures as a National Professor. He was a core team member of the All India Coordinated Project on Medicinal and Aromatic Plants of ICAR. He was also an Honorary Advisor to a private biotech company engaged in commercial research and production in Vadodara. As Head of the Department, he brought many changes in the ambience and the discipline to make the education and research environment more conducive for students and staff, and all his students and colleagues now feel indebted to him for these values he inculcated, though he may have sounded harsh in those days. After his retirement from the University, he worked as a CSIR-Emeritus Scientist at the Department of Plant Breeding and Genetics, Gujarat Agricultural University, Anand. He led to research team to conduct tissue culture work under the Greening of Gujarat and Rajasthan project funded by DBT, New Delhi on indigenous forest species such as guggal (Commiphora*), Acacia and soap nut tree* (Sapindus spp). *Seeds were also sown here by his diversifying interest in employing fundamental techniques of plant tissue culture to commercial micro-propagation of economically important horticulture crops such as banana, while he was Professor Emeritus here.*

He has mentored 25 Ph D students, authored more than 150 research papers in national and international journals of repute, and edited a book with P.N. Bhatt titled 'Handbook of Plant Tissue and Cell Culture'. *He worked on a wide array of topics such as cell differentiation, organ differentiation, the role of polyamines in morphogenesis, plantlet regeneration in tree species like soap nut and recalcitrant species like guggal. He led his team to work on metabolism of many secondary plant products such as diosgenin, guggulsteroids, insect molting hormones,* Lathyrus *neurotoxin and several phenolic compounds in cultured cells. His curiosity for unravelling the physiological and biochemical changes in haploid and diploid tissues of tobacco is exemplary. With his collaboration, stable somaclones of tobacco were selected for resistance to* Fusarium oxysporum. *He remained active till his physical health permitted him to support his family, and provide constant encouragement to his students.*

Apart from his career as a Scientist and Administrator, Mehta was a fond lover of classical music and used to sing songs/ghazals. It used to be a unanimous demand on him to get on stage during the recreational evenings during symposia and scientific gatherings. Known for his cricketing talent, he represented the M.S. University cricket team and also the drama group. He used to be an excellent host along with his wife to all who visited his home. Mehta always attributed his success to his late wife Yashaswini Mehta. He is survived by a son and a daughter, wearing similar feathers.

The plant tissue culture/plant biotechnology fraternity surely miss his dynamic physical presence, but his inspiration, humility and sense of humour, still remains with us.

Pankaj Kumar

Ajay Kumar Thakur

Dinesh Kumar Srivastava

Contents

Contributors

Supriya Aglawe
ICAR-National Research Centre for Grapes
Solapur Road, Pune, Maharashtra, India

Supriya Ambawat
Agriculture University,Jodhpur
Rajasthan, India

Anika
Dr YS Parmar University of Horticulture and Forestry
Nauni, Solan, Himachal Pradesh, India

Jyoti Chaudhary
Devi Ahilya University
Khandwa Road, Indore, MP, India

Shivani Chauhan
Dr YS Parmar University of Horticulture and Forestry
Nauni, Solan, Himachal Pradesh, India

Rajat Chandel
Department of Food Science & Technology
Punjab Agricultural University, Ludhiana, Punjab, India

Atul Dhiman
Dr YS Parmar University of Horticulture and Forestry
Nauni, Solan, Himachal Pradesh, India

Dalamu
ICAR-Central Potato Research Institute
Shimla, HP, India

Namo Dubey
Devi Ahilya University
Khandwa Road, Indore, MP, India

Rakesh Gehlot
Center of Food Science and Technology
CCS-HAU, Hisar, India

Arti Ghabru
Dr YS Parmar University of Horticulture and Forestry
Nauni, Solan, Himachal Pradesh, India

Ramandeep Kaur
Department of Food Science & Technology
Punjab Agricultural University, Ludhiana, Punjab, India

Ankit Kumar
Center of Food Science and Technology
CCS-HAU, Hisar, India

Anshul Kumar
MS Swaminathan School of Agriculture
Shoolini University, Bhajol, Solan, Himachal Pradesh

Amit Kumar
CSIR-Institute of Himalayan Bioresource Technology
Palampur, Himachal Pradesh, India

Dinesh Kumar Srivastava
Dr YS Parmar University of Horticulture and Forestry
Nauni, Solan, Himachal Pradesh, India

Jagesh Kumar Tiwari
ICAR-Indian Institute of Vegetable Research
Varanasi, Uttar Pradesh, India

Pankaj Kumar
Dr YS Parmar University of Horticulture and Forestry
Nauni, Solan, Himachal Pradesh, India

Satish Kumar
Dr YS Parmar University of Horticulture and Forestry
Nauni, Solan, Himachal Pradesh, India

Sujeet Kumar
ASCEN HyVeg Pvt. Ltd.
Sohna, Gurugram, Haryana, India

Vikas Kumar
Department of Food Science & Technology
Punjab Agricultural University, Ludhiana, Punjab, India

S.K. Luthra
ICAR-Central Potato Research Institute Research Station
Modipuram, Uttar Pradesh, India

Vikas Mangal
ICAR-Central Potato Research Institute
Shimla, HP, India

Mukesh
Dr YS Parmar University of Horticulture and Forestry
Nauni, Solan, Himachal Pradesh, India

Sharmistha Naik
ICAR-National Research Centre for Grapes
Solapur Road, Pune, Maharashtra, India

Abhilash Padhan
Dr YS Parmar University of Horticulture and Forestry
Nauni, Solan, Himachal Pradesh, India

Mahinder Partap
CSIR-Institute of Himalayan Bioresource Technology
Palampur, Himachal Pradesh, India

Neerja Rana
Dr YS Parmar University of Horticulture and Forestry
Nauni, Solan, Himachal Pradesh, India

Shiv Rattan
CSIR-Institute of Himalayan Bioresource Technology
Palampur, Himachal Pradesh, India

Roshni R. Samarth
ICAR-National Research Centre for Grapes
Solapur Road, Pune, Maharashtra, India

Raja Shankar
ICAR-Indian Institute of Horticulture Research
Bengaluru, India

Himani Sharma
Dr YS Parmar University of Horticulture and Forestry
Nauni, Solan, Himachal Pradesh, India

Sherly J.
Pandit Jawaharlal Nehru College of Agriculture
Karaikal, Puducherry, India

Subaran Singh
Maharana Pratap Horticultural University
Karnal, Haryana, India

P. Sivasakthivelan
Department of Agricultural Microbiology
Faculty of Agriculture, Annamalai University, Annamalai Nagar
Tamil Nadu, India

Salej Sood
ICAR-Central Potato Research Institute
Shimla, HP, India

Priyanka Thakur
Dr YS Parmar University of Horticulture and Forestry
Nauni, Solan, Himachal Pradesh, India

Ajay Kumar Thakur
ICAR-Central Potato Research Institute
Shimla, HP, India

Komal D. Thorat
ICAR-National Research Centre for Grapes
Solapur Road, Pune, Maharashtra, India

Anuradha Upadhyay
ICAR-National Research Centre for Grapes
Solapur Road, Pune, Maharashtra, India

Kamlesh Verma
Dr YS Parmar University of Horticulture and Forestry
Nauni, Solan, Himachal Pradesh, India

Swati Verma
Dr YS Parmar University of Horticulture and Forestry
Nauni, Solan, Himachal Pradesh, India

Ashish R. Warghat
CSIR-Institute of Himalayan Bioresource Technology
Palampur, Himachal Pradesh, India

Rasna Zinta
ICAR-Central Potato Research Institute
Shimla, HP, India

1

Genomic Approaches in Horticultural Crop Improvement

Subaran Singh, Supriya Ambawat and Sujeet Kumar

1.1 Introduction

Horticulture consisting of vegetables, flowers, fruits, plantation crops, spices, aromatic, and medicinal plants is an important and growing sector of agriculture that contributes about 30.4% of GDP to the agriculture sector. These crops are important dietary nutritional components and sources of medicines and aroma and also have significant aesthetic values for human beings. In recent years, farmers have switched from growing conventional crops to producing horticulture and floriculture. Horticulture crops are important to the global economy because they can provide jobs, supply raw materials for different food processing companies, increase farm profitability because of increased production and produce foreign cash through exports. India has become a top producer of horticultural goods and a dependable source of employment, income generation and promotion of exports. About 33% of India's Gross Value Addition (GDA) is made up of the country's horticulture industry, which generates about 320 million tons of products annually. About 10% of the land in India is used for horticulture, which accounts for 33% of the agricultural value. India ranks first in the production of mango, banana, papaya, okra, lime, and lemon and is the second-largest producer of fruits and vegetables in the world. Flower harvesting, nursery maintenance, hybrid seed production and tissue culture, propagation of fruits and flowers, and food processing are highly remunerative employment options for women in rural areas. Since it requires a lot of manpower, it offers many employment prospects, particularly for those living in rural areas. Flower picking, nursery care, hybrid seed development, tissue culture, fruit and flower propagation, and food processing are among some of the highly lucrative occupations related to horticultural crops for women in rural areas enabling them to sustain their livelihood. Horticulture crops have a great nutritional content, making them highly valuable. They are a vital source of nutrients, including proteins, carbohydrates, minerals, vitamins and organic acids. A considerable

DOI: 10.1201/9781003625674-1

amount of vitamins and minerals, especially from fruits and vegetables, can be obtained. Horticulture has great potential to alleviate the world economy and diversify agriculture, and thus can also solve the global micronutrient crisis, ensuring nutritional security for the people. In the current atmosphere of climate change, we must ensure accelerated growth in agricultural output in order to ensure enough food supply to the ever-increasing human population worldwide. Development of new crop varieties that can withstand unfavorable climatic conditions is necessary as vegetables and fruits are particularly vulnerable to the effects of climate change. Thus, there has been increased research focus towards horticulture crop productivity and quality enhancement. The development of cultivars with improved resistance to biotic and abiotic stresses has become the major priority area of genetic breeding programs. Additionally, the objectives of plant breeding have been widened to enhance fruit quality attributes like flavor and nutraceutical components according to consumer's choices as well as improving fruit shelf-life to reduce post-harvest losses (Hansson et al. 2018; Neves et al. 2018).

Conventional breeding, which uses a succession of backcrossing and selection, for introgressing beneficial traits into elite germplasm has been used widely for crop improvement (Tester and Langridge 2010; Karkute et al. 2017). However, it is a time and resource-consuming approach with a limited range of applications because of the absence of target alleles in the germplasm of a particular species. To overcome these limitations, transgenic technology, which enables the integration of specific genes that may come from a non-crossable plant species or a non-plant organism has also been used as a potential technique for genetic enhancement (Anami et al. 2013). In the past 30 years, many transgenic horticultural crops have been developed with significant modifications (Parmar et al. 2017). However, there are a number of transgenics guidelines and biosafety issues because of public concern. Therefore, in this context, genomic research has great potential to enhance molecular biology research and different advanced genomic techniques can be successfully applied in horticultural crops in several ways. The development of valuable genomic resources is mostly driven by the rising demand for horticultural products in agriculture and their useful genetic variation. Genomic resources are very beneficial and may be used for the welfare of humanity through biotechnological interventions in biological processes. It is possible to isolate and characterize newly beneficial genes and create new DNA markers using the already available genomic and gene sequence data. Horticultural crops have a variety of genetic resources that can be used effectively to exploit recent research in order to develop improved cultivars and define future targets. The combination of the genomic approaches, i.e., starting with genome resource and its sequencing with the discovery of the molecular marker approach to find out the candidate gene for associated character through positional cloning, is very useful. The vast knowledge of genes through microarray studies with comparative genomics approaches and linking traits with useful genes will help in improving breeding programs and developing cultivars with the most desired attributes. Diverse evolutionary studies through molecular, biological and metabolic approaches along with the genomic approaches paved the way for the identification of the better-performing genotype with desired/suitable traits in the horticultural crops for fulfilling the purpose of quality and yield as well. The decoding of plant genomes using high-throughput approaches and recent advancements in automation have played a significant role in advancing genomic research. The technique of generating useful genomic information in horticulture crops has undergone a revolution with the introduction of the genome, transcriptome sequencing and next-generation sequencing. In this context, huge progress has been made through whole-genomic sequencing projects where several genome-sequencing projects were started and completed in horticulture crops, like vegetables, fruits, ornamental plants and trees. Information from these sequencing projects was widely used to analyze and understand genome structures and complexity, functional and comparative genomics and to identify and extract useful molecular markers and genes (Sonah et al. 2011). Additionally, tools like DNA markers are extremely useful and essential and will continue to be important in the future for all types of mapping projects in horticulture crops. Presently, horticulture crops, such as apples, melons, potato, tomato, roses, citrus, pepper, cucumber and others, are being successfully genotyped to generate precise genetic maps using SNP, conserved ortholog set (COS), and SSR markers. Similarly, comparative genetics is also a very helpful technique for understanding the genetics of horticulture crops. These crops showed significant gene conservation both among closely related species and among various dicot species. Nowadays, genome editing has become a very important tool and being widely used as a major technique to improve several crops (Cong et al. 2013). Different genomic resources and tools have been used widely for both applied and basic research, but the progress is still very slow and there is a high need to accelerate it using various advanced genomic tools.

1.2 Various Genomic Approaches for Horticultural Crops

1.2.1 Molecular Markers

Crop improvement relies on the identification of desirable genes and superior genotypes possessing such genes. Molecular markers can highly increase the significance and utility of indirect selection in plant breeding and hence facilitate the method of "conventional" gene transfer. Molecular markers are huge in number and are independent of the environment or stage of development. Saturated linkage maps are very helpful and essential

TABLE 1.1
Molecular Markers Used in Different Horticultural Crops

Crop	Molecular Marker	References
Apple (*Malus domestica*)	AFLP and RAPD	Coart et al. (2003); Botez et al. (2009); Sestras et al. (2009)
Apricot (*Prunus armeniaca*)	EST-SSR	Decroocq et al. (2003)
Avocado (*Persea americana*)	Minisatellite	Ashworth and Cleff (2003)
Banana (*Musa* sp.)	RAPD	Brown et al. (2009)
Citrus (*Citrus sinensis*)	RFLP, EST-SSR	Durham et al. (1992); Chen et al. (2006)
Cashew(*Anacardium occidentale*)	RAPD and ISSR	Thimmappaiah et al. (2009)
Coffee (*Coffea* ssp.)	EST- SSR	Aggarwal et al. (2007)
Faba bean (*Vicia faba*)	EST-SSR	Gong et al. (2010)
Grapes (*Vitis vinifera*)	RFLP, SSR, EST-SSR, SNP	Bourquin et al. (1993); Vezzulli et al. (2008)
Iris (*Iris germanica*)	EST-SSR	Tang et al. (2009)
Lettuce (*Lactuca sativa*)	EST-SSR	Simko (2009)
Litchi (*Litchi chinensis*)	SSR	Ekué et al. (2009)
Mango (*Mangifera indica*)	cpISSR and RAPD	He et al. (2007); Marcela et al. (2009)
Muskmelon (*Cucumis melo*)	SSR	Ritschel et al. (2004)
Mustard (*Brassica* sp.)	SSR, STMS, EST-SSR	Yadava et al. (2009); Singh et al. (2018); Singh et al. (2017)
Pear (*Pyrus communis* L.)	SSR and AFLP	Sisko et al. (2009)
Potato (*Solanum tuberosum*)	SSR	Grover et al. (2009)
Strawberry (*Fragaria vesca*)	EST-SSR	Bassil et al. (2006)
Sunflower (*Helianthus annuus* L.)	SSR, InDels	Heesacker et al. (2008); García-Moreno et al. (2010)
Sweet cherry (*Prunus avium* L.)	SSR, CAPs	Olmstead et al. (2008)
Tomato (*Solanum lycopersicum*)	EST-SSR, SNP	Frary et al. (2005)
White Campion (*Silene latifolia*)	EST-SSR	Moccia et al. (2009)

for marker-assisted selection, gene tagging and map-based gene cloning. Numerous molecular markers, including restriction fragment length polymorphism (RFLP), random amplified polymorphic DNA (RAPD), sequence characterized amplified regions (SCARs), inter-simple sequence repeats (ISSR), simple sequence repeat (SSR), sequence-tagged sites (STS), amplicon length polymorphisms (ALPs), amplified fragment length polymorphic DNA (AFLP), cleaved amplified polymorphic sequence (CAPS), insertion-deletion (InDel), single nucleotide polymorphism (SNP) have been found to play a key role in identification and cloning of disease resistance genes from numerous plants. Different molecular markers used in horticultural crops are listed in Table 1.1. They have also been employed by seed industries to create "fingerprint" patterns for assembling or aggregating complex beneficial traits in vegetable crops. The fingerprinting data is useful for genetic diversity analysis, conservation of germplasm especially the CMS lines and germplasm characterization of different accessions. Tivang et al. (1996) employed RAPD markers to determine variation between and within artichoke breeding populations. Different molecular markers have also been used widely for DNA fingerprinting of cultivars and breeding lines of a number of vegetable crops including beans (Hamann et al. 1995), tomato (Kaemmer et al. 1995), pepper (Prince et al. 1995). The discovery and use of molecular markers have been expedited by technological advancements that allow for the precise identification of genetic variations. It is preferable to use PCR-based markers like RAPD, AFLP and microsatellites as prior knowledge of nucleotides is not essential. SSR markers are most preferred by the researchers because they are relatively easy, highly reproducible, co-dominant and simple to transfer. With the development of microsatellites, cultivars of crops like tomato, cucurbit, potato, spinach, pepper and lettuce were identified with a high degree of reliability. Purity testing of breeding lines, DNA fingerprinting and testing for F_1 progeny of various accessions of tomato was done using microsatellite probes (Kaemmer et al. 1995). Earlier, SSR marker development used to be quite expensive because it required expensive library screening and clone sequencing. However, now huge public SSR databases are available for numerous crops and many of these have been explored and exploited for cross-species transferability studies. SSR markers may also be used to study synteny between species that are not closely related to one another (Howad et al. 2005; Celton et al. 2009) and breeders can use these results for identifying markers for desirable traits in different crops.

SNP, conserved ortholog set (COS) and expressed sequence tag (EST) markers are the most widely used markers developed from genomic resources. ESTs have been utilized to generate SSR and SNP markers in different horticultural crops (Frary et al. 2005; Sargent et al. 2006; Ekué et al. 2009; www.bioinformatics.nl/tools/snpweb/). However, the majority of EST-derived markers are monomorphic limiting their application for developing DNA fingerprinting and linkage mapping. Ortholog-specific EST-SSR markers were used for genotyping and genetic mapping in Iris where they reflected the prevalence and usefulness of SSRs in the transcript assembly. The cross-species usefulness and polymorphisms of *I. brevicaulis–I. fulva* EST-derived SSR markers have also been discussed among wild ecotypes and horticulturally important cultivars (Tang et al. 2009). Gynoecious and hermaphrodite flowers were used to generate more than 8000 cucumber ESTs using FLX 454 pyrosequencing technology, these are now publicly available in the database. Similarly, many species from the *Solanaceae* and *Brassicaceae* families were employed to create EST resources among vegetable crops. Apple and citrus are economically significant fruits and hence being studied widely at the genomic level. Among fruits, grapes (362,193) have the largest number of ESTs in the database followed by apples (324,429) and citrus (208,909). Likewise, 296,963, 277,147 and 149,480 ESTs are available for tomato, potato, and *B. rapa* respectively in the NCBI EST database (dbEST; www.ncbi.nlm.nih.gov/dbEST/). *M. guttatus* has the largest number of ESTs (231,095) among flowers. SNPs and insertion/deletion (InDel) are other marker systems developed from genomic resources. SNPs are easy and simple to develop as they are comprised of only a nucleotide variation at a single locus. SNPs are the most advanced type, which are commercially utilized in genetics or for the development of phylogenetic linkages. SNPs have been developed in many species, including grape (Gilchrist et al. 2006; Hyten et al. 2008). These SNPs were derived from BAC end and EST sequences and employed successfully for the construction of genetic linkage maps (Salmaso et al. 2008; Vezzulli et al. 2008) and anchoring physical maps. SNPs identified from grape gene sequences were also used for linkage disequilibrium and genetic diversity analysis (Le Cunff et al. 2008). Additionally, 1,700,000 SNPs were obtained from decoding the Pinot Noir cultivar's grape genome (Velasco et al. 2007). Similarly, SNPs have also been developed and used in other horticultural crops (Simko et al. 2006; Chagne et al. 2008). Allele-specific markers should be developed for the genes controlling important horticultural traits for their effective use in developing superior cultivars. Using an EST library of about 350,000 sequences, around 71,482 putative SNPs have been found in apple, which opens new opportunities to understand the genetic control of major horticulture traits using linkage disequilibrium or quantitative trait locus (QTL) analysis (Chagne et al. 2008). These can also prove useful and potential markers for aligning genetic and physical maps across the *Rosaceae* family (Chagne et al. 2008; Folta and Gardiner 2009). It was also suggested that SNP markers created in one system may be transferable to other systems. But, now there is a need for the development of different computational methods that can handle and use huge collections of SNP markers datasets available through DNA sequencing and discussions and developments are ongoing in this direction (Hyten et al. 2008).

1.2.2 Marker-Assisted Selection (MAS)

MAS is an indirect selection technique that uses tightly linked molecular markers to select the desired trait. This is more useful than phenotypic selection and has many advantages, such as identifying the desired trait early even in the seedling stage, effective for traits with low heritability, higher selection efficiency than phenotypic selection, and speeds up the breeding cycle. Molecular markers, which are closely linked to genes serve as tags and can be employed in breeding programs to select desired genes indirectly. Thus, MAS facilitates the selection of such genes and genotypes and can be used to enhance breeding effectiveness and lower the costs of breeding programs. Iezzoni et al. (2010) suggested an eight-stage procedure to enhance the regular use of markers for breeding programs. Marker-assisted backcrossing (MABC), marker-assisted recurrent selection (MARS) and genomic selection (GS) are the three main methods typically used to carry out molecular marker-aided selection. MABC is used to integrate large-effect QTLs and oligogenes. In comparison to traditional backcross breeding, it is

TABLE 1.2
MAS for Different Traits in Horticultural Crops

Crop	Trait
Almond	Genetic relatedness among cultivars and breeding lines
Broccoli	Resistance to diamond black moth
Cauliflower	Downy mildew resistance
Cucumber	Yield contributing traits, multiple lateral branching
Peach	Identification of diverse peach cultivars and linkage mapping
Pepper	Tobamo virus resistance, tomato spotted wilt virus resistance, root knot nematode resistance, potyvirus reistance
Potato	Root-knot nematode resistance, potato virus X and Y resistance, root cyst nematode, wart resistance.
Sour cherry	QTL analysis flower and fruit traits
Tomato	Black mold resistance, acyl sugar mediated pest resistance, bacterial spot, fusarium wilt and speck resistance, resistance to nematode and insects, fruit quality

well adapted for recovery of the recurrent parent genome and removal of the donor parent genome, flanking the target gene and minimizing linkage drag. On the other hand, MARS opens the way for population development programs by increasing the frequency of QTLs that have a major impact on character expression. GS enriches the population with QTLs whether they have a major impact on character expression or not and thus elevates minor QTL frequency in the population. MAS has several applications in horticulture crop breeding with different traits being selected in horticultural crops as mentioned in Table 1.2 (Yadav et al. 2015; Sajid and Elçi 2020). Horticultural crop development programs can be definitely improved by marker-assisted breeding using information from model plant species *Arabidopsis*. Ogundiwin et al. (2008) identified a gene encoding the leuco anthocyanidin dioxygenase (*PpLDOX*) enzyme using different advanced genomic tools and data from *Arabidopsis*. This enzyme may be linked to a QTL that regulates the enzymatic browning of the mesocarp of peach fruit, which causes chilling injury. Thus, this molecular data can be used in breeding programs to develop chilling injury-resistant varieties in peach and other stone fruits (Peace et al. 2005; Bliss 2010). Numerous traits have been identified and improved using MAS in horticulture crops (Table 1.3). Tomato leaf curl virus is one of the major and prevalent diseases and six resistant genes (*Ty1, Ty2, Ty3, Ty4, Ty5, Ty6*) have been identified for this trait through MAS. Closely linked P6-25 and SCAR-1 markers were used for foreground selection of the *Ty-3* gene. The *Ty3* gene was introgressed into the genetic background of Pusa Rohini, Pusa-120 and Pusa Ruby, which are known to be the most adaptable and popular varieties among consumers. Three microsatellite markers (CYSTSIN, BVWS02441, MCPI_11) linked to the powdery mildew-resistant gene were also identified on chromosome 2 in water melon, which can be further utilized for indirect

TABLE 1.3
QTLs and Genes Identified for the Main Horticultural Crops

Crop	QTL/Gene	Molecular Identification	References
Apple	Scab resistance gene VF	Leucine rich repeat domain/ Transmembrane domain	Xu and Korban (2003)
Cabbage	Three major QTLs for head splitting resistance (Hsr 3.2, 4.2, 9.2)	–	Su et al. (2015)
Cauliflower	Orange gene (Or) Dna J	Cysteine Rich domian	Lu et al. (2006)
Cucumber	Fusarium wilt resistance (Foc 2.1, fw 2.1 Foc 4) and	–	He et al. (2022)
	Downy mildew resistance *(dm 1.1, dm 2.1, dm 4.1, dm 5.1, dm, 6.1)*		
Papaya	CpCYC-b	Lycopene β-cyclase	Blas et al. (2010)
Pepper	Virus resistance e1F4E gene	Leucine rich repeat	Ruffel et al. (2002)
Potato	Resistance to Ro1Globodera ostochiensis (Gro1-4)	Unknown	Paal et al. (2004)
	Quality trait cold-sweetening (invGE/GF	Invertase	Paal et al. (2004)
	Flavonoid 3,5-hydroxylase (f35h)	–	Jung et al. (2005)
Musk Melon	Fusarium wilt resistance (*Fom-1, Fom*-2)	–	Oumouloud et al. (2013)
Radish	Clubroot resistance (Rsa10025569)	–	Wang et al. (2022)
Sugarbeet	Nematode resistance HS1pro-1	Leucine rich repeat domain	Cai et al. (1997)
Tomato	Fruit shape (Ovate)	Unknown	Liu et al. (2002)
	Fruit sugar content (Brix9-2-5/ Lin5)	Invertase	Fridman et al. (2002)
	Fruit weight (fw2.2)	Unknown	Frary et al. (2000)
	Sw4.1 QTL gene	ABC transporter	Orsi and Tanksley (2009)
Watermelon	Gummy stem blight resistance *(ClGSB3.1, ClGSB5.1, and ClGSB7.1)*	–	Gimode et al. (2021)
	Genetic male sterility (*Cla006625*)	–	Dong et al. (2021)

selection or integration of powdery mildew-resistant gene into the new population (de Souza Gama et al. 2015).

1.2.3 Gene tagging and candidate gene identification

Gene tagging refers to the mapping of genes of economic significance linked to known markers. The candidate gene approach can be used to isolate plant genes underlying specific traits. Molecular biologists can use bioinformatics tools to access publically available genomic resources, identify potential candidate genes in genome sequences and subsequently, clone them from closely related species using PCR amplification and sequencing (Pflieger et al. 2001). The proliferating and arrested inflorescence phenotype of *Arabidopsis thaliana* due to the ap1-1/cal-1 mutant has led to hypotheses that similar genes may be useful to characterize traits in *B. oleracea*. Consequently, using this technique, genes found in *Arabidopsis* were utilized to find similar genes in cauliflower and broccoli (King 2003). Similarly, the *Gro1-4* gene, which provides resistance against pathotype Ro1 of the cyst nematode *Globodera rostochiensis* was cloned from potato using positional cloning and a candidate gene approach (Paal et al. 2004). Although many QTLs have been detected, they are very large and challenging to transfer genetically. However, candidate genes underlying these QTLs can be cloned using advanced molecular tools and genomic resources. Many QTLs were fine mapped and some of them have been successfully cloned using a positional cloning technique. The BAC library serves as the major genetic resource for map-based gene cloning and the development of physical maps (Gaafar et al. 2005). They can also be used to study comparative genome organization and evolution mechanisms among plant species. YAC or BAC libraries have been employed for gene cloning and fine mapping in a number of horticultural crops. A BAC library was constructed from *Rosa rugosa* covering 5.2 genome equivalents (Kaufmann et al. 2003). *R. rugosa* was selected as a source of DNA to construct the library due to its small genome size. A BAC library was built in *Rosa multiflora* hybrids using the information of *R. rugosa* and *Rdr1* locus, which is responsible for black spot resistance, was fine mapped in roses (Biber et al. 2009). Peach has been chosen as the model genome mainly for marker development and execution of gene discovery and genome comparisons in *Rosaceae* (Georgi et al. 2002). With the help of the BAC library, a gene locus for *TERMINAL FLOWER1 (TFL1)* was characterized in peaches and apricots, which will be further useful in understanding the basis of the development of flowers in angiosperms (Liang et al. 2010). Thus, numerous reports have proved the usefulness of the BAC library and are being used for mapping and cloning of genes in several horticultural crops (Cova et al. 2010; Liu et al. 2010). Additionally, sequence data has been generated in many horticultural crops for various fully characterized genes and full-length cDNA clones.

Numerous QTLs have been linked to candidate genes without the use of cloning just based on map information and observations. For example, QTL for fruit shape, sugar content and weight were cloned from tomatoes and functionally validated by transgenic technique and can be further used in the future for other significant horticultural traits (Frary et al. 2000; Liu et al. 2002). Later, marker-assisted selection can also be used to transfer the cloned QTL into cultivated varieties. In addition to the economically significant traits, various studies were performed to improve the nutritional quality of vegetables and fruits. Nutritional genomics should emphasize the cloning of genes associated with nutraceutical, anti-nutritional and medicinal applications in the case of horticultural crops. Production of carotene, which is a precursor of vitamin A, is among the most widely studied areas for gene identification and understanding of related pathways. Horticulture crops like papaya, citrus, and tomato were used to clone numerous genes of carotenoids (Zhang et al. 2009; Blas et al. 2010). A gene isolated from a high-β-carotene orange cauliflower mutant offers a useful resource to develop transgenic in carotenoid-deficient crops (Zhou et al. 2008). Similar to vitamins, antioxidants are of great importance in a number of horticulture crops since they provide defense against several diseases. The majority of widely consumed fruits and vegetables may not contain enough anthocyanin to provide all the advantages. Expression of the two transgenes elevated the hydrophilic antioxidant capacity of tomato three times suggesting that this strategy can be used to develop novel pharmaceuticals (Butelli et al. 2008). Several attempts have been made to make edible vaccines as fruits are most suitable for delivering such vaccines (Rybicki 2010). These studies emphasize how molecular and genomic approaches may be used to manipulate and enhance the nutritional value of various horticulture crops. Using recent and advanced techniques like genome editing, forward- and reverse-genetics, and relevant genetic materials like introgression libraries and panels of unrelated accessions can increase the effectiveness of these efforts. Highly effective bioinformatics tools and vast genome sequences can also be useful in this direction for *in silico* identification of candidate sequences displaying high homology with QTL in positional cloning or association mapping projects.

1.2.4 QTL Mapping

The majority of the significant traits in plants are controlled by QTLs. The development of a linkage map provides a precise depiction of the physical location of a gene on a chromosome. It shows that in addition to oligogenes, minor genes (polygenes) also play a significant role in the inheritance of a trait. The segregation patterns of the specific genes underlying a QTL can be used to characterize and map that QTL. The experiments of Sax (1923) in the common bean (*Phaseolus vulgaris*) revealed that there is a linkage between seed size (a quantitative trait) and seed coat color (a qualitative trait) and thus mapping of polygenes started based on the idea that a genetic marker and a quantitative trait phenotype are associated with each other. Initially, this approach was unable to find the precise location of polygenes due to their large number, dispersed location within the genome, and cumulative effect on character expression. Later, distinct clusters of minor genes were discovered in the genomic areas that were associated with the expression of quantitative traits, and the term QTL was used to describe each such genomic region that consists of a small number of minor genes on different chromosomes. Development of linked molecular markers to a specific gene or QTL is a pre-requisite for MAS. However, the use of bi-parental mapping populations (F_2, backcross, DH, RIL) results in low mapping precision due to the occurrence of relatively few recombination events. These days, NAM (nested association mapping) populations and multi-parent advanced generation intercross (MAGIC) populations are employed as they are relatively better for constructing highly saturated maps. Association mapping is an effective approach that uses historical recombination events to detect QTLs in germplasm collections or natural populations. Extensive efforts have been made for QTL mapping to identify useful QTLs for several important traits in vegetable as well as fruit crops (Kenis et al. 2008; Olmstead et al. 2008; Hatakeyama et al. 2010; Kubo et al. 2010; Zhang et al. 2010). Some QTLs and genes identified in major horticultural crops are enlisted in Table 1.3. QTLs for the late blight disease were found in tomatoes using different mapping populations. Additionally, QTLs for glandular trichomes and yield were also identified using SNP markers in chromosome 1 (Benewitz et al. 2018; Brekketet et al. 2019). Similarly, SSR markers have been used to identify QTLs in cucumber on chromosome 6 for the length of the fruit peduncle and cucumber mosaic virus (Song et al. 2016; Shi et al. 2018).

1.2.5 Genome Sequencing of Different Horticultural Crops

Although the genomic resources for model horticultural crops are growing quickly, many of them are still underutilized. There are many genomic resources already available, including whole-genome sequences, large-insert genomic libraries, expressed sequence tags (ESTs), numerous molecular markers and high-density linkage maps. But, to maintain the development and related value-added opportunities of these crops, genomic resources like high-throughput genome sequences (HTGs) and genomic survey sequences (GSSs) are highly needed. These resources can be used for sequencing and annotation, cloning of genes, mapping for QTL and MAS in important horticultural crops. However, the genomic studies in horticulture crops is relatively slow in comparison to cereals and other crops. Sanger's di-deoxynucleotide chain termination method of DNA sequencing is a conventional method that gives a long read length with a quality sequence but is laborious and time-consuming. On the other hand, advanced genome-sequencing techniques are fast and can be successfully applied to improve horticulture crops (Sonah et al. 2011). After the development and advancement of next-generation sequencing (NGS) technologies such as FLX454 (Roche), SOLiD (Applied Biosystems), Solexa (Illumina), Helicose, Pacific Biosciences for DNA sequencing, there has been a significant increase in the sequence database of many organisms. This has raised hopes that in a few years, genomic resources will be developed for many more horticultural crops, which will help horticulture breeders to use this extensive genomic information in their varietal improvement programs. Large insert genomic libraries generated in bacterial artificial chromosome (BAC), yeast artificial chromosome (YAC), or P artificial chromosome (PAC) vectors were initially used for high-throughput genome sequencing of eukaryotic organisms. Among these vectors, BAC vectors are generally most preferred due to several limitations of other vector systems. BAC-by-BAC sequencing offers a more systematic and effective approach than shotgun sequencing for whole-genome sequencing.

Various projects were started to sequence the genome of horticultural crops and many of them are already publicly available and some of them can be made available in the near future (Sharma et al. 2016; Chen et al. 2019; Sonah et al. 2011). The tomato genome sequencing project (https://solgenomics.sgn.cornell.edu/organism/Solanum_lycopersicum/genome), potato genome sequencing consortium (www.potatogenome.net), grape genome sequencing project (www.vitaceae.org), papaya genome sequencing project, floral genome sequencing project Brassica sequencing (http://brassica.info/genome/sequencing.html) and solanaceous genomes (www.solgenomics.net), etc., are some of the most notable initiatives towards genome sequencing in horticultural crops. Various genome-sequencing information available for different horticultural crops is listed in Table 1.4. To achieve maximum coverage with high-quality sequence and cost-effective methodology, these projects were well-equipped with sophisticated and advanced sequencing tools in addition to conventional methods. Among vegetables, tomato (*Solanum lycopersicum*) and potato (*Solanum tuberosum*) have been employed as scientific models for genomic research in solanaceous crops. The International Solanaceae Genome Project (ISGP) was established in 2003 in order to sequence tomato, which is the first diploid crop among asteroids to be sequenced (Mueller et al. 2009). ISGP distributed 12 tomato chromosomes among 10 international teams for sequencing. The worldwide group has focused on the tomato genome's euchromatic region for decoding. The tomato genome is comprised of about 950 Mb of DNA and annotation of this genome sequence divulged 34,727 protein-coding genes. The majority of region (75%) in the tomato sequence is heterochromatin and largely devoid of genes, hence the euchromatic region of the tomato genome was targeted by the international consortium for decoding. The draft genome sequence of tomato was declared in August, 2010 and is available in the public domain (https://solgenomics.sgn.cornell.edu/). Additionally, the Potato Genome Sequencing Consortium (PGSC; www.potatogenome.net) has announced the draft sequence of the entire potato genome containing nearly 35,004 genes like tomato. These data show that members of the *Solanaceae* family share a great deal of genetic similarities and these sequences will be useful for functional analysis of the numerous *Solanaceae* genes and will serve as important resources for comparative genomic studies among other solanaceous crops. Tomatoes and potatoes are simple to genetically modify and are excellent candidates for biotechnological investigations related to gene function (An

TABLE 1.4

List of Genomes Sequenced for Horticultural Crops

Crop	Scientific Name	Size (Mb)	Chromosomes	Sequencing Method	Center
Apple	*Malus domestica*	800	17	Whole-genome shotgun approach	IASMA research center, The Horticulture and Food Research Institute of New Zealand
Almond	*Prunus dulcis*	300	8	Illumina HiSeq	Beijing Agro-biotechnology Research Center at CUGI
Banana	*Musa acuminata*	600	11	Roche/454 sequencing	Global Musa Genomics Consortium (GMGC)
Brinjal	*Solanum melongena*	1100	12	Illumina paired- end sequencing	Cornell University
Carrot	*Daucus carota*	473	9	454/Illumina/BAC-end sequencing	University of Central Florida, Vegetable Crops Research Unit
Cabbage	*Brassica oleracea*	587	9	Illumina HiSeq/PacBio	TIGR
Coffee	*Coffea arabica*	1300	11	Illumina HiSeq/PacBio	University of Central Florida
Colorado blue columbine	*Aquilegia coerulea*	350	7	–	Joint Genome Institute(JGI)
Cucumber	*Cucumis sativus*	370	7	BAC-by-BAC and next-generation sequencing	The Cucumber Genome Initiative
Date Palm	*Phoenix dactylifera*	605	18	Shotgun library Roche/454 data SOLiD 4 System, BAC-end sequencing	King Abdulaziz City for Science and Technology, Beijing Institute of Genomics
Grape	*Vitis vinifera*	480	19	Sanger shotgun sequencing	International Grape Genome Program, IASMA Research Center
Hot pepper	*Capsicum annuum*	3000	12	Illumina HiSeq	Korea Research Institute of Bioscience and Biotechnology, Seoul National University, Cornell University
Indian Mustard	*Brassica juncea*	922	18	Illumina Hiseq	Multinational Brassica Genome Project
Mango	*Mangifera indica*	439	20	PAC Biosequencing	Mango Genome Consortium
Mustard	*Brassica rapa*	529	10	BAC-by-BAC and next-generation sequencing	Brassica IGF Project, The Multinational Brassica rapa Sequencing Project

(continued)

TABLE 1.4 (Continued)

List of Genomes Sequenced for Horticultural Crops

Crop	Scientific Name	Size (Mb)	Chromosomes	Sequencing Method	Center
Onion	*Allium cepa*	15,000	8	Illumina HiSeq/PacBio	USDA-ARS and University of Wisconsin
Papaya	*Carica papaya*	372	9	Sanger whole-genome shotgun	The Papaya Genome Sequencing Consortium
Peach	*Prunus persica*	230	8	Sanger whole-genome shotgun	Clemson University, CUGI
Pear	*Pyrus bretschneideri*	577	17	BAC-by-BAC and next-generation sequencing	Nanjing Agricultural University, University of Illinois
Pecan	*Carya illinoinensis*	691	16	–	University of Georgia, USA
Potato	*Solanum tuberosum*	840	12	Whole-genome shotgun / Illumina/Sanger sequencing	The Potato Genome Sequencing Consortium, TIGR, The Canadian Potato Genome Project
Red columbine	*Aquilegia formosa*	350	7	–	DOE Joint Genome Institute
Rose	*Rosa multiflora*	750	7	PacBio RS II	Rosaceae International Genomics Initiative (RosIGI)
Siberian apricot	*Prunus armeniaca*	222	8	Illumina HiSeq/ PacBio	Instituto Valencianao de Investigaciones Agrarias, INRA Avignon
Strawberry	*Fragaria × ananassa*	240	28	Roche/454, Illumina/Solexa and Life Technologies/SOLiD platforms	University of Florida University of California, Davis,USDA-ARS
Sugar beet	*Beta vulgaris*	760	9	Roche/454, Illumina and Sanger	Hokkaido University, Japan, Max Planck Institute, Michigan State University
Sweet orange	*Citrus sinensis*	367	9	Whole-genome shotgun paired-end–tag sequence reads, Illumina GAII sequencer	U.S. Citrus Genomics Initiative, University of California
Tomato	*Solanum lycopersicum*	950	12	Sanger and 454/Roche GS FLX reads and SOLiD and Illumina GAIIx	The Tomato Genome Consortium

et al. 1986). Additionally, other sequence resources, such as genomic DNA, cDNA, molecular markers and conserved orthologous genes were also mapped to enable comparative genome analysis between eggplant, potato, pepper, tomato and other *Solanaceae* crops (Pratt et al. 2008). Tomato is a model crop for dicots and its genome sequence information can pave the way for improvement in other plant species, such as apples, strawberries, bananas, melons, and many other fleshy fruits that resemble tomatoes in some ways.

In addition to these, the draft sequence of several other plant species, such as cassava, shepherd's purse, peach, papaya (*Carica papaya*), apple (*Malus x domestica*), grapevine (*Vitis vinifera*), cucumber (*Cucumis sativus*), sorghum (https://phytozome-next.jgi.doe.gov/sorghumpan/), poplar (www.eurekalert.org), etc., are currently available in database (Jaillon et al. 2007; Ming et al. 2022; Huang et al. 2009). Among fruit crops, the International Grape Genome Project (IGGP), a French–Italian partnership, published the first draft of a grapevine sequence with an 8× quality (Jaillon et al. 2007). The cucumber (*Cucumis sativus*), a plant belonging to the *Cucurbitaceae* family that also includes other major crops like squash, melon, pumpkin and watermelon has been sequenced and utilized as a model system for research on plant vascular biology and sex determination. Another significant genome sequencing consortia, the Floral Genome Project (FGP) (www.floralgenome.org/), was also started to explore the origin, conservation and diversification of the genomic architecture of the flowers (www.floralgenome.org/). Additionally, several computational tools for evolutionary and functional genomic studies have also been developed to utilize this vast information. Thus, by utilizing several genomic techniques, the FGP has been investigating the evolution of floral development and characterizing the ancestral floral transcriptome. Comparative and functional genomics techniques are being used to evaluate the sequence data for the characterization of genes and validation of their functions. Thus, many genome sequences are available for horticultural crops and there are lots of opportunities to use the available genomic resources and utilize the genetic potential of these crops in horticulture and also extend them further in the related crops. The development of horticultural genome databases for genome sequencing, data sharing, data quality, data visualization and data storage as well as the advancement of horticultural plant genome sequencing and pan-genome reconstruction will offer useful direction for horticultural plant genomic studies. Several high-throughput techniques, genomic platforms, and approaches are presently accessible to the plant researchers for using genomics in crop improvement. This massive sequence data from numerous species makes comparative genomics research easier in plants. But, effective bioinformatics tools and trained computational biology personnel are needed to use this information effectively.

1.2.6 Comparative Genomics in Horticultural Crops

The study of the relationships between the structural and functional aspects of genomes from various species is known as comparative genomics. Sequence data of several horticultural crops have been deposited in public databases, which can be used for both applied and basic research in related genera. It will be useful to elucidate the evolutionary relationships between various species and improve phylogenetic classification. In horticultural crops, comparative genomic studies have been applied mainly to families like *Rosaceae* (Dirlewanger et al. 2004), *Solanaceae* (Mueller et al. 2009), *Fabaceae* (Zhu et al. 2005; Cronk et al. 2006), *Asteraceae* (The Compositae Genome Project: https://compgenomics.ucdavis.edu/compositae_resources.php) and *Brassicaceae* (Hall et al. 2002; Lysak and Lexer 2006). These investigations have demonstrated that the majority of chromosomal rearrangements that have occurred during genome evolution have resulted in syntenic co-linearity between sizable chromosomal areas within a family. For example, the genomes of *Malus* and *Pyrus* were found to exhibit a high level of co-linearity in the *Maloideae* family (Yamamoto et al. 2004). This was further revealed by the first genetic linkage map of loquat (*Eriobotrya japonica*), which was developed using SSR markers derived from *Malus* (Gisbert et al. 2009). As a result, this co-linearity can be used further for comparative mapping research with other *Maloideae* species like loquat and quince. Comparative plant genomics started in 1988 in the *Solanaceae* family when a genetic linkage map of potato chromosomes was constructed on the basis of genomic and cDNA clones from tomato (Bonierbale et al. 1988) and at that time genomic information of tomato was also used in pepper (Tanksley et al. 1988). Later, genomic data from tomato was used to compare several different *Solanaceae* crops. *Arabidopsis* genome sequence can prove to be very useful for structural and functional analysis of *Brassica* as *Arabidopsis* and *Brassica* genes share 87% sequence similarity (Cavell et al. 1998). This facilitates the use of *Arabidopsis* microarray for expression analysis in *Brassica* (Fu et al. 2009). Several studies revealed that *Arabidopsis* genome can be used for fine mapping of QTLs and cloning of genes in *Brassica* (Formanová et al. 2010; Zhao et al. 2010). *Brassica* unigene-derived markers have been also anchored on the *A. thaliana* genome (Parida et al. 2010). This study showed that unigene-derived markers can be used to establish synteny between related genera, which may be further used to understand complex chromosomal rearrangements like tandem and segmental duplications, inversions, and insertions/deletions occurring in different species. Several online databases and tools are also available to access data on various horticulture crops. For example, a Genome Database for Rosaceae (GDR) is an integrated web-based relational database that provides centralized access to *Rosaceae* genomics and genetics data and its analytical tools can be utilized to study cross-species associations (Jung et al. 2008; www.rosaceae.org). Mimulus is phylogenetically well positioned for comparative genome analysis as it has close relationships with different well-developed genomic resource crop plants like *Arabidopsis*, tomato, sunflower, asterids, and lettuce, and its flower structure can be also used as a model for developmental studies.

1.2.7 DNA Microarray and Eco-TILLING

DNA microarray is a useful tool to analyze the overall expression of genes in an organism. Among the different techniques available for rapid identification and genotyping, specifically engineered DNA chip arrays are the most promising approach for detecting genetic differences in horticultural crops. However, well-characterized cDNA or EST sequences are required to develop these microarray chips and thus generation of large EST collections is important for large-scale gene discovery. Oligonucleotide arrays are commercially accessible for numerous crop species like *Solanum lycopersicon* and *Vitis vinifera*. cDNA arrays are also now being used extensively for many horticultural crops, such as citrus (Forment et al. 2005), potato (Rensink et al. 2005), *B. oleracea* (Soeda et al. 2005), strawberry (Aharoni et al. 2004) and gerbera (Laitinen et al. 2005). The Institute for Genomic Research (TIGR) created a cDNA potato microarray using a non-redundant set of 10,000 ESTs and made it inexpensively available to the different researchers. They also provided a transcription profiling service to many users in order to evaluate these arrays in various *Solanaceous* species, resulting in the development of huge microarray data. The Agilent "44K feature platform" system, which became accessible for use in gene discovery in 2006, has been chosen by the potato oligo chip initiative (POCI). The flexibility of this system allows redesigning the arrays if more information about gene sequences becomes available in the future. Thus, array-based SNP technologies have been employed to analyze the potato genome (Rickert et al. 2005), which can be surely applied to many other species. Tomato sequencing efforts have produced results of translational value as data from publicly available databases have been used to identify candidate genes and molecular markers to be used in genetics and breeding research (Yang et al. 2004; Frary et al. 2005). Eco-TILLING can be used as a powerful tool to identify InDels and SNPs inside a 1-kb window in a set of genotypes, which further helps to identify diverse haplotypes on target loci and make them accessible for association mapping (Comai et al. 2004). It has been used effectively for identifying genetic variations in tomato (Stergiopoulos et al. 2007) and melon ecotypes (Nieto et al. 2007). It is also very useful to identify allelic variations within natural populations and is very helpful for quick gene-discovery techniques between species. The use of a microarray for discovering polymorphism between large numbers of species is very expensive. But, with the fast advancement of cost-effective tools, the use of a microarray for different applications like diversity analysis, eco-TILLING, and genetic mapping among various horticultural crops is expected in the future.

1.2.8 Genome Editing

Although conventional breeding is a good strategy, crop improvement is confined by a waning genetic base that is becoming more dependent on natural allelic variations and crossing barriers between wild and cultivated species. To overcome these crossing barriers, the transgenic technique is a good alternative for plant breeding as it does not use the native genetic stock of the plant to generate traits of interest but instead relies on a foreign source for the gene of interest. However, a major concern is that transgenic crops cannot be commercialized because of public acceptance. In recent years, genome editing using engineered nucleases, such as clustered regularly interspaced short palindromic repeats/CRISPR associated 9 (CRISPR/Cas9) (Cong et al. 2013) or RNA-guided engineered nucleases (RGEN), transcription activator-like effector nucleases (TALENs) (Mahfouz et al. 2011; Li et al. 2012) and zinc finger nucleases (ZFNs) (Carroll 2011), have proved to be useful tools for editing target sites in DNA sequences. These genome-editing approaches are being widely used in several crops to exploit the native genomic stock of plants in a precise manner. ZFNs and TALENs are constrained by their technical complexity and ineffectiveness, whereas CRISPR/Cas9 is a simple and highly efficient genome-editing tool that allows very precise and targeted genetic variation and is hence highly preferred by breeders. It uses a guide RNA, which triggers a nuclease to create the double-strand break (DSB) at the target site in DNA and activates the cell repair mechanisms, which ultimately results in nucleotide deletions or insertions. Therefore, CRISPR/Cas is a toolbox to accomplish various objectives like basic science research, the development of crops with good agronomic traits, introduction of novel ideas to food production. Off-target cleavage by Cas9 is the main issue that can be evaded by choosing a specific target site. CRISPR/Cas9 can be used to modify and improve several desired traits in horticultural crops and thus can help enhance their production and productivity in the future (Liu et al. 2022; Rukavtsova et al. 2023). This technology has been effectively used for improvement in many horticultural crops like tomato (Brooks et al. 2014), orange (Jia and Wang 2014), apple (Nishitani et al. 2016), grape (Ren et al. 2022), kiwifruit (Wang et al. 2018), cacao (Fister et al. 2018), pear (Charrier et al. 2019), yam (Syombua et al. 2020), rubber tree (Dai et al. 2021), citrus (Alquézar et al. 2022), pear (Ming et al. 2022), duncan grape fruit (Jia et al. 2022). Some other studies reported for use of CRISPR/Cas9 genome-editing technique in horticultural crops for improving different traits are listed in Table 1.5. Increased yield, abiotic stress tolerance, disease and pest resistance, processing quality, increased shelf life, improved nutrition and aesthetic value, etc., are some of the important traits that can be focused on for improvement in horticultural crops using genome editing. Some of these traits are monogenic, whereas others are complex, polygenic, labor-intensive, time-consuming and difficult to improve by conventional breeding. But, multiple genes can be modified in the plant genome using CRISPR/Cas9 and

TABLE 1.5

CRISPR/Cas-Mediated Genome Editing in Some Horticultural Plants

Crop	Trait	References
Apple	Resistance to fire blight disease	Pompili et al. (2020)
Banana	Development of semi-dwarf banana cultivar	Shao et al. (2020)
	Resistance to banana streak virus	Tripathi et al. (2019)
	Fortification of β-carotene in banana fruit	Kaur et al. (2020)
	Increased shelf life	Hu et al. (2021)
Cabbage	Dwarf phenotype	Lawrenson et al. (2015)
Carrot	Blockage of the anthocyanin biosynthesis in purple-colored carrot	Klimek-Chodacka et al. (2018)
Citrus	Resistance to citrus canker disease	Jia et al. (2017); Peng et al. (2017)
Cucumber	Resistance to Cucumber vein yellowing virus; Potyviruses Zucchini yellow mosaic virus; Papaya ring spot mosaic virus	Chandrasekaran et al. (2016)
Potato	Amylose content & glycoalkaloids accumulation	Andersson et al. (2018)
Strawberry	Increasing of auxin biosynthesis leading to faster growth of seedlings	Zhou et al. (2018)
Tomato	Color pigmentation	Ito et al. (2015)
	Production of seedless fruits	Klap et al. (2017)
	Resistance to powdery mildew	Nekrasov et al. (2017)
	Fruits with ripening, ethylene production, lycopene accumulation repressed	Li et al. (2018)
	Longer fruit shelf life	Wang et al. (2019); Yu et al. (2017)
Watermelon	Albino phenotype, seed size and promoted seed germination	Tian et al. (2017); Wang et al. (2021); Liu et al. (2022)
Strawberry	Improvement in the quality of berry	Zhou et al. (2018); Martin-Pizarro et al. (2019)

they can also be added through homologous recombination. Several tools can be used to choose the specific target site, including CRISPR-Plant (Xie et al. 2014), GT-scan (O'Brien and Bailey 2014), CRISPR direct (Naito et al. 2015), etc., but CRISPR direct is most suitable among these techniques for horticultural crops. Genes like *Botrytis cinerea75, Xanthomonas citri73, 74* that provide resistance against plant pathogens have been edited in grapes, citrus and apple.

The majority of the CRISPR/Cas9 research was conducted mainly on tomato plant, where 18 distinct genes were targeted independently. First of all, CRISPR/Cas9-mediated genome-editing experiments were done in 2014 in vegetable crop tomatoes, where the *ARGONAUTE7 (SlAGO7)* gene, which plays an important role in leaf development was targeted (Brooks et al. 2014). Since the loss of function mutation of the *SlAGO7* gene resulted in needle or wiry-like leaves, it was chosen as a target for easy identification of edited plants. Developmental genes such as *BLADE-ON-PETIOLE*

(*SlBOP*; inflorescence; Xu et al. 2016); *SHORTROOT (SHR*; root development; Ron et al. 2014); *SELF PRUNING* and *SELF PRUNING 5G (SP5G)* (SP; plant development) were also modified for functional validation in tomato. It was observed that tomato plants modified with both *sp* and *sp5G* genes were early flowering and very compact. This experiment proved the efficacy of CRISPR/Cas9 in improving vegetable crops. *Anthocyanin 1 (ANT1)*, which is involved in the biosynthesis of anthocyanins phytoene desaturase (*SlPDS*), phytochrome interacting factor (*SlPIF4*), and phytoene synthase (*PSY1*), which is involved in the biosynthesis of carotenoids were among some of the other genes edited by CRISPR/Cas9 in tomato (Cermak et al. 2015; Hayut et al. 2017). Hayut et al. (2017) exhibited targeted recombination in heterozygotes by crossing two tomato plants that were stably expressing Cas9 and gRNA independently. They also demonstrated effective shuffling of chromosomal regions between homologous chromosomes in somatic cells. Thus, with this experiment, the inaccessible regions of a chromosome could be effectively shuffled by targeted crossover to use genes present in such locations.

The application of CRISPR/Cas9 for actual vegetable crop improvement was reported in 2017. Two distinct groups simultaneously developed parthenocarpic tomato fruits that are in high demand in the food processing industry. Klap et al. (2017) performed gene deletion of the *Slagamous-like 6 (SlAGL6)* gene resulting in mutant plants that may produce parthenocarpic fruits under heat-stress conditions that would normally substantially impair fertilization-dependent fruit set. Other vegetables like *Lactuca sativa* and *Brassica oleracea* were also used to target developmental genes using CRISPR/Cas9 (Lawrenson et al. 2015; Woo et al. 2015). The conversion of sugar to starch in pea seeds causes them to lose their sweetness with maturity. But this process can be stopped by altering the starch branching enzyme 1 (*SBE1*) gene to develop the "staysweet" trait in peas (Bhattacharyya et al. 1990). Cucurbitacin causes bitterness in cucumbers, which can be targeted by knocking down the genes responsible for cucurbitacin synthesis (Shang et al. 2014). Breeders work to create cucumber types devoid of the bitter cucurbitacin. Cucumbers without bitterness can result from cucurbitacin production gene knockdown and cucumbers lacking bitterness can be developed (Shang et al. 2014). Similarly, anti-nutritious factors can be eliminated and nutritional factors can be improved in the case of soybeans and leafy vegetables by altering their biosynthetic pathway. Oxidation of phenolic compounds by polyphenol oxidase (PPO) causes browning in fruits and vegetables leading to a detrimental impact on color, flavor, taste and nutritional content, which can be resolved by manipulating the *PPO* gene as done in non-browning mushrooms. Fruit crops are a vital source of nutrition but their perennial nature makes it a time-consuming breeding process and hence little attention has been paid to improving them through genome editing. Hence, transgenic or genetic engineering techniques are best suited for their improvement. Genome editing in fruit crops using CRISPR/Cas9 is a promising area and some studies have been reported for these (Tian et al. 2017; Martin-Pizarro et al. 2019; Kaur et al. 2020; Pompili et al. 2020; Shao et al. 2020). CRISPR/Cas9 was used in citrus for editing the phytoene desaturase (*CsPDS*) gene of the carotenoid biosynthesis pathway (Jia and Wang 2014). Citrus canker-resistant plants were also developed using CRISPR/Cas9 by deleting the susceptibility gene *CsLOB1* (Jia et al. 2017). Citrus has several alleles of the *CsLOB1* gene containing effector-binding element (*EBEPthA4*) that is recognized by the primary *Xcc* effector *PthA4* to activate *CsLOB1* expression. Citrus canker-resistant plants were developed by targeting this effector-binding region in the *CsLOB1* promoter (Peng et al. 2017). This technique was also used for targeting the phytoene desaturase gene in both watermelons and grapes (Nakajima et al. 2017; Tian et al. 2017). Moreover, target site databases for CRISPR/Cas9-based genome editing are being created, and "Grape-CRISPR," is one such grape-specific database that is made publically available (Wang et al. 2016). Similarly, apple has been modified for resistance against fire blight disease using CRISPR/cas9 technology (Pompili et al. 2020). Recently, banana has also been used for genome editing to fortify β-carotene, development of semi-dwarf banana variety and resistance to banana streak virus (Tripathi et al. 2019; Kaur et al. 2020; Shao et al. 2020).

In addition to fruits and vegetables, CRISPR/Cas9 genome editing has recently been used in other horticultural crops like ornamental and medicinal plants. A model legume called *Lotus japonicus* was the first ornamental plant to undergo CRISPR/Cas9 genome editing. Leghemoglobin loci (*LjLb1*, *LjLb2* and *LjLb3*) and other symbiotic nitrogen fixation-related genes like symbiosis receptor-like kinase were targeted in this plant and single as well as multiple gene-mutant plants were developed (Wang et al. 2016). Additionally, it was observed that the nodule-specific *LjLb2* promoter is also equally effective for gene editing in nodules as the constitutively active CaMV35S promoter. *Chrysanthemum morifolium* (chrysanthemum), an ornamental plant that stably expresses yellowish-green fluorescent protein (*CpYGFP*) gene was also utilized for genome editing where this gene was deleted by two sgRNAs targeted at different positions (Kishi-Kaboshi et al. 2017). Further, it was also proposed that the *PcUbi* promoter for Cas9 and the *AtU6* promoter for sgRNA are appropriate for genome editing in chrysanthemum. Chrysanthemum is also a natural resource for the development of insecticides and successful genome editing in this plant has opened the doors for altering the biosynthetic pathways resulting in the development of new, broad-spectrum and more efficient insecticides. In addition to the ornamental plants, the Chinese medicinal plant *Salvia miltiorrhiza* was also modified using CRISPR/Cas9 to delete the diterpene synthase gene (*SmCPS1*) engaged in the biosynthesis of tanshinones. This technique led to the successful development of plants devoid of tanshinones without affecting other phenolic metabolites (Li et al. 2017). Therefore, genome editing in medicinal plants will be useful to study secondary metabolite biosynthetic pathways as well as the enhancement of their yield and quality.

1.3 Challenges and Future Prospects

Horticultural plants differ from agricultural plants in a number of ways, such as the need for controlled environments and specialized techniques or facilities for plant growth, the need for grafting, post-harvest care, and a protracted juvenile phase as well as the fact that they typically reproduce asexually and have a special metabolism. Although genomic tools have several benefits over traditional crop breeding, there are still certain difficulties in applying them to horticultural crops. It is challenging to conduct molecular and genetic investigations on horticultural crops, which makes it difficult to identify the genes underlying desirable traits. Thus, it will be useful to sequence the genomes of important horticultural crops in order to find the genes responsible for the desired traits. Lot of data has been generated due to advanced genomic tools. Integrating massive amounts of data is the ultimate objective of functional genomics and bioinformatics in order to comprehend the molecular mechanisms behind different developmental stages. Thus, now there is a need to integrate and analyze this data for horticultural plants and this has become a significant challenge for many horticultural biologists and needs collaborative efforts of different scientists from different disciplines. Contrary to this, the availability of genomic resources for many economically important horticulture crops is either low or non-existent, which needs top priority. The lack of well-established molecular genetic linkage maps is another big issue in the case of many fruit species. Due to their perennial nature, these species are not easy to study at the genetic and molecular levels and thus development of mapping population and genetic linkage map for these is challenging, which needs to be addressed. Hence, advancements in genomics and strong statistical tools are highly needed to develop markers linked to a specific trait. Moreover, the non-availability of ESTs or other genomic resources in the case of woody plants is also another big concern in comprehending their genome structure. Due to their complicated genomes, it would be challenging to use NGS technologies to sequence these species from scratch as one can generate enough sequences from any species, but *de novo* genome assembly is a major problem. Therefore, it would be ideal to first estimate the genome size of each species and develop enough DNA markers from ESTs or other sources. Due to their non-amountability to tissue culture methods, trait-specific genes would be difficult to functionally validate even after being identified from genomic resources. Therefore, alternative approaches should be followed for functional analysis of identified and cloned genes in the majority of these crops.

1.4 Conclusion

Molecular markers have already proved to be an effective tool to support conventional plant breeding in improving different crops. They can be straight away applied in basic research and advanced breeding programs to improve horticultural crop yields. Additionally, the use of next-generation sequencing technology and other advanced genomic approaches will greatly help in developing high-density linkage maps ensuring the identification of QTLs with high accuracy. Thus, the identification of closely linked molecular markers can make marker-assisted selection more efficient in improving the desired traits. MAS makes it easier to identify progeny containing the desired traits in an early generation and accelerates the selection process in breeding programs. The genomic tools are extremely useful and efficient for quickly understanding the basic genetic mechanism and genome organization at the molecular level, which can be further utilized by a variety of genetic resources to develop superior cultivars. Moreover, the use of these recent and upcoming advanced genomic tools can lead to the next green revolution in the future. Despite the ban on genetically modified organisms, there are a number of conventional and advanced breeding and genomic tools and techniques that can be applied for crop improvement in organic vegetables. However, the accomplishment of DNA markers and genomic techniques to improve horticultural crops will depend on close and strong collaboration between plant breeders and biotechnologists, availability of skilled technicians, and significant financial investment in research. Functional genomics has undergone a revolutionary change due to the development of new sequencing technologies. On the other hand, the introduction of genome-editing technology can play a significant role in crop improvement by producing unique allelic series of mutants. Genome-editing technology like CRISPR/Cas9 can be used to improve nutrient content, yield, abiotic stress resistance, host–pathogen interaction, etc., in different horticultural crops. Hence, more research efforts should be put into transcriptomics and genomic analysis of orphan horticultural crops allowing for the deployment of new technologies for their improvement in the future.

References

Aggarwal RK, Hendre PS, Varshney RK, Bhat PR, Singh VKL. Identification, characterization and utilization of EST-derived genic microsatellite markers for genome analyses of coffee and related species. *Theor Appl Genet* 2007;114:359–372.

Aharoni A, Giri AP, Verstappen FWA, Bertea CM, Sevenier R, Sun Z, et al. Gain and loss of fruit flavor compounds produced by wild and cultivated strawberry species. *Plant Cell* 2004;16:3110–3131.

Alquézar B, Bennici S, Carmona L, Gentile A, Peña L. Generation of transfer-DNA-free base-edited citrus plants. *Front Plant Sci* 2022;13:835282.

An G, Watson BD, Chiang CC. Transformation of tobacco, tomato, potato, and Arabidopsis thaliana using a binary Ti vector system. *Plant Physiol* 1986;81:301–305.

Anami S, Njuguna E, Coussens G, Aesaert S, Van Lijsebettens M. Higher plant transformation: principles and molecular tools. *Int J Dev Biol* 2013;57:483–494.

Andersson M, Turesson H, Olsson N, Fält As, Ohlsson P,; Gonzalez Mn, Samuelsson M, Hofvander P. Genome editing in potato via CRISPR/Cas9 ribonucleoprotein delivery. *Physiol Plant* 2018;164:119–240.

Ashworth VE, Clegg MT. Microsatellite markers in Avocado (Persea mericana Mill.): genealogical relationships among cultivated Avocado Genotypes. *J Hered* 2003;94:407–415.

Bassil NV, Gunn M, Folta K, Lewers K. Microsatellite markers for fragaria from 'strawberry festival' expressed sequence tags. *Mol Ecol Notes* 2006;6(2):473–476.

Bennewitz S, Bergau N, Tissier A. QTL mapping of shape of type VI glandular trichomes in tomato. *Front Plant Sci* 2018;9:1421.

Bhattacharyya MK, Smith AM, Ellis TH, Hedley C, Martin C. The wrinkled-seed character of pea described by Mendel is caused by a transposon-like insertion in a gene encoding starch-branching enzyme. *Cell* 1990;60:115–122.

Biber A, Kaufmann H, Linde M, Spiller M, Terefe D, Debener T. Molecular markers from a BAC contig spanning the Rdr1 locus: a tool for marker-assisted selection in roses. *Theor Appl Genet* 2009;120(4):765–773.

Blas AL, Ming R, Liu Z, Veatch OJ, Paull RE, Moore PH, Yu Q. Cloning of the papaya chromoplast-specific lycopene beta-cyclase, CpCYC-b, controlling fruit flesh color reveals conserved microsynteny and a recombination hot spot. *Plant Physiol* 2010;152(4):2013–2022.

Bliss FA. Marker assisted breeding in horticulture crops. In: Bassil NV, Martin R, editors. *Proc.IS on molecular markers in horticulture,* 859. *Acta hort*; 2010. p. 339–350.

Bonierbale MW, Plaisted RL, Tanksley SD. RFLP maps based on a common set of clones reveal modes of chromosomal evolution in potato and tomato. *Genetics* 1988;120:1095–1103.

Botez C, Sestras R, Ardelean M, Pamfil D, Patrascu B, Sestras A. . Phenotypic Selection Assisted by Molecular Markers for Scab Resistance in Apple. *Acta Hort* 2009;814:771–776.

Bourquin JC, Sonko A, Otten L, Walter B. . RFLP molecular taxonomy in Vitis vinifera. *Theor Appl Genet* 1993;87:431–38.

Brekketet TD, Stroud JA, Shaw DS, Crawford S, Steele KA. QTL mapping in salad tomatoes. *Euphytica*. 2019; 215:115.

Brooks C, Nekrasov V, Lippman ZB, Van Eck J. Efficient gene editing in tomato in the first generation using the clust ered regularly interspaced short palindromic repeats/CRISPR-associated9 system. *Plant Physiol* 2014;166:1292–1297.

Brown N, Venkatasamy S, Khittoo G, Bahorun T, Jawaheer S. Evaluation of genetic diversity between 27 banana cultivars (Musa spp.) in Mauritius using RAPD markers. *Afr J Biotech* 2009;8:1834–1840.

Butelli E, Titta L, Giorgio M, Mock HP, Matros A, Peterek S, et al. Enrichment of tomato fruit with health-promoting anthocyanins by expression of select transcription factors. *Nat Biotechnol* 2008;26(11):1301–1308.

Cai D, Kleine M, Kifle S, Harloff H, Sandal NN, Marcker KA, et al. Positional cloning of a gene for nematode resistance in sugar beet. *Science* 1997;7(275):832–834.
Carroll D. Genome engineering with zinc-finger nucleases. *Genetics* 2011;188: 773–782.
Cavell AC, Lydiate D, Parkin IAP, Dean C, Trick M. Collinearity between a 30-centimorgan segment in Arabidopsis thaliana chromosome 4 and duplicated regions within the Brassica napus genome. *Genome* 1998;41:62–9.
Celton JM, Tustin DS, Chagné D, Gardiner SE. Construction of a dense genetic linkage map for apple rootstocks using SSRs developed from Malus ESTs and Pyrus genomic sequences. *Tree Genet Genomes* 2009;5:93–107.
Cermak T, Baltes NJ, Cegan R, Zhang Y, Voytas DF. High-frequency, precise modification of the tomato genome. *Genome Biol* 2015;16:232.
Chagne D, Gasic K, Crowhurst RN, Han Y, Bassett HC, Bowatte DR, et al. Development of a set of SNP markers present in expressed genes of the apple. *Genomics* 2008;92:353–8.
Chandrasekaran J, BruminM, Wolf D, Leibman D, Klap C, Pearlsman M, Sherman A, Arazi T, Gal-On A. Development of broad virus resistance in non-transgenic cucumber using CRISPR-Cas9 technology. *Mol Plant Pathol* 2016;17:1140–1153.
Charrier A,Vergne E, Dousset N, Richer A, Petiteau A, Chevreau E. Efficient targeted mutagenesis in apple and first time edition of pear using the CRISPR-Cas9 system. *Front Plant Sci* 2019;10:40.
Chen C, Zhou P, Choi YA, Huang S, Gmitter FG. Mining and characterizing from citrus ESTs. *Theor Appl Genet* 2006;112:1248–57.
Chen F, SongY, Li X, Chen J, Mo L, Zhang X, Lin Z, Zhang L. Genome sequences of horticultural plants: past, present, and future. *Hortic Res* 2019;6:112.
Coart E, Vekemans X, Smulders MJM, Wagner I, Van Huylenbroeck J, Van Bockstaele E, Roldán-Ruiz I. Genetic variation in the endangered wild apple (Malus sylvestris L. Mill.) in Belgium as revealed by amplified fragment length polymorphism and microsatellite markers. *Mol Ecol* 2003;12:845–857.
Comai L, Young K, Till BJ, Reynolds SH, Greene EA, Codomo CA, et al. Efficient discovery of DNA polymorphisms in natural populations by EcoTILLING. *Plant J* 2004;37:778–86.
Cong L, Ran FA, CoxD Lin S, Barretto R, Habib N, et al. Multiplex genome engineering using CRISPR/Cas systems. *Science* 2013;339:819–823.
Cova V, Paris R, Passerotti S, Zini E, Gessler C, Pertot I, et al. Mapping and functional analysis of four apple receptor-like protein kinases related to LRPKm1 in HcrVf2-transgenic and wild-type apple plants. *Tree Genet Genomes* 2010;6:389–403.
Cronk Q, Ojeda I, Pennington RT. Legume comparative genomics: progress in phylogenetics and phylogenomics. *Curr Opin Plant Biol* 2006;9:99–103.
Dai X,Yang X,Wang C, Fan Y, Xin S, Hua Y, Wang K, Huang H. CRISPR/Cas9-mediated genome editing in Hevea brasiliensis. *Ind Crops Prod* 2021;164:113418.
de Souza Gama RNC, Santos CAF, Dias RCS, de Souza RRC, de Queiróz MA. Microsatellite markers linked to powdery mildew resistance locus in watermelon. *Aust J Crop Sci* 2015;9(1):92–97.
Decroocq V, Fave MG, Hagen L, Bordenave L, Decroocq S. Development and transferability of apricot and grape EST microsatellite markers across taxa. *Theor Appl Genet* 2003;106:912–22.
Dirlewanger E, Graziano E, Joobeur T, Garriga-Calderé F, Cosson P, Howad W, Arús P. Comparative mapping and marker-assisted selection in Rosaceae fruit crops. *Proc Natl Acad Sci USA* 2004;101(26):9891–6.

Dong W, Wu D, Yan C, Wu D. Mapping and analysis of a Novel Genic Male Sterility Gene in Watermelon (*Citrullus lanatus*). *Front Plant Sci* 2021;12:639431.

Durham RE, Liou PC, Gmitter FG, Moore GA. Linkage of restriction fragment length polymorphism and isozymes in Citrus. *Theor Appl Genet* 1992;84:39–48.

Ekué MRM, Gailing O, Finkeldey R. Transferability of simple sequence repeat (SSR) markers developed in Litchi chinensis to Blighia sapida (Sapindaceae). *Plant Mol Biol Rep* 2009;27:570–4.

Fister AS, Landherr L, Maximova SN, Guiltinan MJ. Transient expression of CRISPR/Cas9 machinery targeting TcNPR3 enhances defense response in Theobroma cacao. *Front Plant Sci*, 2018;9:268.

Folta KM, Gardiner SE. *Genetics and genomics of Rosaceae*. In: Jorgensen R, editor. Series Ed. New York: Springer; 2009. p. 600.

Formanová N, Stollar R, Geddy R, Mahé L, Laforest M, Landry BS, Brown GG. High-resolution mapping of the Brassica napus Rfp restorer locus using Arabidopsis-derived molecular markers. *Theor Appl Genet* 2010;120:843–851.

Forment J, Gadea J, Huerta L, Abizanda L, Agusti J, Alamar S, et al. Development of a citrus genome-wide EST collection and cDNA microarray as resources for genomic studies. *Plant Mol Biol* 2005;57:375–91.

Frary A, Nesbitt TC, Frary A, Grandillo S, van der Knaap E, Cong B, et al. fw2.2: a quantitative trait locus key to the evolution of tomato fruit size. *Science* 2000;7:85–88.

Frary A, Xu Y, Liu J, Mitchell S, Tedeshi E, Tanksley S. Development of a set of PCR-based anchor markers encompassing the tomato genome and evaluation of their usefulness for genetics and breeding experiments. *Theor Appl Genet* 2005;111:291–312.

Fridman E, Liu YS, Carmel-Goren L, Gur A, Shoresh M, Pleban T, et al. Two tightly linked QTL modify tomato sugar content via different physiological pathways. *Mol Genet Genomics* 2002;266:821–826.

Fu SX, Cheng H, Qi C. Microarray analysis of gene expression in seeds of Brassica napus planted in Nanjing (altitude: 8.9 m), Xining (altitude: 2261.2 m) and Lhasa (altitude: 3658 m) with different oil content. *Mol Biol Rep* 2009;36(8):2375–2386.

Gaafar RM, Hohmann U, Jung C. Bacterial artificial chromosome-derived molecular markers for early bolting in sugar beet. *Theor Appl Genet* 2005;110:1027–1037.

García-Moreno MJ, Velasco L, Pérez-Vich B. Transferability of non-genic microsatellite and gene-based sunflower markers to safflower. *Euphytica* 2010;175(2):145–150.

Georgi L, Wang Y, Yvergniaux D, Ormsbee T, Iñigo M, Reighard G, et al. Construction of a BAC library and its application to the identification of simple sequence repeats in peach [Prunus persica (L.) Batsch]. *Theor Appl Genet* 2002;105(8):1151–1158.

Gilchrist EJ, Haughn GW, Ying CC, Otto SP, Zhuang J, Cheung D, et al. Use of Ecotilling as an efficient SNP discovery tool to survey genetic variation in wild populations of Populus trichocarpa. *Mol Ecol* 2006;15:1367–1378.

Gimode, W., Bao, K., Fei, Z. et al. QTL associated with gummy stem blight resistance in watermelon. *Theor Appl Genet* 2021;134: 573–584. https://doi.org/10.1007/s00122-020-03715-9

Gisbert AD, Martinez-Calvo J, Llacer G, Badenes ML, Romero C. Development of two loquat [Eriobotrya japonica (Thunb.) Lindl.] linkage maps based on AFLPs and SSR markers from different Rosaceae species. *Mol Breed* 2009;23:523–538.

Gong YM, Xu SC, Mao WH, Hu QZ, Zhang GW, Ding J, et al. Generation and characterization of 11 novel est derived microsatellites from Vicia faba (Fabaceae). *Am J Bot* 2010;97:e69–71.

Grover A, Ramesh B, Sharma PC. Development of microsatellite markers in potato and their transferability in some members of Solanaceae. *Physiol Mol Biol Plants* 2009;15:343–358.

Hall AE, Fiebig A, Preuss D. Beyond the Arabidopsis genome: opportunities for comparative genomics. *Plant Physiol* 2002;129(4):1439–1447.

Hamann A, et al. Microsatellite fingerprinting in the genus Phaseolus. *Genome* 1995;38:507–515.

Hansson S, Aman P, Becker W, Koning D, Lagerkvist C, Larsson I, Lehrman A, Risérus U, Stymne S. Breeding for public health: a strategy. *Trends Food Sci Technol* 2018;76:481–486.

Hatakeyama K, Horisaki A, Niikura S, Narusaka Y, Abe H, Yoshiaki H, et al. Mapping of quantitative trait loci for high level of self-incompatibility in Brassica rapa L. *Genome* 2010;53(4):257–265.

Hayut SF, Bessudo CM, Levy AA. Targeted recombination between homologous chromosomes for precise breeding in tomato. *Nat Commun* 2017;8:15605.

He X, Yong-zeb G, Yang-ruia L, Shi-jinb O. Assessment of the genetic relationship and diversity of mango and its relatives by cpISSRMarker. *Agric Sci China* 2007;6:137–142.

He Y, Wei M, Yan Y, Yu C, Cheng S, Sun Y, Zhu X, Wei L, Wang H , Miao L. Research advances in genetic mechanisms of major Cucumber diseases resistance. *Front Plant Sci* 2022;13:862486. doi: 10.3389/fpls.2022.862486

Heesacker A, Kishore VK, Gao W, Tang S, Kolkman JM, Gingle A, et al. SSRs and INDELs mined from the sunflower EST database: abundance, polymorphisms, and cross-taxa utility. *Theor Appl Genet* 2008;117(7):1021–1029.

Howad W, Yamamoto T, Dirlewanger E, Testolin R, Cosson P, Cipriani G, et al. Mapping with a few plants: using selective mapping for microsatellite saturation of the Prunus reference map. *Genetics* 2005;171:1305–1309.

Hu C, Sheng O, Deng G, He W, Dong T, Yang Q, DouT, Li C, Gao H, Liu S. CRISPR/Cas9-mediated genome editing ofMaACO1 (aminocyclopropane-1-carboxylate oxidase1) promotes the shelf life of banana fruit. *Plant Biotechnol J* 2021;19:654–656.

Huang S, Li R, Zhang Z, Li L, Gu X, Fan W, et al. The genome of the cucumber, Cucumis sativus L. *Nat Genet* 2009;41:1275–1281.

Hyten DL, Song Q, Choi IY, Yoon MS, Specht JE, Matukumalli LK, et al. High-throughput genotyping with the Golden Gate assay in the complex genome of soybean. *Theor Appl Genet* 2008;16:945–952.

Iezzoni A, Weebadde C, Luby J, Yue C, van de Weg E, Fazio G, et al. RosBREED: enabling marker-assisted breeding in Rosaceae. In: Bassil NV, Martin R, editors. *Proc. IS on molecular markers in horticulture*, 859. *Acta Hort*; 2010. p. 385–389.

Ito Y, Yokoi N, Endo M, Mikami M, Toki S. CRISPR/Cas9-mediated mutagenesis of the RIN locus that regulates tomato fruit ripening. *Biochem Biophys Res Commun* 2015;6:76–82.

Jaillon O, Aury JM, Noel B, Policriti A, Clepet C, Casagrande A, et al. The grapevine genome sequence suggests ancestral hexaploidization in major angiosperm phyla. *Nature* 2007; 449:463–467.

Jia H, Wang N. Targeted genome editing of sweet orange using Cas9/sgRNA. *PLoS One* 2014;9:e93806.

Jia H, Zhang Y, Orbovic V, Xu J, White FF, Jones JB., et al. Genome editing of the disease susceptibility gene CsLOB1 in citrus confers resistance to citrus canker. *Plant Biotechnol J* 2017;15:817–823.

Jia H, Omar A, Orbovi´c V, Wang N. Biallelic editing of the LOB1 promoter via CRISPR/Cas9 creates canker-resistant 'Duncan'grapefruit. *Phytopathology* 2022;112:308–314.

Jung CS, Griffiths HM, De Jong DM, Cheng S, Bodis M, De Jong WS. The potato P locus codes for flavonoid 3, 5-hydroxylase. *Theor Appl Genet* 2005;110:269–275.

Jung S, Staton M, Lee T, Blenda A, Svancara R, Abbott A, Main D. GDR (Genome Database for Rosaceae): integrated web-database for Rosaceae genomics and genetics data. *Nucleic Acids Res* 2008;36(Database issue):D1034–40.

Kaemmer D, et al. Oligonucleotide fingerprinting of tomato DNA. *Plant Breed* 1995;114:12–17.

Karkute SG, Singh AK, Gupta OP, Singh PM, Singh B. CRISPR/Cas9 Mediated Genome Engineering for Improvement of Horticultural Crops. *Front Plant Sci* 2017;8:1635.

Kaufmann H, Mattiesch L, Lörz H, Debener T. Construction of a BAC library of Rosa rugosa Thunb. and assembly of a contig spanning Rdr1, a gene that confers resistance to blackspot. *Mol Genet Genom* 2003; 268:666–674.

Kaur N, et al. CRISPR/Cas9 directed editing of lycopene epsilon-cyclasemodulates metabolic flux for β-carotene biosynthesis in banana fruit. *Metab Eng* 2020;59:76–86.

Kenis K, Keulemans J, Davey WM. Identification and stability of QTLs for fruit quality traits in apple. *Tree Genet Genomes* 2008;4:647–661.

King GJ. Using molecular allelic variation to understand domestication processes and conserve diversity in brassica crops. In: Düzyaman E, Tüzel Y, editors. *International symposium on sustainable use of plant biodiversity to promote new opportunities for horticultural production development*, 598. *Acta Hort (ISHS)*; 2003. p. 181–186.

Kishi-Kaboshi M, Aida R, Sasaki K. Generation of gene-edited Chrysanthemum morifolium using multicopy transgenes as targets and markers. *Plant Cell Physiol* 2017;58:216–226.

Klap C, Yeshayahou E, Bolger A, Arazi T, Gupta SK, Shabtai S, Usadel B, Salts Y, BargR. Tomato facultative parthenocarpy results from SIAGAMOUS-LIKE 6 loss of function. *Plant Biotechnol J* 2017;15:634–647.

Klimek-Chodacka M, OleszkiewiczT, Lowder LG, Qi Y, BaranskiR. Efficient CRISPR/Cas9-based genome editing in carrot cells. *Plant Cell Rep* 2018;37:575–586.

Kubo N, Saito M, Tsukazaki H, Kondo T, Matsumoto S, Hirai M. Detection of quantitative trait loci controlling morphological traits in Brassica rapa L. *Breeding Sci* 2010;60:164–171.

Laitinen RA, Immanen J, Auvinen P, Rudd S, Alatalo E, Paulin L, et al. Analysis of the floral transcriptome uncovers new regulators of organ determination and gene families related to flower organ differentiation in Gerbera hybrida (Asteraceae). *Genome Res* 2005;15:475–486.

Lawrenson T, Shorinola O, Stacey N, Li C, Ostergaard L, Patron N, et al. Induction of targeted, heritable mutations in barley and Brassica oleracea using RNA-guided Cas9 nuclease. *Genome Biol* 2015;16:258.

Le Cunff L, Fournier-Level A, Laucou V, Vezzulli S, Lacombe T, Adam-Blondon AF, et al. This P. construction of nested genetic core collections to optimize the exploitation of natural diversity in Vitis vinifera L. subsp sativa. *BMC Plant Biol* 2008;8:31.

Li B, CuiG, ShenG, Zhan Z, Huang L, Chen J, et al. Targeted mutagenesis in the medicinal plant Salvia miltiorrhiza. *Sci Rep* 2017;7:43320.

Li R, Fu D, Zhu B, Luo Y, Zhu H. CRISPR/Cas9-mediated mutagenesis of lncRNA1459 alters tomato fruit ripening. *Plant J* 2018;94:513–524.

Li T, Liu B, Spalding MH, Weeks DP, YangB. High-efficiency TALEN-based gene editing produces disease-resistant rice. *Nat Biotechnol* 2012;30:390–392.

Liang H, Zhebentyayeva T, Olukolu B, Wilde D, Reighard GL, Abbott A. Comparison of gene order in the chromosome region containing a TERMINAL FLOWER 1 homolog in apricot and peach reveals microsynteny across angiosperms. *Plant Sci* 2010;179:390–398.

Liu J, Van Eck J, Cong B, Tanksley SD. A new class of regulatory genes underlying the cause of pear-shaped tomato fruit. *Proc Natl Acad Sci USA* 2002;99:13,302–306.

Liu SY, Yu K, Huffner M, Park SJ, Banik M, Pauls KP, Crosby W. Construction of a BAC library and a physical map of a major QTL for CBB resistance of common bean (*Phaseolus vulgaris* L.). *Genetica* 2010;138:709–716.

Liu Y, Zhang C, Wang X, Li X, You C. CRISPR/Cas9 technology and its application in horticultural crops. *Hortic Plant J* 2022;8(4):395–407.

Lu S, Van Eck J, Zhou X, Lopez AB, O'Halloran DM, Cosman KM, et al. The cauliflower or gene encodes a DnaJ cysteine-rich domain-containing protein that mediates high levels of beta-carotene accumulation. *Plant Cell* 2006;18(12):3594–3605.

Lysak M, Lexer C. Towards the era of comparative evolutionary genomics in Brassicaceae. *Plant Syst Evol* 2006;259:175–98.

Mahfouz MM, Li L, Shamimuzzaman M, WibowoA, FangX, Zhu JK. De novo-engineered transcription activator-like effector (TALE) hybrid nuclease with novel DNA binding specificity creates double-strand breaks. *Proc Natl Acad Sci USA* 2011;108:2623–2628.

Marcela DM, Ingrid SG, Manuel RGE, Jaramillo H. Analysis of diversity among six populations of Colombian mango (Mangifera indica L. cv. Hilacha) using RAPDs markers. *Electron J Biotech* 2009;12:1–6.

Martín- Pizarro C, Triviño JC, Posé D. Functional analysis of the *TM6* MADS-box gene in the octoploid strawberry by CRISPR/Cas9-directed mutagenesis. *J Exp Bot* 2019;70:885–895.

Ming M, Long H, Ye Z, Pan C, ChenJ, Tian R, Sun C, Xue Y, ZhangY, Li J, et al. Highly efficient CRISPR systems for loss-of-function and gain-of-function research in pear calli. *Hortic Res* 2022;9:148.

Moccia M, Oger-Desfeux C, Marais GA, Widmer A. A White Campion (Silene latifolia) floral expressed sequence tag (EST) library: annotation, EST-SSR characterization, transferability, and utility for comparative mapping. *BMC Genomics* 2009;10:243.

Mueller LA, Lankhorst RK, Tanksley SD, Giovannoni JJ, White R, Vrebalov J, et al. A snapshot of the emerging tomato genome sequence. *Plant Genome* 2009;2:78–92.

Naito Y, Hino K, BonoH, Ui-Tei K. CRISPRdirect: software for designing CRISPR/Cas guide RNA with reduced off-target sites. *Bioinformatics* 2015;31:1120–1123.

Nakajima I, Ban Y, AzumaA, Onoue N, Moriguchi T, Yamamoto T, et al. CRISPR/Cas9-mediated targeted mutagenesis in grape. *PLoS One* 2017;12:e0177966.

Nekrasov V, Wang C, Win J, Lanz C, Weigel D, KamounS. Rapid generation of a transgene-free powdery mildew resistant tomato by genome deletion. *Sci Rep* 2017;7:482.

Neves CG, Amaral DOJ, PaulaMFB, NascimentoGC, Passos OS, Santos MA, OllitraultP, Gesteira AS, Luro F, Micheli F. Characterization of tropical mandarin collection: implications for breeding related to fruit quality. *Sci Hortic* 2018;239:289–299.

Nieto C, Piron F, Dalmais M, Marco CF, Moriones E, Gómez-Guillamón ML, et al. EcoTILLING for the identification of allelic variants of melon eIF4E, a factor that controls virus susceptibility. *BMC Plant Biol* 2007;7:34.

Nishitani C, Hirai N, Komori S, Wada M, OkadaK, Osakabe K,Yamamoto T, Osakabe Y.. Efficient genome editing in apple using a CRISPR/Cas9 system. *Sci Rep,* 2016;6:31481.

O'Brien A, Bailey TL. GT-Scan: identifying unique genomic targets. *Bioinformatics* 2014;30: 2673–2675. doi: 10.1093/bioinformatics/btu354

Ogundiwin EA, Peace CP, Nicolet CM, Rashbrook VK, Gradziel TM, Bliss FA, et al. Leucoanthocyanidin dioxygenase gene (PpLDOX): a potential functional marker for cold storage browning in peach. *Tree Genet Genomes* 2008;4:543–554.

Olmstead JW, Sebolt AM, Cabrera A, Sooriyapathirana SS, Hammar S, Iriarte G, et al. Construction of an intra-specific sweet cherry (Prunus avium L.) genetic linkage map and synteny analysis with the Prunus reference map. *Tree Genet Genomes* 2008;4:897–910.

Orsi CH, Tanksley SD. Natural variation in an ABC transporter gene associated with seed size evolution in tomato species. *PLoS Genet* 2009;5(1):e1000347.

Oumouloud A, El-Otmani M, Chikh-Rouhou H, Claver AG, Torres RG, Perl-Treves R., Álvarez JM. Breeding melon for resistance to Fusarium wilt: recent developments. *Euphytica* 2013;192:155–169.

Paal J, Henselewski H, Muth J, Meksem K, Menéndez CM, Salamini F, et al. Molecular cloning of the potato Gro1-4 gene conferring resistance to pathotype Ro1 of the root cyst nematode Globodera rostochiensis, based on a candidate gene approach. *Plant J* 2004;38(2):285–297.

Parida SK, Yadava DK, Mohapatra T. Microsatellites in Brassica unigenes: relative abundance, marker design, and use in comparative physical mapping and genome analysis. *Genome* 2010;53(1):55–67.

Parmar N, Singh KH, Sharma D, Singh L, Kumar P, Nanjundan J, Khan YJ, Chauhan DK, Thakur AK. Genetic engineering strategies for biotic and abiotic stress tolerance and quality enhancement in horticultural crops: a comprehensive review. *3 Biotech* 2017;7:239.

Peace CP, Crisosto CH, Gradziel TM. Endopolygalacturonase: a candidate gene for freestone and melting flesh in peach. *Mol Breed* 2005;15:420–427.

Peng, A. et al. Engineering canker-resistant plants through CRISPR/Cas9 targeted editing of the susceptibility gene CsLOB1 promoter in citrus. *Plant Biotechnol J* 2017;15:1509–1519

Pflieger S, Lefebvre V, Causse M. The candidate gene approach in plant genetics: a review. *Mol Breed* 2001;7:275–291.

Pompili V, Dalla Costa L, Piazza S, Pindo M, Malnoy M. Reduced fireblight susceptibility in apple cultivars using a high-efficiency CRISPR/Cas9-FLP/FRT-based gene editing system. *Plant Biotechnol J* 2020;18:845–858.

Pratt RC, Francis DM, Barrero Meneses LS. Genomics of tropical Solanaceous species: established and emerging crops, plant genetics and genomics. In: Moore

PH, Ming R, editors. *Genomics of tropical crop plants*, Vol. 1. New York: Springer; 2008. p. 453–67.

Prince JP. et al. A survey of DNA polymorphism within the genus Capsicum and the fingerprinting of pepper cultivars. *Genome* 1995;38:224–231.

Ren C, Lin Y, Liang Z. CRISPR/Cas genome editing in grapevine: recent advances, challenges and future prospects. *Fruit Res* 2022;2:7.

Rensink WA, Lee Y, Liu J, Iobst S, Ouyang S, Buell CR. Comparative analyses of six solanaceous transcriptomes reveal a high degree of sequence conservation and species-specific transcripts. *BMC Genomics* 2005;14(6):124.

Rickert AM, Ballvora A, Matzner U, Klemm M, Gebhardt C. Quantitative genotyping of single-nucleotide polymorphisms by allele-specific oligonucleotide hybridization on DNA microarrays. *Biotechnol Appl Biochem* 2005;42:93–96.

Ritschel PS, Lins TC, Tristan RL, Buso GS, Buso JA, Ferreira ME. Development of microsatellite markers from an enriched genomic library for genetic analysis of melon (Cucumis melo L.). *BMC Plant Biol* 2004;18(4):9.

Ron M, Kajala K, Pauluzzi G, Wang D, Reynoso MA, Zumstein K, et al. Hairy root transformation using Agrobacterium rhizogenes as a tool for exploring cell type-specific gene expression and function using tomato as a model. *Plant Physiol* 2014;166:455–469.

Ruffel S, lène Dussault M, Palloix A, Moury B, Bendahmane A, Robaglia C, et al. A natural recessive resistance gene against potato virus Y in pepper corresponds to the eukaryotic initiation factor 4E (eIF4E). *Plant J* 2002;32:1067–1075.

Rukavtsova EB, Zakharchenko NS, Lebedev VG, Shestibratov KA. CRISPR-Cas genome editing for horticultural crops improvement: advantages and prospects. *Horticulturae* 2023;9:38.

Rybicki EP. Plant-made vaccines for humans and animals. *Plant Biotechnol J* 2010;8(5):620–637.

Sajid QUA, Elçi E. Recent development in molecular markers system for disease resistance in Tomato. *Res J Biotechnol* 2020;141(12):140–151.

Salmaso M, Malacarne G, Troggio M, Faes G, Stefanini M, Grando MS, et al. A grapevine (Vitis vinifera L.) genetic map integrating the position of 139 expressed genes. *Theor Appl Genet* 2008;116:1129–1143.

Sargent DJ, Clarke J, Simpson DW, Tobutt KR, Arus P, Monfort A, et al. An enhanced microsatellite map of diploid Fragaria. *Theor Appl Genet* 2006;112:1349–1359.

Sax K. The association of size differences with seed-coat pattern and pigmentation in Phaseolus vulgaris. *Genetics* 1923;8:552–560.

Sestras R, Pamfil D, Ardelean M, Botez C, Sestras A, Dan C, Mihalte LUse of phenotypic and MAS selection based on bulk segregant analysis for study of genetic variability induced by artificial hybridization on Apple. *Not Bot Horti Agrobo* 2009;37(1):273–277.

Shang Y, Ma Y, Zhou Y, Zhang H, Duan L, Chen H, et al. Biosynthesis, regulation, and domestication of bitterness in cucumber. *Science* 2014;346:1084–1088. doi: 10.1126/science.1259215

Shao X, et al. Using CRISPR/Cas9 genome editing system to createMaGA20ox2 gene-modified semi-dwarf banana. *Plant Biotechnol J* 2020;18:17–19.

Sharma N, Singh SK, Lal S, Singh NK. Genome sequence information in fruit crops: current status. *Transcriptomics* 2016;4(1):130.

Shi L, Yang Y, Xie Q, Miao H, Bo K, Song Z et al.Inheritance and QTL mapping of cucumber mosaic virus resistance in cucumber. *PLoS One* 2018;13:7.

Simko I, Haynes KG, Jones RW. Assessment of linkage disequilibrium in potato genome with single nucleotide polymorphism markers genetics. *Genetics* 2006;173:2237–2245.

Simko I. Development of EST-SSR markers for the study of population structure in lettuce (Lactuca sativa L.). *J Hered* 2009;100(2):256–262.

Singh BK, Choudhary SB, Yadav S, Malhotra EV, Rani R, Supriya A, et al. Genetic structure identification and assessment of interrelationships between Brassica and allied genera using newly developed genic-SSRs of Indian mustard (*Brassica juncea* L.). *Ind Crop Prod* 2018;113:111–120.

Singh S, Singh VV, Supriya A, Dubey M, Singh D. Screening and estimation of allelic differentiation in Indian Mustard (*Brassica juncea* L.) genotypes based on SSR Markers for background selection. *Int J Curr Microbiol App Sci* 2017;6(9):2506–2516.

Sisko M, Javornik B, Siftar A, Ivancic A. . Genetic relationships among Slovenian pears assessed by molecular markers. *J Am Soc Hortic Sci* 2009;134:97–108.

Soeda Y, Konings MCJM, Vorst O, van Houwelingen AMM, Stoopen GM, Maliepaard CA, et al. Gene expression programs during Brassica oleracea seed maturation osmopriming, and germination are indicators of progression of the germination process and the stress tolerance level. *Plant Physiol* 2005;137:354–368.

Sonah H, Deshmukh RK, SinghVP, Gupta DK, Singh NK, Sharma TR. Genomic resources in horticultural crops: status, utility and challenges. *Biotechnol Adv* 2011;29:199–209.

Song, ZC, Miao H, Zhang S, Wang Y, Zhang SP, Gu XF. Genetic analysis and QTL mapping of fruit peduncle length in cucumber. *PLoS One* 2016;11:167–845.

Stergiopoulos I, De Kock MJD, Lindhout P, De Wit PJGM. Allelic variation in the effector genes of the tomato pathogen Cladosporium fulvum reveals different modes of adaptive evolution. *Am Phytopathological Soc MPMI* 2007;20:1271–1283.

Su Y, Liu Y, Li Z, Fang Z, Yang L, Zhuang M, et al. QTL Analysis of head splitting resistance in Cabbage (Brassica oleracea L. var.capitata) using SSR and InDel makers based on whole-genome re-sequencing. *PLoS One* 2015;10(9):e0138073. doi:10.1371/journal.pone.0138073

Syombua ED, Zhang Z, Tripathi JN, Ntui VO, Kang M, George OO, Edward NK, Wang K, Yang B, Tripathi L. A CRISPR/Cas9-based genome-editing system for yam (Dioscorea spp.). *Plant Biotechnol J* 2020;19:1–3.

Tang S, Okashah RA, Cordonnier-Pratt MM, Pratt LH, Johnson Virgil Ed, Taylor CA. EST and EST-SSR marker resources for Iris. *BMC Plant Biol* 2009;9:72.

Tanksley SD, Bernatzky R, Lapitan NL, Prince JP. Conservation of gene repertoire but not gene order in pepper and tomato. *Proc Natl Acad Sci USA* 1988;85:6419–6423.

Tester M, Langridge P. Breeding technologies to increase crop production in a changing world. *Science* 2010;327:818–822.

Thimmappaiah WG, Santhosh Shobha D, Melwyn GS. Assessment of genetic diversity in cashew germplasm using RAPD and ISSR markers. *Sci Hortic* 2009;120:411–417.

Tian S, Jiang L, Gao Q, Zhang J, Zong M, Zhang H, Ren Y, Guo S, Gong G, Liu F, Xu Y. Efficient CRISPR/Cas9-based gene knockout in watermelon. *Plant Cell Rep* 2017;36:399–406.

Tivang J, et al. Randomly amplified polymorphic DNA (RAPD) variation among and within artichoke (Cynara scolymus L.) cul-tivars and breeding populations. *J Am Soc Hortic Sci* 1996;121:783–788.

Tripathi JN, et al. CRISPR/Cas9 editing of endogenous banana streak virus inthe B genome of Musa spp. overcomes a major challenge in bananabreeding. *Commun Biol* 2019;2:1–11.

Velasco R, Zharkikh A, Troggio M, Cartwright DA, Cestaro A, Pruss D, et al. High quality draft consensus sequence of the genome of a heterozygous grapevine variety. *PLoS One* 2007; 2:1326.

Vezzulli S, Troggio M, Coppola G, Jermakow A, Cartwright D, Stefanini M, et al. 2008. A functional integrated map for cultivated grapevine (*Vitis vinifera* L.) from three pedigrees, based on 283 SSR and 501 SNP markers. *Theor Appl Genet* 117:499–511.

Wang Q, Wang Y, Qian H, Zhang Z, Zhang L. Evaluation of germplasm and development of markers for resistance to *Plasmodiophora brassicae* in Radish (*Raphanus sativus* L.). *Agronomy* 2022;12:554. https://doi.org/10.3390/agronomy12030554

Wang T, ZhangH, ZhuH. CRISPR technology is revolutionizing theimprovement of tomato and other fruit crops. *Hortic Res* 2019;6:77.

Wang Y, Liu X, Ren C, Zhong GY, Yang L, Li S, et al. Identification of genomic sites for CRISPR/Cas9-based genome editing in the Vitis vinifera genome. *BMC Plant Biol* 2016;16:96.

Wang Y, Wang J, Guo S, Tian S, Zhang J, Ren Y, Li M, Gong G, Zhang H, Xu Y. CRISPR/Cas9-mediated mutagenesis of ClBG1 decreased seed size and promoted seed germination in watermelon. *Hortic Res* 2021;8:70.

Wang Z, Wang S, Li D, Zhang Q, Li L, Zhong C, Liu Y, Huang H. Optimized paired-sgRNA/Cas9 cloning and expression cassette triggers high-efficiency multiplex genome editing in kiwifruit. *Plant Biotechnol J* 2018;16:1424–1433.

Woo JW, Kim J, Kwon SI, Corvalan C, Cho SW, Kim H, et al. DNA-free genome editing in plants with preassembled CRISPR-Cas9 ribonucleoproteins. *Nat Biotechnol* 2015;33:1162–1164.

Xie K, Zhang J, Yang Y. Genome-wide prediction of highly specific guide RNA spacers for the CRISPR-Cas9 mediated genome editing in model plants and major crops. *Mol Plant* 2014;7:923–926.

Xu C, Park SJ, Van Eck J, Lippman ZB. Control of inflorescence architecture in tomato by BTB/POZ transcriptional regulators. *Genes Dev* 2016;30:2048–2061.

Xu M, Korban SS. Positional cloning of the apple scab resistance gene vf. *Acta Hort ISHS* 2003;625:79–87.

Yadav S, Kumar S, Savita K, Singh AK, Singh R. Genomics for Crop Improvement-an Indian perspective. *Biotech Today* 2015;5(1):45–50.

Yadava DK, Parida SK, Dwivedi VK, Varshney A, Ghazi IA, Sujata V, Mohapatra T. Cross-transferability and polymorphic potential of genomic STMS markers of Brassica species. *J Plant Biochem Biotech* 2009;18:29–36.

Yamamoto T, Kimura T, Soejima J, Sanada T, Ban Y, Hayashi T. Identification of quince varieties using SSR markers developed from pear and apple. *Breed Sci* 2004;54(3):239–244.

Yang W, Bai X, Kabelka E, Eatonr C, Kamoun S, van der E, et al. Discovery of single nucleotide polymorphisms in Lycopersicon esculentum by computer aided analysis of expressed sequence tags. *Mol Breed* 2004;14:21–34.

Yu QH, Wang B, Li N, Tang Y, Yang S, Yang T, Xu J, Guo C, Yan P, Wang Q, Asmutola P. CRISPR/Cas9-induced targeted mutagenesis and gene replacement to generate long-shelf-life tomato lines. *Sci Rep* 2017;7:11874.

Zhang G, Sebolt AM, Sooriyapathirana SS, Wang D, Bink MCAM, Olmstead JW, et al. Fruit size QTL analysis of an F1 population derived from a cross between a domesticated sweet cherry cultivar and a wild forest sweet cherry. *Tree Genet Genomes* 2010;6:25–36.

Zhang J, Tao N, Xu Q, Zhou W, Cao H, Xu J, Deng X. Functional characterization of Citrus PSY gene in Hongkong kumquat (Fortunella hindsii Swingle). *Plant Cell Rep* 2009;28(11):1737–1746.

Zhao J, Kulkarni V, Liu N, Del Carpio DP, Bucher J, Bonnema G. BrFLC2 (FLOWERING LOCUS C) as a candidate gene for a vernalization response QTL in Brassica rapa. *J Exp Bot* 2010;61(6):1817–1825.

Zhou J, Wang G, Liu Z. Efficient genome-editing of wild strawberry genes, vector development, and validation. *Plant Biotechnol J* 2018;16:1868–1877.

Zhou X, Van Eck J, Li L. Use of the cauliflower Or gene for improving crop nutritional quality. *Biotechnol Annu Rev* 2008;14:171–190.

Zhu H, Choi HK, Cook DR, Shoemaker RC. Bridging model and crop legumes through comparative genomics. *Plant Physiol* 2005;137(4):1189–1196.

2

Apple Breeding Through Biotechnological Interventions

Anshul Kumar, Kamlesh Verma, Abhilash Padhan, Pankaj Kumar and Dinesh Kumar Srivastava

2.1 Introduction

Apple (*Malus* × *domestica* Borkh.) is an important member of the *Rosaceae* family. It is one of the most consumed fruits globally. Apple is an allopolyploid ($2n = 2x = 34$), heterozygous, cross-pollinated and self-incompatible horticultural crop. It has been cultivated from the ancient past around 8000 to 2000 BCE and originated from Central Asia (Forsline et al., 2003), including Kazakhstan, Kyrgyzstan and Tajikistan (Dzhangaliev, 2003). Botanically it is a pome fruit. It is a small to medium (2–10 m) deciduous single-trunk tree with purplish-brown densely hairy buds. The leaves are dark green and alternately arranged, and the flowers have five sepals and five petals that range from 3 to 4 cm in diameter and are white to pink in color (Kumar et al., 2022a). Apple is a popular fruit with many health benefits. It has been shown to minimize the risk of lung cancer, asthma, diabetes, thrombotic stroke, obesity, cardiovascular disease and proliferation activities (Kumar et al., 2022a). It has a high nutritional value in terms of water content, which is above 80%, organic acids (0.2–0.8%), sugars, vitamins (primarily vit. C 2.3–31.1 mg/100 g DM), 0.34%–1.23% minerals, dietary fibers and pectin (Kiczorowska and Kiczorowski, 2005; Musacchi and Serra, 2018). Furthermore, they are known to have high antioxidant polyphenol content and contain phenolic compounds, such as epicatechin, phloretin, chlorogenic acid, procyanidins and quercetin, which depends on variety, stage of maturation, environment conditions, cultural practices and storage time (Pissard et al., 2013).

Apple needs extended periods of chilling to break bud dormancy, which is problematic in this era of global warming and climate change. Plant breeders are primarily concerned with breeding for fruit quality and horticulturally important key attributes. However, conventional apple breeding or genetic improvement is a difficult and time-consuming procedure as apple has a long juvenile phase, perennial nature, heterozygosity and sexual

DOI: 10.1201/9781003625674-2

incompatibilities (Kumar et al., 2022a; Kumar et al., 2022b). Applications of advanced biotechnological tools have fastened apple breeding for different traits. The popular biotechnological strategies for crop improvement of apple include tissue culture, genetic engineering, molecular breeding and several omic approaches. In order to accelerate the rate of crop improvement, Watson et al. (2018) recently concentrated on integrating speed breeding with other modern crop-breeding methods, such as genome editing, genomic selection and high-throughput genotyping. Gene pyramiding, marker-assisted backcrossing, mapping for linked targeted traits by particular genes or QTLs, fine mapping of the QTL region, etc., are all examples of MAS (Adlak et al., 2019). To screen for genetic diversity, a variety of DNA markers, including RAPD, SSR and RFLP are available (Sharma and Modgil, 2020). The CRISPR-associated system (Cas) provides appealing DNA/RNA editing vectors for plant virus resistance. These methods offer enormous versatility in viral or host gene editing based on the RNA or DNA genome (Pandey et al., 2023).

2.2 Various Biotechnological Approaches Used for Apple Crop Improvement

2.2.1 Plant Tissue Culture and Genetic Stability Studies

In vitro tissue culture is a technique of culturing plant cells, tissues or organs on nutrient media under controlled conditions to produce a large number of plantlets that are genetically identical to the mother plant in a relatively short period of time. Since the beginning of apple tissue culture, it has been used to genetically improve cultivars (scions) and rootstocks and is employed for the quick production of plants. By developing improved procedures, it is feasible to provide disease-free propagation material for growers, novel cultivars with desirable features, and rapid multiplication of superior cultivars (Kabylbekova et al., 2019). Totipotency is a unique characteristic of plant cells that allows them to retain the latent ability to develop an entire plant (Reinert and Backs, 1968; Vasil and Vasil, 1972). Micropropagation consists of four different stages, i.e., stage 1 – establishment of *in vitro* cultures from *ex vitro* plants, stage 2 – shoot multiplication, stage 3 – root regeneration in microshoots and stage 4 – hardening and acclimatization of *in vitro* plantlets. A high ability to regenerate shoots is a crucial pre-requisite for developing an effective transformation system. Tis108, a strigolactone-biosynthesis inhibitor, was successfully employed to promote the shoot regeneration of woody plants. The *MdCCD7* strigolactone-biosynthesis gene could be a target gene for biotechnological enhancements in shoot regeneration capacity (Asghar et al., 2022). A rise in synthesized IAA concentration increased callus development at the base of shoots. Acclimatization of "Marubakaido" shoots rooted *in vitro* was effective. Both IAA concentration and source had an effect on plantlet survival during acclimation. Plant survival reduced from 96% to 66% during the acclimation stage as the concentration of synthetic IAA increased. "Marubakaido" inoculated on MS media enriched with bacterial IAA produced thinner roots with no callus formation, leading to greater (up to 98%) acclimation survival (Souza et al., 2022). Plants grown from *in vitro* propagation of apples may exhibit genetic or somaclonal variation (Modgil et al., 2017). Thus, it is essential to use molecular markers to evaluate the clonal uniformity early in the micropropagation process and afterwards, at regular intervals of subsequent subcultures (Modgil et al., 2017). Using ISSR markers to examine genetic stability in the apple rootstock MM106 and the genetic instability was discovered in tissue culture plants when they were multiplied for an extended period of time (Pathak and Dhawan, 2011). In contrast, using the ISSR marker, apple rootstock MM111 and Merton 793 showed no genetic variation between *in vitro* cultures (Pathak and Dhawan, 2010, 2012). Using SSR markers, Jin et al. (2014) studied the genetic fidelity of regenerated plantlets of apple rootstock "Pingyitiancha" and produced genetically uniform plants as a result. *In vitro* culture of rootstock M7 was examined using RAPD primers by Modgil et al. (2017), who observed that there was no difference in the banding pattern.

2.2.2 Double Haploids

Fruit tree crops cannot produce genetically pure lines via selfing because of their prolonged reproductive cycles and frequent self-incompatibility. They are propagated vegetatively; hence homozygosity is not necessary for the propagation. Moreover, the problem of fruit crop genomes with large levels of heterozygosity can be resolved using DH lines, speeding the development of homozygous lines and removing a barrier to accurate genetic analysis of F1 populations due to the range of alleles to evaluate (Kunihisa et al., 2019). In order to induce ploidy level, zygosity state utilizing isozyme, flower, tree and fruit quality, as well as SSR analysis, Hofer et al. (2008) employed *in vitro* androgenesis and *in situ* parthenogenesis followed by embryo culture. They found that using homozygous lines is not a practical strategy due to low efficiency in *in vitro* androgenesis and *in situ* parthenogenesis, as well as low fertility. For the construction of the genetic map of "Prima," Kunihisa et al. (2019) used double haploid lines for QTL analyses using the F1 population from "Prima" "Apple Chukanbohon 95P6." They concluded that apple fruit skin splitting has recessive inheritance and that the allele for splitting is closely associated with high acidity and low juice browning in "Prima."

2.2.3 Virus Elimination

Viruses in apple can be eliminated successfully through *in vitro* culture. In order to eradicate viruses, a variety of techniques are used, such as thermotherapy (Hu et al., 2015; Wang et al., 2018), cryotherapy (Li et al., 2016), chemotherapy (Paprštein et al., 2013; Hu et al., 2015), meristem culture or shoot tip culture (Wang et al., 2016; Vivek and Modgil, 2018), combined thermotherapy and cryotherapy (Zhao et al., 2018), combined dose-dense (dd) thermotherapy and chemotherapy (Hu et al., 2015) and thermotherapy and combined meristem culture (Vivek and Modgil, 2018). High virus eradication efficiency of 95% in apple cv. "Xinhongjiangjun" infected with apple chlorotic leaf spot virus (ACLSV), apple stem pitting virus (ASPV) and apple stem grooving virus (ASGV) was achieved by Hu et al. (2015) by combining thermotherapy with chemotherapy. They found that ACLSV and ASGV were more difficult to eradicate than ASPV. Using shoot tips from adventitious shoots grown from apple cv. "Gala" leaf explants, Wang et al. (2016) completely eradicated the ASPV virus. Vivek and Modgil (2018) successfully eliminated a virus from an apple cv. Oregon Spur-II infected with ACLSV, ApMV, ASGV and ASPV viruses using a combination of thermotherapy and meristem culture. By combining cryotherapy and thermotherapy, Zhao et al. (2018) successfully eradicated the apple stem grooving virus (ASGV) virus from *in vitro* apple plants of the "Gala" variety.

2.2.4 Abiotic and Biotic Stress Tolerance

Apple *in vitro* culture has been used to evaluate its tolerance to biotic and abiotic stress. *Vfa 1, 2,* and *4* genes were used to alter the apple variety "McIntosh" and transgenic lines expressing *Vfa 1* and *2* displayed greater resistance to *Venturia inaequalis*, whereas plants expressing the *Vfa 4* gene were shown to be susceptible (Malnoy et al., 2005). According to Bahmani et al. (2012), a low NaCl concentration (20 mM) significantly increased shoot length and fresh weight, while a high salt concentration (above 100 mM NaCl) significantly inhibited the growth of *in vitro* MM106 shoots and reduced the number of regenerated shoots, rooting percentage and root number. Nucleotide-binding/leucine-rich-repeat (NLR) gene, *MdNLR16* gene overexpression in apple "Greensleeves" antisense lines resulted in improved resistance to *A. alternata* (Meng et al., 2018). The *Md-miRln20* gene was down-regulated, which decreased the up-regulation of *Md-TN1-GLS* and disease incidence against glomerella leaf spots in apple (Zhang et al., 2019a). The calmodulin gene of *Marssonina coronaria* displayed partial/complete resistance to Marssonina blotch in the apple cultivar "Red Chief," using an RNAi construct (Verma et al., 2021).

2.2.5 Genome Editing

Zinc-finger nucleases, transcription activator-like effector nucleases (TALEN), and the clustered regulatory interspaced short palindromic repeats/CRISPR-associated protein (Cas) system (CRISPR/Cas) are some of the tools available for editing the genome (Chen and Gao, 2014). Because of its easy construction and high specific efficiency, CRISPR/Cas9 has emerged as a strong technology for genome editing when compared to alternative methods. In a number of crop species, including apple, CRISPR-Cas9 genome-editing technologies have been successfully used. For a single gRNA, Nishitani et al. (2016) used the CRISPR-Cas9 system to knockout the phytoene desaturase (*PDS*) gene in apple, resulting in clear and partial albino phenotypes in 31.8% of regenerated plantlets. In the *Rosaceae* family, this is the first report of genome editing. The CRISPR-Cas9 genome-editing technique was used by Osakabe et al. (2018) to accomplish DNA-free targeted mutations in apple and grapevine using both direct delivery of CRISPR-Cas9 ribonuclease proteins (RNPs) and plasmid-mediated genome editing. Using *A. tumefaciens* genetic transformation, Pompili et al. (2020) used the CRISPR/Cas9-FLP/FRT gene-editing technique for the first time to achieve 75% editing efficiency in two susceptible apple cultivars for *MdDIPM4* knockout. It is a very effective site-directed mutagenesis approach that is DNA-free. For Xinjiang wild apple genome, Zhang et al. (2021) used the CRISPR-Cas9 genome-editing technique. Compared to a few buds with edited genes, they were able to easily obtain edited calli. Using CRISPR-Cas9 single-guide RNA (sgRNA) targeted trihydroxynaphthalene reductase (*THN*) gene of *Venturia inaequalis*, Rocafort et al. (2022) produced six of the 36 stable transformants of apple exhibiting a light-brown phenotype, indicating 16.7% gene inactivation efficiency. They also reported high-throughput and low-cost methods by providing a high-resolution melting (HRM) curve procedure for the quick identification of CRISPR-Cas9 gene-editing mutant of this fungus. Several tropical fruit varieties with improved disease resistance, fruit quality, productivity and shelf life have been produced using CRISPR-Cas9 technology and recently it has emerged as a revolutionary tool for improving sustainable agriculture (Irfan et al., 2023). The growing popularity of plant gene editing has sparked research into consumer perspectives as gene-edited food are frequently better acceptable than genetically engineered food (Strobbe et al., 2023).

2.2.6 Marker-Assisted Selection

A new field of molecular breeding known as "marker-assisted selection" uses linked DNA markers to evaluate traits based on their banding patterns. The three major classifications of markers are morphological markers, biochemical markers (proteins, isoenzymes, etc.) and molecular markers. Morphological markers can be scored visually, which typically include qualitative traits. Sturtevant (1913) created a genetic linkage map for *Drosophila melanogaster* with six morphological traits. On the other hand, biochemical markers are based on variations in particular banding patterns of proteins separated in native/SDS PAGE. The use of isozymes as genetic markers for analyzing genetic variation is restricted by the developmentally stage-dependent expression of isozyme loci. Molecular markers are pieces of DNA that encode particular traits and their inheritance could be easily detected. They are divided into hybridization-based types or non-PCR-based markers (RFLP), PCR-based (RAPD, SSR, AFLP, SCAR, ISSR, CAPS), and sequence-based markers (SNP) (Gupta et al., 1999; Ramesh et al., 2020). First, a non-PCR-based or hybridization-based marker, i.e., restriction fragment length polymorphism (RFLP), is used for human linkage mapping (Botstein et al. 1980). Restriction fragment length polymorphism (RFLP) (Botstein et al., 1980), amplified fragment length polymorphism (AFLP) (Zabeau and Vos, 1993), randomly amplified polymorphic DNA (RAPD) (Williams et al., 1990), expressed sequence tags (ESTs) (Adams et al., 1991), simple sequence repeats (SSRs) or microsatellites or variable number tandem repeats (VNTR) (Weber and May, 1989; Litt and Lutty, 1989), inter-simple sequences repeats (ISSR) (Zietkiewicz et al., 1994), short tandem repeats (Edwards et al., 1991) and simple sequence length polymorphism (SSLPs) (Tautz, 1989), sequence-characterized amplified regions (SCARs) derived from RAPDs (Paran and Michelmore, 1993) or AFLPs (Ashutosh et al., 2007) and single nucleotide polymorphisms (SNPs) (Taillon-Miller et al., 1998; Picoult-Newberg et al., 1999) are some of the DNA markers that have been developed and are used for diversity studies in plants (Joshi et al., 1999, Adlak et al., 2019).

Conventional breeding strategies and crop improvement programs have been revolutionized by next-generation sequencing (NGS) technologies (Kilian and Graner, 2012; Ramesh et al., 2020). In plant-breeding programs, particularly for fruit crops, marker-assisted selection (MAS) is a useful method. Molecular markers are used for gene pyramiding, QTL mapping, marker-assisted foreground and background selection, gene tagging, association mapping, fine mapping, TILLING, and eco-TILLING, etc. In addition to these, restriction site-associated DNA (RAD) sequencing, genotyping by sequencing (GBS), single-cell-based whole-genome sequencing (SCWGS), whole-genome resequencing (WGRS) techniques, ddRAD sequencing *de novo* sequencing, transcriptome and metagenome analysis are some of the methods that have been widely employed for crop improvement (Varshney et al., 2009). An apple reference genome and thousands of SNPs were produced by next-generation sequencing (NGS) with the price of NGS is falling continuously (Velasco et al., 2010; Bianco et al., 2014; Bianco et al., 2016; Daccord et al., 2017). Using next-generation sequencing technology, the whole chloroplast genomes of 24 different apple types were retrieved and compared with the genomes of the wild species. The practical application of MAS in breeding programs for perennial plants, such as apples, is still challenging (Kenis et al., 2008; Migicovsky et al., 2016; Peace, 2017; Jia, 2018).

Forecasting the probability of storage disorder development using biomarkers at or before harvest is in its early stages and necessitates a much better understanding of the maturity, environmental and stress factors involved (Gapper et al., 2023). On apple chromosome 09, the main quantitative trait loci (QTLs) related to fruit cover color have been routinely mapped using either pedigree-based or genome-wide association studies (GWAS) (Chagné et al., 2016; McClure et al., 2018). In addition, genomic selection (GS), is a recently established method of genome-assisted breeding that varies from conventional molecular marker-assisted selection method in that it estimates the genetic impact of whole-genome markers (Muranty et al., 2015; McClure et al., 2018). Although GS only examines the effects of genome-wide markers and excludes the importance of individual markers, numerous efforts have been undertaken to date for QTL mapping and the development of molecular markers for a variety of traits of cultivated plants (Werner et al., 2018). Yet, when QTL-based markers were added to the GS marker panel or used as fixed markers, the prediction accuracy significantly improved (Minamikawa et al., 2017; Werner et al., 2018).

Zhang et al. (2019b) performed a comparative genomic analysis with the available apple genomes and discovered that a long terminal repeat (LTR) retrotransposon insertion upstream of the *MdMYB1* gene, a key transcriptional activator of anthocyanin production, is connected with the red-skinned phenotype. In addition to this, pedigree-based quantitative trait locus (QTL) mapping and genome-wide association studies are alternative techniques for mapping both qualitative and quantitative features (Hamblin et al., 2011; Ye et al., 2017; Jia, 2018). Recently, the next-generation sequencing (NGS)-based QTL mapping technique known as bulk segregant analysis (BSA)-seq, which isolates and genotypes two phenotypically different subpopulations (bulks) (BSA extremity bulks) of recombinant progeny (segregants), has been successful in quickly mapping genes (Takagi et al., 2013; Huo et al., 2016; Dougherty et al., 2018). Exploitation of data from high-throughput whole-genome sequencing (WGS) and transcriptomes (WTS), this method has had a revolutionary influence. It is hoped that it would aid in the development of new biotechnological solutions in molecular diagnostics as well as the biocontrol of *Neofabraea* spp. in apples, which causes postharvest bull's eye rot (Oszust et al., 2023). Jia (2018) used MapQTL and BSA-seq (bulk segregant analysis utilizing next-generation sequencing) with two separate pedigree-based populations to identify four key genome-wide QTLs responsible for apple fruit acidity. They discovered that qtl08.2 was identified as a significant QTL in Jonathan's 9 "Golden Delicious," but Zisai Pearl' 9 "Red Fuji" did not contain it. Cleaved amplified polymorphic sequence (CAPS) markers were developed to detect single nucleotide polymorphisms (SNPs) on the *MdMYB1* gene linked to anthocyanin synthesis (Takos et al., 2006; Chagné et al., 2016). In order to identify the recently discovered sugar transporters known as SWEETs, simple sequence repeat (SSR) or cleaved amplified polymorphism sequence (CAPS) markers were developed. These markers also suggest that the duplication of SWEET genes is connected to whole-genome duplication (WGD) (Zhen et al., 2018).

2.3 Various Omics Approaches in Apple Improvement

2.3.1 Genomics

Throughout the course of the past two decades, "omics" and advanced biotechnological technologies have enabled the decoding of crop plant genomes, the assigning of roles to previously unrecognized genes, and the development of genome-wide DNA markers (Bashir et al., 2023). The Malus genome measures 742.3 Mbp/1C, which is comparable to the length of the tomato genome (Shulaev et al., 2008). Several reference maps of the apple genome are displayed in the Genome Database of Rosaceae and the database developed from Europen Malus projects like HiDRAS, allowing users to focus on specific regions of interest or align maps for comparison (Jung et al., 2008). The first *Malus × domestica* physical maps were created using BAC-libraries made for "florina" and for M. *floribunda* 821, which were based on partial sequencing of areas influencing the resistance to apple scab (Patocchi et al., 1999a, 1999b). The efforts of MAS in apple breeding have been significantly boosted by the publication of an apple reference genome (Velasco et al., 2010). SNP discovery utilizing the next-generation sequencing genotyping-by-sequencing (GBS) methodology and SNP arrays (Bianco et al., 2014, 2016) are two methods for producing genetic marker data (Elshire et al., 2011). The apple REFPOP, is an apple reference population, a large collection of 534 genotypes planted in six European countries, a unique tool to accelerate the breeding of apples. The suitability of the apple REFPOP design was confirmed for genomics-assisted breeding using related germplasm, which plays an advantageous role in collaboration among the countries for customized apple-breeding methods in the current era of genomics (Jung et al., 2020). The online database Apple *MDO* was created for cross-referencing and study of potential novel functions of apple genes through collaborative analysis of various omics datasets. Additionally, data on gene annotation and toolkits for functional support were offered. The database may make it simple for researchers to learn more about how genes associated with significant agronomic qualities function and may be used as a resource for other fruit plants (Da et al., 2019).

Apple possesses two FT-like genes (*MdFT1* and *MdFT2*) and on two different linkage groups (LGs) with partial homology, *LG 12* and *LG 4*, *MdFT1* and *MdFT2* were mapped, respectively. The research could be used to pinpoint the molecular basis for flowering and create strategies for reducing the juvenile stage in a variety of fruit crops, including apple (Kotoda et al., 2010). A high-quality draft genome sequence of the domesticated apple was reported, which showed that a relatively recent genome-wide duplication (GWD) resulted in the transition from nine ancestral chromosomes to 17 chromosomes in the Pyreae (Velasco et al., 2010). The genome version 2.0 of the diploid Golden Delicious is now used as a model for plant genome alignment and functional gene annotation (Velasco et al., 2010).

2.3.2 Transcriptomics

Transcriptomic analysis using RNA-seq technology was conducted in red and white fleshed apples and discovered a transcription factor belonging to the WRKY family known as *MdWRKY11* and the expression of *F3H, FLS, DFR, ANS,* and *UFGT* were encouraged by the overexpression of *MdWRKY11*, which also led to an increase in the accumulation of flavonoids and anthocyanin (Wang et al., 2018). The research examined the unique function of MdWRKY11 in flavonoid biosynthesis and suggested several additional genes that may also play a role. To breed superior red-fleshed apples, this information on flavonoid synthesis is very useful (Wang et al., 2018). Using the Illumina (Hi Seq) platform, a comparative transcriptome analysis was conducted to mine new scab resistance genes in three scab resistant (Florina, Prima and White Dotted Red) and three susceptible (Ambri, Vista Bell and Red Delicious) apple genotypes resulting in the identification of 822 differentially expressed genes. Such findings provide the framework for further research into the genetics of apple scab resistance and the defense pathways that may be responsible for the apple's scab resistance to *V. inaequalis* (Masoodi et al., 2022). A bud sport was identified from apple cv. Oregon Spur II showing reduced red coloration. Using it as plant material, a high-quality haplotype reference genome consisting of 649.61 Mb sequences with a N50 value of 2.04 Mb was assembled. As per the transcriptome studies, anthocyanin content was found to be highly linked with 69 differently expressed genes between the original type and mutant fruit skin (Liu et al., 2022). The study conducted on the green skin color apple variety "white winter pearmain," which was treated with ultraviolet-B (UV-B) exhibited red skin color and marked anthocyanin accumulation. To confirm the variation in gene expression in green apples, transcriptomic analysis revealed 37,110 and 37,709 differentially expressed genes, respectively, as well as 382 and 475 transcription factors (TFs) (Ding et al., 2021). With a focus on the earliest stages of fruit initiation, Li et al. (2022) have used comparative transcriptomics to investigate the molecular events that correlate with the development of fleshy pome (apple) and drupe (peach) fruit. Regardless of the type of fruit, the team discovered the MADS-box genes *PI* and *TM6*, whose expression negatively correlates with fruit flesh-forming tissues. The work offers significant insights into fruit type specification and draws on the vast transcriptomic datasets at the early stages of pome and drupe fruit development (Li et al., 2022).

2.3.3 Proteomics

To separate the total proteins from apple fruit at various phases of ripening and senescence, a total of 2340 sites were subjected to imaging and statistical analysis, and 316 spots, or roughly 13.5% of the total protein population were observed to have undergone considerable alteration (Zheng et al., 2013) and in response to ethylene treatment among the 316 proteins, 219 spots were only visible at a certain stage of fruit ripening and 97 spots differed substantially across the course of fruit ripening. Further investigation and protein identification revealed that ethylene synthesis, antioxidation and redox, glucose metabolism, oxidative stress, energy production as well as defense response are only a few of the proteins whose abundance increases as apple fruit ripens (Zheng et al., 2013). Several pathogenesis-related (PR) proteins including beta 1,3-glucanase, ascorbate peroxidase (APX), glutathione peroxidase (GPX), and *mald1* were found in susceptible as well as resistant hosts against *A. alternata* (Zhang et al., 2015). A bud sport mutant with red skin as well as increased anthocyanin concentration was identified in apple var. "Fuji." A total of 451 differently expressed proteins were identified in apple skin as a result of quantitative SWATH-MS (sequential window acquisition of all theoretical spectra mass spectrometry) proteomics investigations. These investigations indicated proteome modifications in apple red skin bud mutants (Chen et al., 2019). To examine the apple defense responses 48 hours after inoculation, infection of *C. gloeosporioides* was done on the leaves of apple varieties "Fuji" (resistant) and Gala (sensitive) and led to the identification of 42 distinct proteins expressed at 24 and 48 hours after inoculation. The majority of these proteins were found to be directly connected to defense responses (Rockenbach et al., 2015). Parallel reaction monitoring (PRM) technology was utilized for differential proteomic analysis in *P. expansum* infected and uninfected apple samples and 343 differentially expressed proteins were identified, most of which were connected to stress responses, oxidative phosphorylation, metal ion binding and defensive responses (Xu et al., 2022). Fruit organoleptic quality is mostly determined by soluble sugars and organic acids. A total of 13,036 unique peptides and 1079 differentially expressed proteins were identified by Ma et al. (2019) on the fruits of two apple species, *M. sargentii* and *M. niedzwetzkyana*, which showed significant differences in soluble sugar and organic acid contents of fruits. Zhang et al. (2017) found that the blue mold degradation of apples caused by *Penicillium expansum* was greatly prevented by *Yarrowia lipolytica*. As per proteomic and transcriptomic analysis, a total of 35 proteins were found to be up- and down-regulated in response to the *Y. lipolytica* stimulation. Results showed that the activities of polyphenol oxidase, peroxidase, chitinase and L-phenylalanine ammonia-lyase involved in increasing defense responses in apple fruit tissue were enhanced by *Y. lipolytica*. Cao et al. (2008) conducted a study to identify the mechanism of apple flower bud differentiation, shorten juvenile period and accelerate its breeding process. When compared to the leaf bud, the expression of 283 protein locations both in terms of number and quality have changed significantly in the flower bud. Among the 283 protein spots, four (16.4, 30.2, 40.3 and 65.1 kD) protein spots of the flower bud in the archae stage at the start of flower bud differentiation, three (39.3, 60.2 and 66.3 kD) in the post-stage and one (77.1 kD) in the sepal stage on 2-DE maps were indicative (Cao et al., 2008).

2.3.4 Metabolomics

A metabolomic analysis in a study links individual polyphenolic compounds in apple pulp with quality traits like antioxidant activity and the presence of phenolic compounds and anthocyanins in apple skin (Ceci et al., 2021). By using spectrometric assays, the antioxidant activity, phenolic content and anthocyanins on the sunny and shaded sides of apple skin were assessed. The measured parameters in the older apple types were greater than in comparison to the red-fleshed and commercial cultivars. These findings demonstrated the potential to identify cultivars with attractive health characteristics based on their level of polyphenols appropriate for breeding or commercialization (Ceci et al., 2021). The study conducted on the green skin color apple variety "white winter pearmain," which was treated with ultraviolet-B (UV-B) exhibited red skin color and marked anthocyanin accumulation. To confirm the variation in metabolites, metabolic analysis showed that the visible light and UV-B treated green apples had 152 and 178 significantly altered metabolites compared to the control respectively with flavone, flavanol and anthocyanin being the most significantly enhanced (Ding et al., 2021). A comprehensive proteome research report of apple fruit growth over five different growth stages starting from the juvenile period to maturity along with metabolomic profiling led to the identification and quantification of 7098 and 6247 proteins, respectively using a tandem mass tag (TMT) based comparative proteomics technique. This extensive proteome dataset offers a worldwide perspective of the crucial metabolic and developmental pathways involved in fruit growth (Li et al., 2016). Commisso et al. (2021) using metabolic analysis (NMR and HPLC-DAD/MS), profiled the metabolome of six different apple cultivars and discovered that the ancient Italian apple "Pom Prussian" was enriched with chlorogenic acid as well as more distinctive polyphenols (Phloretin derivatives) with high *in vitro* antioxidant activity. They also observed the cultivar-specific accumulation of sugars, amino acids, malic acid and various polyphenols, which was also related to the growing season for some cultivars (Commisso et al., 2021). Superficial scald is a physiological condition of apple fruit characterized by sunken, necrotic lesions that after prolonged cold storage. Together with the oxidation of α-farnesene and eventually, the generation of methanol and the emergence of symptoms, overall metabolic and transcriptome changes that represented numerous pathways and processes took place. According to Gapper et al. (2017), rather than being caused by physical injury, this type of peel necrosis is the result of an active metabolic change involving several pathways that are sparked by cooling temperatures at the beginning of cold storage. Among many other pathways, the peel that is exhibiting scald symptoms akin to damage brought on by mechanical stress and herbivory in other plants has elevated methanol and methyl ester levels as well as increased pectin methylesterases (Gapper et al., 2017).

Zhang et al. (2023) investigated the impact of 1-methylcyclopropene (1-MCP) on apple fruit quality under prolonged cold storage using integrative metabolomics and transcriptomic studies. The findings demonstrated that 1-MCP (1 l/l) treatment could preserve fruit apparent quality (such as external color and firmness), inhibit the growth of rot rate and soluble solids content/titrable acid ratio, reduce ethylene release and respiratory intensity during cold storage and increase shelf life. Moreover, 1-MCP demonstrated long-lasting effects on the development of numerous fruit attributes associated with metabolite and gene expression. The use of 1-MCP altered the glycolysis and tricarboxylic acid cycle pathway genes, which are involved in metabolism at the early stage of storage. An improved understanding of how fruit quality is formed can be obtained by examining metabolic changes that occur during apple fruit development. In order to evaluate the profound metabolic changes occurring during the development and ripening of the "Pinova" apple, Xu et al. (2020) used a broadly targeted LC-MS-based metabolomics technique with large-scale detection, identification and quantification. The findings demonstrated that throughout the four stages of fruit growth, organic acid, including intermediates of the tricarboxylic acid cycle (TCA) exhibited a clear lowering tendency, whereas the majority of sugars showed increasing accumulation levels. To better understand the biochemical mechanisms underlying internal browning by examining the effects of pre-harvest applications of calcium, potassium and triazole fungicides on the post-harvest primary metabolites of "Braeburn" fruit a study was carried out by Hatoum et al. (2014). They found that the fruit began to brown early on during storage and the severity increased with time. This was associated with a group of main metabolites whose concentrations changed with time, either increasing (e.g., alanine, galactose, mannitol, sorbitol, xylose) or decreasing (e.g., malate, sucrose). By one week of the start of storage, there was a metabolomic difference between untreated controls and fruit treated with DPA or 1-MCP and after 4 weeks of storage, there was a metabolic divergence between controls and DPA-treated fruit and this occurred 2 months before scald symptoms appeared (Rudell et al., 2009). These findings showed that considerable metabolomic changes related to scald occur before the onset of symptoms. Bilbrey et al. (2021) have created a high-throughput integration technique for genomic and multiplatform metabolomics data in order to comprehend genotype-phytochemical connections in apple. This platform showed the value of integrating multiple genomics to promote data-driven, phytochemical-based plant breeding. Wang et al. (2021) reported that bagging of fruit altered the microenvironment of the fruit, causing cell structure to be destroyed, flavonoids to decrease and triterpenoids to increase, which are the main causes of the browning of the peel. Metabolites directly correlate with quality traits like antioxidant activity, total phenolics and total anthocyanins, which are crucial criteria in the breeding (Cuthbertson et al., 2012).

2.3.5 Epigenetics

DNA methylation is one of the most important epigenetic alternations and to describe the degree and pattern of DNA methylation at the two various ploidies, a diploid apple cultivar, and its derived autotetraploid, polyploidization probably had little impact on the overall level of DNA methylation. However, in terms of the methylation pattern, the tetraploid had a higher frequency of hemi-methylated sites (He et al., 2017). Daccord et al. (2017) have created a high-quality *de novo* assembly of the apple genome using the most recent sequencing and optical mapping technology and genome-wide DNA methylation suggests that epigenetic markers may have a role in the formation of agronomically important traits in apple. Cho et al. (2020) conducted a study on the apple cultivar "Fuji" and its mutant "Beni Shogun" with enhanced skin color and characterized at five developmental stages. As per the data from L^* and a^*/b^* values, the blushed skin of "Beni Shogun" fruits is linked to a higher anthocyanin accumulation than the striped skin of "Fuji" fruits. The expression of structural genes encoding enzymes involved in the anthocyanin biosynthesis pathway is activated by a higher level of *MdMYB10* transcript in "Beni Shogun" (Cho et al., 2020). To examine the various methylation patterns associated with anthocyanin concentrations, Jiang et al. (2019) used RNA-Seq and bisulfite sequencing to analyze three deep-red apple mutants and their lighter-skinned parents and discovered that there is no relationship between DNA methylation and gene expression, and increased expression of ANS (MD06G1071600) and F3H (MD05G1074200) genes in apple mutants led to higher anthocyanin concentrations.

Histone lysine methylation is crucial for plant development and stress responses. A class of SET-domain group proteins catalyzes the methylation of the histone lysine residue (SDGs). A total of 67 SDG members were discovered in the apple genome and most of the Md SDG duplicated gene pairs were connected to a recent apple genome duplication event (Li et al., 2021). In a study, the team of researchers mapped cytosine methylation in the commercial apple genome at a single base resolution and examined alternations in methylation patterns linked to water deficit in representative cultivars that are both drought-sensitive and tolerant. Xu et al. (2018) discovered that the methylation levels at the *CG*, *CHG* and *CHH* sequence contexts in the apple genome are 54%, 38% and 8.5%, respectively and these findings will be useful for comprehending potential relationships between DNA methylation and gene expression in plants that are exposed to abiotic and biotic challenges while developing in natural habitats (Xu et al., 2018). An experiment was conducted on deep red apple mutants and their lighter-skinned parents to explore the various changes in methylation patterns associated with anthocyanin concentrations using bisulfite sequencing (BS-Seq) and RNA-sequencing (RNA-Seq.). From this experiment, it was discovered that 13,405, 13,384 and 10,925 differentially methylated regions (DMRs) and 1987, 956 and 1180 differentially expressed genes (DEGs) were present, respectively. In apple mutants, several genes had an increased expression, and variations in the methylation patterns of their promoters were seen (Jiang et al., 2019).

Perrin et al. (2020) chose apple as a model plant to examine the transmission of DNA methylation patterns in a perennial plant. The initial inquiry compared the DNA methylation patterns of mature trees to young seedlings produced through selfing to explore the inheritance of DNA methylation patterns during sexual reproduction in apples. They discovered that the majority of DNA methylation patterns from the fully developed donor tree are transmitted to the newly grafted plants. Epigenomic and transcriptomic data demonstrated that grafted plants are at a transitional stage between an adult tree and a seedling, inheriting some of the donor tree's epigenomic history (Perrin et al., 2020). Many variations in the "Fuji" apple line share the same genetic background and consistently trigger conditions that led to epigenetic occurrence, making it a prime target for epigenetic study. Du et al. (2020) utilized 91 mutants of "Fuji" and the findings revealed that the average worldwide genomic DNA methylation level of the 91 bud mutants of "Fuji" ranged from 29.12% to 45.048%, with an average of 35.91%. To investigate a way of non-different polyphenol oxidase (PPO) browning, multi-omics (methylomics, transcriptomics and proteomics) methodologies were used for the global profiles of DNA methylation and gene and protein expression. Wang et al. (2022) selected two apple varieties with natural bud mutations and discovered an association between methylation (*MdCMT3* and *MdCMT3c*) and browning index (BI) as well as a negative correlation between BI and demethylation (*MdROS1* and *MdDME*). The 5-azacytidine DNA methylation inhibitor can stop apples from becoming brown. Further research revealed that *OMT1* and methylated-*NCA1* dramatically enhanced during apple browning. Methylated-NCA1 might suppress the production of the *NCA1* gene, which would then decrease catalyze activity and significantly increase apple browning. These results provide new information on the landscape and mechanism through which DNA hypermethylation significantly increased browning in "Fuji" apples (Wang et al., 2022). Although paper-bagging is an efficient method to maximize apple skin color, it has not been shown to have any effect on fruit skin pattern. Cho et al. (2022) observed that increased anthocyanin concentration in the cultivar "Fuji" fruit's red-stripped skin turned reddish blushed after being exposed to light again was connected with higher *MdMYB10* expression and it may be governed by DNA methyltransferases like *MdMET*, *MdCMT* and *MdDRM* having their expression down-regulated. Histone deacetylase enzymes have a role in the control of a variety of developmental processes in plants. Hu et al. (2022) found possible downstream target genes controlled by *MdHDA19* and directly inhibits *MdACS3a* expression by promoting *H3K9* deacetylation and forming a complex with *MdERF4-MdTPL4* to reduce the level of acetylation, which has an impact on ethylene synthesis. These findings showed that the *MdERF4-MdTPL-MdHDA19* repressor complex is involved in the epigenetic control of apple fruit ripening. As anthocyanin level in apple skin influences its red coloration. *MdMYB90*-like expression was 21.3 times greater in apple bud sport mutations and reported reduced levels of DNA methylation, which may account for the gene's increased expression and the mutant apple skin's higher anthocyanin content (Sun et al., 2021). According to genome-wide DNA methylation data, epigenetic markings may contribute to agronomically important traits (Daccord et al., 2017).

2.4 Future Perspective and Conclusion

Apple holds significant economic importance within the fruit industry, and breeders across the globe are consistently engaged in the development of novel cultivars characterized by enhanced horticultural traits. However, the long juvenile phase and substantial spatial requirements of apple populations pose substantial limitations to efficient breeding endeavors. Recent advancements in biotechnology have substantially alleviated these constraints, thereby streamlining apple-breeding processes. Emerging biotechnological approaches, such as tissue culture, genome editing, marker-assisted selection and various omics methodologies, have exhibited substantial potential for their application in apple-breeding programs. Epigenetic modifications have furnished scientists with a powerful tool for deciphering the intricate relationship between genotype and phenotype, a pivotal factor in crop improvement efforts. To further expand the scope of fruit production, concerted breeding efforts will be indispensable in the future. The judicious utilization of these biotechnological tools holds the promise of augmenting both the qualitative and quantitative attributes of apples. Moreover, these tools are instrumental in enhancing resistance to biotic and abiotic stressors and improving other traits of horticultural significance. Furthermore, the establishment of a comprehensive regulatory network in this domain necessitates an exhaustive exploration of the hierarchical organization of transcription factors and a thorough investigation of the molecular mechanisms governing fruit ripening. This chapter provides a comprehensive review of the latest advancements in apple breeding through the application of biotechnological tools. The advent of these novel breeding techniques has ushered in promising strategies for crop improvement in the apple industry.

References

Adams, M. D., Kelley, J. M., Gocayne, J. D., Dubnick, M., Polymeropoulos, M. H., Xiao, H., ... & Venter, J. C. (1991). Complementary DNA sequencing: Expressed sequence tags and human genome project. *Science, 252*(5013), 1651–1656.

Adlak, T., Tiwari, S., Tripathi, M. K., Gupta, N., Sahu, V. K., Bhawar, P., & Kandalkar, V. S. (2019). Biotechnology: An advanced tool for crop improvement. *Current Journal of Applied Science and Technology, 33*(1), 1–11.

Asghar, S., Xiong, Y., Che, M., Fan, X., Li, H., Wang, Y., ... & Han, Z. (2022). Transcriptome analysis reveals the effects of strigolactone on shoot regeneration of apple. *Plant Cell Reports, 41*(7), 1613–1626.

Ashutosh, Sharma, P. C., Prakash, S., & Bhat, S. R. (2007). Identification of AFLP markers linked to the male fertility restorer gene of CMS (Moricandia arvensis) Brassica juncea and conversion to SCAR marker. *Theoretical and Applied Genetics, 114*, 385–392.

Bahmani, R., Gholami, M., Mozafari, A. A., & Alivaisi, R. (2012). Effects of salinity on in vitro shoot proliferation and rooting of apple rootstock MM. 106. *World Applied Sciences Journal, 17*(3), 292–295.

Bashir, T., Ul Haq, S. A., Masoom, S., Ibdah, M., & Husaini, A. M. (2023). Quality trait improvement in horticultural crops: OMICS and modern biotechnological approaches. *Molecular Biology Reports, 50*(10), 8729–8742.

Bianco, L., Cestaro, A., Linsmith, G., Muranty, H., Denancé, C., Théron, A., & Troggio, M. (2016). Development and validation of the Axiom® Apple480K SNP genotyping array. *The Plant Journal, 86*(1), 62–74.

Bianco, L., Cestaro, A., Sargent, D. J., Banchi, E., Derdak, S., Di Guardo, M., & Troggio, M. (2014). Development and validation of a 20K single nucleotide polymorphism (SNP) whole genome genotyping array for apple (Malus× domestica Borkh). *PLoS One, 9*(10), e110377.

Bilbrey, E. A., Williamson, K., Hatzakis, E., Miller, D. D., Fresnedo-Ramírez, J., & Cooperstone, J. L. (2021). Integrating genomics and multiplatform metabolomics enables metabolite quantitative trait loci detection in breeding-relevant apple germplasm. *New Phytologist, 232*(5), 1944–1958.

Botstein, D., White, R. L., Skolnick, M., & Davis, R. W. (1980). Construction of a genetic linkage map in man using restriction fragment length polymorphisms. *American Journal of Human Genetics, 32*(3), 314.

Cao, S., Zhang, Q., Zhu, Z., Guo, J., Chen, Y., & Xue, H. (2008). Preliminary proteomics analysis of the total proteins of flower bud induction of apple trees. *Frontiers of Agriculture in China, 2*, 467–473.

Ceci, A. T., Bassi, M., Guerra, W., Oberhuber, M., Robatscher, P., Mattivi, F., & Franceschi, P. (2021). Metabolomic characterization of commercial, old, and Red-Fleshed apple varieties. *Metabolites, 11*(6), 378.

Chagné, D., Kirk, C., How, N., Whitworth, C., Fontic, C., Reig, G., ... & Iglesias, I. (2016). A functional genetic marker for apple red skin coloration across different environments. *Tree Genetics & Genomes, 12*, 1–9.

Chen, K., & Gao, C. (2014). Targeted genome modification technologies and their applications in crop improvements. *Plant Cell Reports, 33*, 575–583.

Chen, M. X., Sun, C., Zhang, K. L., Song, Y. C., Tian, Y., Chen, X., ... & Zhu, F. Y. (2019). SWATH-MS-facilitated proteomic profiling of fruit skin between Fuji apple and a red skin bud sport mutant. *BMC Plant Biology, 19*(1), 1–13.

Cho, H. J., Han, A. R., & Choi, C. (2022). Effect of paper-bagging on apple skin patterning associated with MdMYB10 promoter methylation. *International Journal of Molecular Sciences, 23*(6), 3319.

Cho, H. J., Kim, G. H., & Choi, C. (2020). Differential gene expression and epigenetic analyses between striped and blushed skinned sports of 'Fuji'apple. *Scientia Horticulturae, 261*, 108944.

Commisso, M., Bianconi, M., Poletti, S., Negri, S., Munari, F., Ceoldo, S., & Guzzo, F. (2021). Metabolomic profiling and antioxidant activity of fruits representing diverse apple and pear cultivars. *Biology, 10*(5), 380.

Cuthbertson, D., Andrews, P. K., Reganold, J. P., Davies, N. M., & Lange, B. M. (2012). Utility of metabolomics toward assessing the metabolic basis of quality traits in apple fruit with an emphasis on antioxidants. *Journal of Agricultural and Food Chemistry, 60*(35), 8552–8560.

Da, L., Liu, Y., Yang, J., Tian, T., She, J., Ma, X., & Su, Z. (2019). AppleMDO: A multi-dimensional omics database for apple co-expression networks and chromatin states. *Frontiers in Plant Science, 10*, 1333.

Daccord, N., Celton, J. M., Linsmith, G., Becker, C., Choisne, N., Schijlen, E., & Bucher, E. (2017). High-quality de novo assembly of the apple genome and methylome dynamics of early fruit development. *Nature Genetics, 49*(7), 1099–1106.

Ding, R., Che, X., Shen, Z., & Zhang, Y. (2021). Metabolome and transcriptome profiling provide insights into green apple peel reveals light-and UV-B-responsive pathway in anthocyanins accumulation. *BMC Plant Biology, 21*(1), 1–15.

Dougherty, L., Singh, R., Brown, S., Dardick, C., & Xu, K. (2018). Exploring DNA variant segregation types in pooled genome sequencing enables effective mapping of weeping trait in Malus. *Journal of Experimental Botany, 69*(7), 1499–1516.

Du, X., Wang, Y., Liu, M., Liu, X., Jiang, Z., Zhao, L., ... & Song, L. (2020). The assessment of epigenetic diversity, differentiation, and structure in the 'Fuji'mutation line implicates roles of epigenetic modification in the occurrence of different mutant groups as well as spontaneous mutants. *PLoS One, 15*(6), e0235073.

Dzhangaliev, A. (2003). The wild apple tree of Kazakhstan. *Horticultural Reviews-Westport then New York-, 29*, 63–304.

Edwards, A., Civitello, A., Hammond, H. A., & Caskey, C. T. (1991). DNA typing and genetic mapping with trimeric and tetrameric tandem repeats. *American journal of human genetics, 49*(4), 746.

Elshire, R. J., Glaubitz, J. C., Sun, Q., Poland, J. A., Kawamoto, K., Buckler, E. S., & Mitchell, S. E. (2011). A robust, simple genotyping-by-sequencing (GBS) approach for high diversity species. *PloS one, 6*(5), e19379.

Forsline, P. L., Aldwinckle, H. S., Dickson, E. E., Luby, J. J., & Hokanson, S. C. (2003). Collection, maintenance, characterization, and utilization of wild apples of Central Asia. *Horticultural Reviews-Westport Then New York-, 29*, 1–62.

Gapper, N. E., Bowen, J. K., & Brummell, D. A. (2023). Biotechnological approaches for predicting and controlling apple storage disorders. *Current Opinion in Biotechnology, 79*, 102851.

Gapper, N. E., Hertog, M. L., Lee, J., Buchanan, D. A., Leisso, R. S., Fei, Z., ... & Rudell, D. R. (2017). Delayed response to cold stress is characterized by successive metabolic shifts culminating in apple fruit peel necrosis. *BMC Plant Biology, 17*(1), 1–18.

Gupta, P. K., Varshney, R. K., Sharma, P. C., & Ramesh, B. J. P. B. (1999). Molecular markers and their applications in wheat breeding. *Plant Breeding, 118*(5), 369–390.

Hamblin, M. T., Buckler, E. S., & Jannink, J. L. (2011). Population genetics of genomics-based crop improvement methods. *Trends in Genetics, 27*(3), 98–106.

Hatoum, D., Annaratone, C., Hertog, M. L. A. T. M., Geeraerd, A. H., & Nicolai, B. M. (2014). Targeted metabolomics study of 'Braeburn'apples during long-term storage. *Postharvest Biology and Technology, 96*, 33–41.

He, P., Cheng, L., Li, H., Wang, H., & Li, L. (2017). A comparative analysis of DNA methylation in diploid and tetraploid apple (Malus× domestica Borkh.). *Czech Journal of Genetics and Plant Breeding, 53*(2), 63–68.

Höfer, M., Grafe, C., Boudichevskaja, A., Lopez, A., Bueno, M. A., & Roen, D. (2008). Characterization of plant material obtained by in vitro androgenesis and in situ parthenogenesis in apple. *Scientia Horticulturae, 117*(3), 203–211.

Hu, G. J., Zhang, Z. P., Dong, Y. F., Fan, X. D., Ren, F., Zhu, H. J. (2015). Efficiency of virus elimination from potted apple plants by thermotherapy coupled with shoot-tip grafting. *Australasian Plant Pathology, 44*, 167–173.

Hu, Y., Han, Z., Wang, T., Li, H., Li, Q., Wang, S., ... & Wu, T. (2022). Ethylene response factor MdERF4 and histone deacetylase MdHDA19 suppress apple fruit ripening through histone deacetylation of ripening-related genes. *Plant Physiology, 188*(4), 2166–2181.

Huo, H., Henry, I. M., Coppoolse, E. R., Verhoef-Post, M., Schut, J. W., de Rooij, H., ... & Bradford, K. J. (2016). Rapid identification of lettuce seed germination mutants by bulked segregant analysis and whole genome sequencing. *The Plant Journal, 88*(3), 345–360.

Irfan, M., Kumar, P., Ahmad, M. F., & Siddiqui, M. W. (2023). Biotechnological interventions in reducing losses of tropical fruits and vegetables. *Current Opinion in Biotechnology, 79*, 102850.

Jia, D., Shen, F., Wang, Y., Wu, T., Xu, X., Zhang, X., & Han, Z. (2018). Apple fruit acidity is genetically diversified by natural variations in three hierarchical epistatic genes: MdSAUR37, MdPP2CH and MdALMTII. *The Plant Journal, 95*(3), 427–443.

Jiang, S. H., Sun, Q. G., Chen, M., Wang, N., Xu, H. F., Fang, H. C., ... & Chen, X. S. (2019). Methylome and transcriptome analyses of apple fruit somatic mutations reveal the difference of red phenotype. *BMC Genomics, 20*(1), 1–13.

Jin, W., Wang, Y., & Wang, H. (2014). Adventitious shoot regeneration from leaves of apple rootstock 'Pingyitiancha'(Malus hupehensis var. pinyiensis) and genetic fidelity of regenerated plantlets using SSR markers. *Canadian Journal of Plant Science, 94*(8), 1345–1354.

Joshi, S. P., Ranjekar, P. K., & Gupta, V. S. (1999). Molecular markers in plant genome analysis. *Current Science, 77*(2), 230–240.

Jung, M., Rotha, M., Aranzana, M. J., Auwerkerken, A., Bink, M., Denance, C., Dujak, C., Durel, C-E., Forcada, C. F. I..... Muranty, H. (2020). The apple REFPOP—a reference population for genomics-assisted breeding in apple. *Horticulture Research, 7*, 189.

Jung, S., Staton, M., Lee, T., Blenda, A., Svancara, R., Abbott, A., & Main, D. (2008). GDR (Genome Database for Rosaceae): Integrated web-database for Rosaceae genomics and genetics data. *Nucleic Acids Research, 36* (suppl_1), D1034–D1040.

Kabylbekova, B., Turdiev, T., Chukanova, N., & Kovalchuk, I. (2019). Optimization of in vitro cloning of different genotypes of fruit and berry crops. *Journal of Biotechnology, 305*, S59.

Kenis, K., Keulemans, J., & Davey, M. W. (2008). Identification and stability of QTLs for fruit quality traits in apple. *Tree Genetics & Genomes, 4*, 647–661.

Kiczorowska, B., & Kiczorowski, P. (2005). Comparison of some chemical components in apples Szampion variety produced in the Lublin province. *Acta Scientiarum Polonorum Technologia Alimentaria, 4*(2), 37–45.

Kilian, B., & Graner, A. (2012). NGS technologies for analyzing germplasm diversity in genebanks. *Briefings in Functional Genomics, 11*(1), 38–50.

Kotoda, N., Hayashi, H., Suzuki, M., Igarashi, M., Hatsuyama, Y., Kidou, S. I., ... & Abe, K. (2010). Molecular characterization of FLOWERING LOCUS T-like genes of apple (Malus× domestica Borkh.). *Plant and Cell Physiology, 51*(4), 561–575.

Kumar, A., Sharma, D. P., Dogra, R. K., Singh, G., Sharma, R., & Kumar, P. (2022b). Identification and characterization of Apple (*Malus* × *domestica* Borkh.) bud sports mutations in the apple growing Northwestern Himalayan region. *Scientia Horticulturae, 304*, 111308.

Kumar, A., Sharma, D. P., Kumar, P., Sharma, G., & Suprun, I. I. (2022a). Comprehensive insights on Apple (*Malus* × *domestica* Borkh.) bud sport mutations and epigenetic regulations. *Scientia Horticulturae, 297*, 110979.

Kunihisa, M., Takita, Y., Yamaguchi, N., Okada, H., Sato, M., Komori, S., ... & Yamamoto, T. (2019). The use of a fertile doubled haploid apple line for QTL analysis of fruit traits. *Breeding Science, 69*(3), 410–419.

Li, B. Q., Feng, C. H., Hu, L. Y., Wang, M. R., & Wang, Q. C. (2016). Shoot tip culture and cryopreservation for eradication of Apple stem pitting virus (ASPV) and Apple stem grooving virus (ASGV) from apple rootstocks 'M9'and 'M26'. *Annals of Applied Biology, 168*(1), 142–150.

Li, M., Galimba, K., Xiao, Y., Dardick, C., Mount, S. M., Callahan, A., & Liu, Z. (2022). Comparative transcriptomic analysis of apple and peach fruits: Insights into fruit type specification. *The Plant Journal, 109*(6), 1614–1629.

Li, M., Li, D., Feng, F., Zhang, S., Ma, F., & Cheng, L. (2016). Proteomic analysis reveals dynamic regulation of fruit development and sugar and acid accumulation in apple. *Journal of Experimental Botany, 67*(17), 5145–5157.

Li, W., Yan, J., Wang, S., Wang, Q., Wang, C., Li, Z., ... & Xu, J. (2021). Genome-wide analysis of SET-domain group histone methyltransferases in apple reveals their role in development and stress responses. *BMC genomics, 22*, 1–17.

Litt, M., & Luty, J. A. (1989). A hypervariable microsatellite revealed by in vitro amplification of a dinucleotide repeat within the cardiac muscle actin gene. *American Journal of Human Genetics, 44*(3), 397.

Liu, Y., Gao, X. H., Tong, L., Liu, M. Z., Zhou, X. K., Tahir, M. M., ... & Zhang, D. (2022). Multi-omics analyses reveal MdMYB10 hypermethylation being responsible for a bud sport of apple fruit color. *Horticulture Research, 9*, uhac179.

Ma, C., Zhang, J., Yuan, J. L., Guo, J. R., Xiong, Y., & Feng, Y. L. (2019). Differential expression of microRNAs are responsive to drought stress and exogenous methyl jasmonate in wheat (*Triticum aestivum*). *International Journal of Agriculture and Biology*, 22(3), 475–486.

Malnoy, M., Xu, M., Borejsza-Wysocka, E. E., Korban, S. S., & Aldwinckle, H. S. (2007). The role of VFA1, VFA2 and VFA4 genes in resistance to Venturia inaequalis in transgenic 'McIntosh' apple plants. *Acta Horticulturae, 738*, 323–326

Masoodi, K. Z., Ahmed, N., Mir, M. A., Bhat, B., Shafi, A., Mansoor, S., ... & Ganai, N. A. (2022). Comparative transcriptomics unravels new genes imparting scab resistance in apple (Malus x domestica Borkh.). *Functional & Integrative Genomics, 22*(6), 1315–1330.

McClure, K. A., Gardner, K. M., Douglas, G. M., Song, J., Forney, C. F., DeLong, J., ... & Myles, S. (2018). A genome-wide association study of apple quality and scab resistance. *The Plant Genome, 11*(1), 170075.

Meng, D., Li, C., Park, H. J., González, J., Wang, J., Dandekar, A. M., ... & Cheng, L. (2018). Sorbitol modulates resistance to Alternaria alternata by regulating the expression of an NLR resistance gene in apple. *The Plant Cell, 30*(7), 1562–1581.

Migicovsky, Z., Gardner, K. M., Money, D., Sawler, J., Bloom, J. S., Moffett, P., ... & Myles, S. (2016). Genome to phenome mapping in apple using historical data. *The Plant Genome, 9*(2), plantgenome2015–11.

Minamikawa, M. F., Nonaka, K., Kaminuma, E., Kajiya-Kanegae, H., Onogi, A., Goto, S., ... & Iwata, H. (2017). Genome-wide association study and genomic prediction in citrus: Potential of genomics-assisted breeding for fruit quality traits. *Scientific Reports, 7*(1), 4721.

Modgil, M., Parmar, S., & Negi, N. P. (2017). RAPD analysis of long term micropropagated rootstock plants of apple Malling 7. *Indian Journal of Experimental Biology, 55*, 178–183

Muranty, H., Troggio, M., Ben Sadok, I., Al Rifai, M., Auwerkerken, A., Banchi, E., ... Bink, M. C. A. M. (2015). Accuracy and responses of genomic selection on key traits in apple breeding. *Horticulture Research, 2*, 15060.

Musacchi, S., & Serra, S. (2018). Apple fruit quality: Overview on pre-harvest factors. *Scientia Horticulturae, 234*, 409–430.

Nishitani, C., Hirai, N., Komori, S., Wada, M., Okada, K., Osakabe, K., ... & Osakabe, Y. (2016). Efficient genome editing in apple using a CRISPR/Cas9 system. *Scientific Reports, 6*(1), 31481.

Osakabe, Y., Liang, Z., Ren, C., Nishitani, C., Osakabe, K., Wada, M., ... & Nagamangala Kanchiswamy, C. (2018). CRISPR–Cas9-mediated genome editing in apple and grapevine. *Nature Protocols, 13*(12), 2844–2863.

Oszust, K., Szpilska, K., Gryta, A., Panek, J., Pylak, M., Lipa, T., & Frąc, M. (2023). New biotechnological solutions in biocontrol and molecular diagnostics of Neofabraea spp. in apples–A review. *Postharvest Biology and Technology, 204*, 112442.

Pandey, S., & Mishra, S. (2023). CRISPR/Cas-Based Strategies for Resistance Against Plant Viruses. In Pandita, D., & Pandita, A. (eds), *CRISPR/Cas-Mediated Genome Editing in Plants* (pp. 391–414). Apple Academic Press.

Paran, I., & Michelmore, R. W. (1993). Development of reliable PCR-based markers linked to downy mildew resistance genes in lettuce. *Theoretical and Applied Genetics, 85*, 985–993.

Parštein, F., Sedlák, J., Svobodová, L., Polák, J., & Gadiou, S. (2013). Results of in vitro chemotherapy of apple cv. Fragrance-Short communication. *Horticultural Science, 40*(4), 186–190.

Pathak, H., Dhawan, V. (2010). Molecular analysis of micropropagated apple rootstock MM111 using ISSR markers for ascertaining clonal fdelity. *Acta Hortic, 865*, 73–80.

Pathak, H., & Dhawan, V. (2011). Evaluation of genetic fidelity of in vitro propagated apple (Malus× domestica Borkh.) rootstock MM 106 using ISSR markers. *Acta Horticulturae, 961*, 303–310.

Pathak, H., & Dhawan, V. (2012). ISSR assay for ascertaining genetic fidelity of micropropagated plants of apple rootstock Merton 793. *In Vitro Cellular & Developmental Biology-Plant, 48*, 137–143.

Patocchi, A., Gianfranceschi, L., & Gessler, C. (1999a). Towards the map-based cloning of Vf: Fine and physical mapping of the Vf region. *Theoretical and Applied Genetics, 99*, 1012–1017.

Patocchi, A., Vinatzer, B. A., Gianfranceschi, L., Tartarini, S., Zhang, H. B., Sansavini, S., & Gessler, C. (1999b). Construction of a 550 kb BAC contig spanning the genomic region containing the apple scab resistance gene Vf. *Molecular and General Genetics MGG, 262*, 884–891.

Peace, C. P. (2017). DNA-informed breeding of rosaceous crops: Promises, progress and prospects. *Horticulture Research, 4*, 17006.

Perrin, A., Daccord, N., Roquis, D., Celton, J. M., Vergne, E., & Bucher, E. (2020). Divergent DNA methylation signatures of juvenile seedlings, grafts and adult apple trees. *Epigenomes, 4*(1), 4.

Picoult-Newberg, L., Ideker, T. E., Pohl, M. G., Taylor, S. L., Donaldson, M. A., Nickerson, D. A., & Boyce-Jacino, M. (1999). Mining SNPs from EST databases. *Genome Research, 9*(2), 167–174.

Pissard, A., Fernández Pierna, J. A., Baeten, V., Sinnaeve, G., Lognay, G., Mouteau, A., ... & Lateur, M. (2013). Non-destructive measurement of vitamin C, total polyphenol and sugar content in apples using near-infrared spectroscopy. *Journal of the Science of Food and Agriculture, 93*(2), 238–244.

Pompili, V., Dalla Costa, L., Piazza, S., Pindo, M., & Malnoy, M. (2020). Reduced fire blight susceptibility in apple cultivars using a high-efficiency CRISPR/Cas9-FLP/FRT-based gene editing system. *Plant Biotechnology Journal, 18*(3), 845–858.

Ramesh, P., Mallikarjuna, G., Sameena, S., Kumar, A., Gurulakshmi, K., Reddy, B. V., ... & Sekhar, A. C. (2020). Advancements in molecular marker technologies and their applications in diversity studies. *Journal of Biosciences, 45*, 1–15.

Reinert, J., & Backs, D. (1968). Control of totipotency in plant cells growing in vitro. *Nature, 220*(5174), 1340–1341.

Rocafort, M., Arshed, S., Hudson, D., Sidhu, J. S., Bowen, J. K., Plummer, K. M., ... & Mesarich, C. H. (2022). CRISPR-Cas9 gene editing and rapid detection of gene-edited mutants using high-resolution melting in the apple scab fungus, Venturia inaequalis. *Fungal Biology, 126*(1), 35–46.

Rockenbach, M. F., Boneti, J. I., Cangahuala-Inocente, G. C., Gavioli-Nascimento, M. C. A., & Guerra, M. P. (2015). Histological and proteomics analysis of apple defense responses to the development of Colletotrichum gloeosporioides on leaves. *Physiological and Molecular Plant Pathology, 89*, 97–107.

Rudell, D. R., Mattheis, J. P., & Hertog, M. L. (2009). Metabolomic change precedes apple superficial scald symptoms. *Journal of Agricultural and Food Chemistry, 57*(18), 8459–8466.

Sharma, M., & Modgil, M. (2020). A reproducible protocol for adventitious shoot regeneration from leaves of apple rootstock Merton793 and assessment of genetic stability. *Indian Academy of Horticultural Sciences (IAHS), 77*(3), 418–425.

Shulaev, V., Korban, S. S., Sosinski, B., Abbott, A. G., Aldwinckle, H. S., Folta, K. M., ... & Veilleux, R. E. (2008). Multiple models for Rosaceae genomics. *Plant Physiology, 147*(3), 985–1003.

Souza, J. A., Bettoni, J. C., Costa, M. D., Baldissera, T. C., Passos, J. F. M. D., & Primieri, S. (2022). 3-IAA on in vitro rooting and acclimatization of apple rootstock. *Communications, 12*, 016–023

Sturtevant, A. H. (1913). The linear arrangement of six sex? linked factors in Drosophila, as shown by their mode of association. *Journal of Experimental Zoology, 14*(1), 43–59.

Strobbe, S., Wesana, J., Van Der Straeten, D., & De Steur, H. (2023). Public acceptance and stakeholder views of gene edited foods: A global overview. *Trends in Biotechnology, 41*(6), 736–740.

Sun, C., Wang, C., Zhang, W., Liu, S., Wang, W., Yu, X., ... & Qu, S. (2021). The R2R3-type MYB transcription factor MdMYB90-like is responsible for the enhanced skin color of an apple bud sport mutant. *Horticulture Research, 8*, 156.

Taillon-Miller, P., Gu, Z., Li, Q., Hillier, L., & Kwok, P. Y. (1998). Overlapping genomic sequences: A treasure trove of single-nucleotide polymorphisms. *Genome research, 8*(7), 748–754.

Takagi, H., Abe, A., Yoshida, K., Kosugi, S., Natsume, S., Mitsuoka, C., ... & Terauchi, R. (2013). QTL-seq: Rapid mapping of quantitative trait loci in rice by whole genome resequencing of DNA from two bulked populations. *The Plant Journal, 74*(1), 174–183.

Takos, A. M., Jaffé, F. W., Jacob, S. R., Bogs, J., Robinson, S. P., & Walker, A. R. (2006). Light-induced expression of a MYB gene regulates anthocyanin biosynthesis in red apples. *Plant Physiology, 142*(3), 1216–1232.

Tautz, D. (1989). Hypervariability of simple sequences as a general source for polymorphic DNA markers. *Nucleic Acids Research, 17*(16), 6463–6471.

Varshney, R. K., Nayak, S. N., May, G. D., & Jackson, S. A. (2009). Next-generation sequencing technologies and their implications for crop genetics and breeding. *Trends in Biotechnology, 27*(9), 522–530.

Vasil, I. K., & Vasil, V. (1972). Totipotency and embryogenesis in plant cell and tissue cultures. *In Vitro, 8*(3), 117–125.

Velasco, R., Zharkikh, A., Affourtit, J., Dhingra, A., Cestaro, A., Kalyanaraman, A., ... & Viola, R. (2010). The genome of the domesticated apple (Malus× domestica Borkh.). *Nature Genetics, 42*(10), 833–839.

Verma, S., Modgil, M., & Chauhan, A. (2021). Deciphering the role of fungal calmodulin gene using Host-mediated gene silencing approach in apple. *Research Square, 1*, 1–34.

Vivek, M., & Modgil, M. (2018). Elimination of viruses through thermotherapy and meristem culture in apple cultivar 'Oregon Spur-II'. *Virus Disease, 29*, 75–82.

Wang, H., Wang, S., Fan, M. M., Zhang, S. H., Sun, L. L., & Zhao, Z. Y. (2021). Metabolomic insights into the browning of the peel of bagging 'Rui Xue'apple fruit. *BMC Plant Biology*, *21*(1), 209.

Wang, L., Tang, T., Wang, W., Zhang, J., Wang, Z., & Wang, F. (2022). Multi-Omics Landscape of DNA Methylation Regulates Browning in "Fuji" Apple. *Frontiers in Nutrition*, *8*, 800489.

Wang, M. R., Cui, Z. H., Li, J. W., Hao, X. Y., Zhao, L., & Wang, Q. C. (2018). In vitro thermotherapy-based methods for plant virus eradication. *Plant Methods*, *14*(1), 1–18.

Wang, M. R., Li, B. Q., Feng, C. H., & Wang, Q. C. (2016). Culture of shoot tips from adventitious shoots can eradicate Apple stem pitting virus but fails in Apple stem grooving virus. *Plant Cell, Tissue and Organ Culture (PCTOC)*, *125*, 283–291.

Wang, N., Liu, W., Zhang, T., Jiang, S., Xu, H., Wang, Y., ... & Chen, X. (2018). Transcriptomic analysis of red-fleshed apples reveals the novel role of MdWRKY11 in flavonoid and anthocyanin biosynthesis. *Journal of Agricultural and Food Chemistry*, *66*(27), 7076–7086.

Watson, A., Ghosh, S., Williams, M. J., Cuddy, W. S., Simmonds, J., Rey, M. D., Asyraf Md Hatta, M., Hinchliffe, A., Steed, A., Reynolds, D., & Adamski, N. M. (2018). Speed breeding is a powerful tool to accelerate crop research and breeding. *Nature Plants*, *4*(1), 23–29.

Weber, J. L., & May, P. E. (1989). Abundant class of human DNA polymorphisms which can be typed using the polymerase chain reaction. *American Journal of Human Genetics*, *44*(3), 388.

Werner, C. R., Voss-Fels, K. P., Miller, C. N., Qian, W., Hua, W., Guan, C. Y., ... & Qian, L. (2018). Effective genomic selection in a narrow-genepool crop with low-density markers: Asian rapeseed as an example. *The Plant Genome*, *11*(2), 170084.

Williams, J. G., Kubelik, A. R., Livak, K. J., Rafalski, J. A., & Tingey, S. V. (1990). DNA polymorphisms amplified by arbitrary primers are useful as genetic markers. *Nucleic Acids Research*, *18*(22), 6531–6535.

Xu, J., Yan, J., Li, W., Wang, Q., Wang, C., Guo, J., ... & Ma, F. (2020). Integrative analyses of widely targeted metabolic profiling and transcriptome data reveals molecular insight into metabolomic variations during apple (Malus domestica) fruit development and ripening. *International Journal of Molecular Sciences*, *21*(13), 4797.

Xu, J., Zhou, S., Gong, X., Song, Y., van Nocker, S., Ma, F., & Guan, Q. (2018). Single-base methylome analysis reveals dynamic epigenomic differences associated with water deficit in apple. *Plant Biotechnology Journal*, *16*(2), 672–687.

Xu, M., Wang, K., Li, J., Tan, Z., Godana, E. A., & Zhang, H. (2022). Proteomic analysis of apple response to penicillium expansum infection based on label-free and parallel reaction monitoring techniques. *Journal of Fungi*, *8*(12), 1273.

Ye, J., Wang, X., Hu, T., Zhang, F., Wang, B., Li, C., ... & Ye, Z. (2017). An InDel in the promoter of Al-ACTIVATED MALATE TRANSPORTER9 selected during tomato domestication determines fruit malate contents and aluminum tolerance. *The Plant Cell*, *29*(9), 2249–2268.

Zabeau, M., & Vos, P. (1993). Selective restriction fragment amplification, a general method for DNA fingerprinting, (European Patent Application Number 0534858 A1).

Zhang, C. X., Tian, Y., & Cong, P. H. (2015). Proteome analysis of pathogen-responsive proteins from apple leaves induced by the alternaria blotch Alternaria alternata. *PLoS One, 10*(6), e0122233.

Zhang, H., Chen, L., Sun, Y., Zhao, L., Zheng, X., Yang, Q., & Zhang, X. (2017). Investigating proteome and transcriptome defense response of apples induced by Yarrowia lipolytica. *Molecular Plant-Microbe Interactions, 30*(4), 301–311.

Zhang, L., Hu, J., Han, X., Li, J., Gao, Y., Richards, C. M., ... & Cong, P. (2019b). A high-quality apple genome assembly reveals the association of a retrotransposon and red fruit colour. *Nature Communications, 10*(1), 1494.

Zhang, L., Zhang, P., Gao, M., Zhao, Y., Zhang, C., & Zhu, H. (2023). Integrative metabolomics and transcriptomics analyses reveal pivotal regulatory mechanisms of 1-methylcyclopropene in maintaining postharvest storage quality of 'Fuji' apples. *Food Quality and Safety, 7*, fyac063.

Zhang, Y., Zhang, Q., Hao, L., Wang, S., Wang, S., Zhang, W., ... & Li, T. (2019a). A novel miRNA negatively regulates resistance to Glomerella leaf spot by suppressing expression of an NBS gene in apple. *Horticulture Research, 6*, 93.

Zhang, Y., Zhou, P., Bozorov, T. A., & Zhang, D. (2021). Application of CRISPR/Cas9 technology in wild apple (Malus sieverii) for paired sites gene editing. *Plant Methods, 17*(1), 1–9.

Zhao, L., Wang, M. R., Cui, Z. H., Chen, L., Volk, G. M., & Wang, Q. C. (2018). Combining thermotherapy with cryotherapy for efficient eradication of Apple stem grooving virus from infected in-vitro-cultured apple shoots. *Plant Disease, 102*(8), 1574–1580.

Zhen, Q., Fang, T., Peng, Q., Liao, L., Zhao, L., Owiti, A., & Han, Y. (2018). Developing gene-tagged molecular markers for evaluation of genetic association of apple SWEET genes with fruit sugar accumulation. *Horticulture Research, 5*, 14.

Zheng, Q., Songa, J., Campbell-Palmera, L., Thompsona, K., Walkera, B., Cuic, Y., & Lib, X. (2013). A proteomic investigation of apple fruit during ripening and in response to ethylene treatment. *Journal of Proteomics, 93*, 276–294.

Zietkiewicz, E., Rafalski, A., & Labuda, D. (1994). Genome fingerprinting by simple sequence repeat (SSR)-anchored polymerase chain reaction amplification. *Genomics, 20*(2), 176–183.

3

Comprehensive Insight on Root Rot Disease of Apple

Mukesh, Anshul Kumar and Pankaj Kumar

3.1 Introduction

Root rot is a disease primarily caused by a soil-borne plant pathogen that attacks the roots of trees growing in wet or damp soil, causing symptoms such as poor growth, wilted leaves, early leaf drop, branch dieback and crop plant death. It is a serious problem in apple plantations that can cause crop failure and complete plant death during favorable climatic conditions. Oomycetes and fungi have been discovered to be the most commonly prevalent root rot pathogens and are highly influenced by the environment, with a diverse range of hosts, hidden underground symptoms and overwinter structures of many root-rot pathogens. The disease control and management are complex and difficult to achieve (Bodah 2017). *Phytophthora cactorum* has been identified as the causative agent of over 200 plant species diseases ranging from temperate to subtropical regions (Garrido et al. 2016). It causes significant production losses by killing both young and mature trees.

An outbreak of *Phytophthora* causes complete bark-girdling near the ground after bark lesions frequently extend up the trunk in narrow tongues, rarely spreading more than one-third around it (Sewell and Wilson 1963). It kills both young and mature trees, resulting in significant production losses. The fungus survives in soil due to its soil-borne nature, and pathogen zoospores can swim easily in free moisture, so wet soil conditions can cause complete orchard destruction. The genus *Rosellinia necatrix* Prills (1902) having an alternate state (Anamorph) *Dematophora necatrix* R. Hartig (1883) live as saprobes, some as endophytes and a few as root pathogens of economically important plant species, growing mainly on deciduous woods of dicotyledonous plants in temperate and tropical regions (Cline and Farr 2006). It is a soil-borne pathogen that causes sudden death by forming a white cottony mycelium and mycelia strands at the base of seedling stems or on the main roots of older plants.

Disease identification begins with visual symptoms on infested plants in the field. Pathogen identification in a pathology laboratory can be accomplished

DOI: 10.1201/9781003625674-3

by isolating the pathogen on PDA and other specific media and observing its morphology under a microscope. Recent molecular approaches for plant pathogen identification can result in accurate pathogen identification.The relationship between soil-borne microbes and tree crops are delicate and complex, with benefits and drawbacks for the host. Integrated "omics" analyses combined with meta-genomics, meta-transcriptomics, meta-proteomics and meta-bolomics, give a more accurate picture of the activities and physiological potential of the microbiota associated with below-ground plants (Mercado-Blanco et al. 2018). Artificial intelligence and sensor-based image analysis, such as visible spectrum, hyper-spectral and multi-spectral sensor-based technologies, may help overcome issues with automating severity measurement in hyper-variable field conditions (Bock et al. 2020).

The progression of root rot disease infection is determined by host–pathogen–environment interactions, which are influenced by environmental conditions, such as moderate to high soil moisture, the optimal temperature for pathogen growth, soil compaction, poor drainage, continuous or frequent cropping, and other factors that contribute to plant stress (Bodah 2017). The root rot pathogen causes the continuous destruction of apple trees in an orchard at regular time intervals because the pathogen keeps moving in the soil if not treated. It is difficult to manage apple root rot disease in field conditions because of its complex soil-borne nature, therefore, preventing water stagnation in orchards or tree basins, weed management of the orchard, and clean-basin management can reduce the severity of the disease to a certain extent. Using the bio-agent formulation of *Trichoderma*, *Bacillus* and arbuscular mycorrhizal fungi (AMF) can also be helpful in managing soil-borne disease in field conditions. Because collar and white root rot pathogens belong to different genera, the selection of fungicide will be the key step for disease management because certain fungicides are ineffective against *Phytophthora*. Identification of the pathogen is very important; if collar rot is detected then the recommended fungicides are Ridomil, Alliete and Infinito, etc., and if the pathogen is identified as *Dematophora* then Carbendazim and other systemic fungicides can be used for drenching the tree basin at recommended doses (0.25%).

3.2 Etiology

3.2.1 Collar Rot

Phytophthora cactorum (domain: eukaryote, kingdom: fungi, phylum: Ascomycota, subphylum: pezizomycotina, class: sordariomycetes) and *Dematophora necatrix* (domain: Eukaryote; kingdom: Chromista, phylum: Oomycote, class: Oomycetes; order: Peronosporales) are the major pathogens associated with collar rot. *P. cactorum* is homothallic, with spherical to irregularly shaped antheridia (13–15 μm), spherical or tapering base oogonia (25–32 μm) and aplerotic oospore (20–26 μm diameter and 2 μm thick). Sporangia are papillate ellipsoidal, ovoid and spherical in shape, with each sporangium containing up to 50 zoospores. The genome of *Phytophthora cactorum*, a homothallic oomycete pathogen with a genome assembly of 121.5 Mb using third-generation single-molecule real-time (SMRT) sequencing technology, is the second-largest genome sequenced so far in the *Phytophthora* genera, with 27,981 protein-coding genes. The use of high-level detoxification enzymes and hydrolase activity-associated genes after ginsenoside exposure plays a critical role in the pathogen's rapid adaptation to host plant defense compounds and fungicides (Yang et al. 2018).

3.2.2 White Root Rot

Dematophora necatrix attacks approximately 170 plant species in 63 genera and 30 families, including deciduous trees (apple, pear, plum, and almond), olive trees, citrus rootstock (*Citrus aurantium* and *Troyer citrange*), grape rootstocks (Richer-110, Salt Creek), avocado, mango, macadamia, field crops (cotton, alfalfa, and bean), and weeds (*Prosopsisfarcta*) (Sztejnberg and Madar 1980). *Rosellinia necatrix*s clerotia are black, hard and spherical nodules, several millimeters in diameter, located mainly on invaded roots and connected from their base with the subcortical mycelium possibly related to its survival in the soil (Perez-Jimenez 2006). Based on the morphological and pathological characteristics, the fungus was identified as *Dematophoranecatrix*, the conidial state of the white root rot pathogen. Conidiophores are erect, brown, branched, 500 μm tall, 2.5–2.8 μm wide, bearing 2–30 conidia in two rows on the conidiogenous cells of the apical branches of the conidiophores. Conidia are hyaline or subhyaline, one-celled, elliptical, ovate or cylindrical, 3.7–5–2–2.2 μm (Watanabe 1992). The transcriptomes analysis of *R. necatrix* revealed 12,104 transcripts; among which 1937 were differentially expressed genes (DEG), 137 exclusively expressed in *R. necatrix* colonizing avocado roots (RGA), and 160 in *R. necatrix* growing on potato dextrose agar media (RGPDA). The production of different compounds, which singly or in combination interfere with defense or signaling mechanisms, was shared among distinct plant families (Zumaquero et al. 2019).

3.3 Disease Detection and Identification

3.3.1 Collar Rot

The disease is evident on the collar portion near the soil level or the graft union. Initially, small necrotic areas are produced, which turn dark brown and have a characteristic alcoholic odor, and wood below such lesions turns dark brown and the staining extends beyond the edge of the lesion. These lesions girdle the tree, which may produce a variety of above-ground symptoms, and have a direct correlation with the underground infection, leaves and fruits are small, foliage is sparse and chlorotic in spring and bronze in the rainy season, which is typical of autumn color (Gupta 1999). *Phytophthora* was previously assigned to the Pythiales, but more recent phylogenetic analysis using large (LSU) or small (SSU) ribosomal DNA (rDNA) sequences or the *cox2* gene has revealed a closer relationship with downy mildew and white rusts (*Albugo* spp.) in the Peronosporales (Martin et al. 2012). Pathogens can be identified from soil samples using the floating biological bait method, followed by pathogen isolation on a semi-selective nutrient medium – P5ARP, from which DNA can be isolated. Baiting techniques for detecting and isolating *Phytophthora* spp. involves the principle of providing an environment that allows the development of a fresh lesion on the susceptible host-bait by *Phytophthora* zoospores and/or other propagules present in a plant sample or soil facilitate easier isolation. While selective media contain several antimicrobial agents, which selectively inhibit the development of undesired secondary fungi and bacteria associated with decayed tissues but is non-inhibitory to *Phytophthora*, suppress the secondary microflora, allow the germination and development of the dormant *Phytophthora* propagules, the emergence of recognizable *Phytophthora hyphae* and colonies (Tsao 1990).

Because the nucleotide sequence of the *Ypt1* gene differs enough between *Phytophthora* species, it has been widely used as a target for molecular detection of *Phytophthora* pathogens (König et al. 2015). Based on the sequence of RAPD, generated fragments form a clear band with diluted DNA as a template, Lilja et al. (2006) developed specific primers *PcactSF* and *PcactSR* from *RAMS* gene markers for the detection of *P. cactorum*. Sequencing after PCR analysis using *YPh1F/YPh2R* and *ITS4/ITS5* primers confirmed the morphological characteristics of *P. cactorum* (Surina et al. 2020). Whole-genome phylogenetic analysis of *P. cactorum* revealed distinct lineages with well-supported non-recombining clades for strawberry CR and apple infecting isolates specialized to strawberry crowns and apple tissue (Nellist et al. 2021). The genome of *Phytophthora cactorum*, a homothallic oomycete pathogen with a genome assembly of 121.5 Mb using third-generation single-molecule real-time (SMRT) sequencing technology, is the second-largest genome sequenced so far in the *Phytophthora* genera, with 27,981 protein-coding genes. The use of high-level detoxification enzymes and hydrolase activity-associated genes after ginsenoside exposure plays a critical role in the pathogen's rapid adaptation to host plant defense compounds and fungicides (Yang et al. 2018). The transcriptome and effector arsenal of *P. cactorum* provide valuable data for understanding the pathogenicity basis of this multi-host pathogen. Single nucleotide polymorphisms for *RXLR* effectors were discovered in a collection of *P. cactorum* isolates from various countries and hosts. The *RXLR*, elicitin and NLP effectors are highly conserved across oomycete species (Chen et al. 2014).

3.3.2 White Root Rot

Visual observation of the apple plant root for the presence of white cottony mycelium and mycelial strands in between the root, bark and wood shows the fungus invades the entire root, causing general rotting and the development of typical mycelium fans (Pliego et al. 2012). The plant usually suffers from wet rot during Dematophora infestation. The pathogen can be isolated using PDA or a specific bait medium from the soil and identified under the microscope through observing its morphology. Arjona-López et al. (2019) demonstrated effective, sensitive and reliable molecular detection and quantification method for *R. necatrix* in soil samples using the duplex TaqMan qPCR protocol and accurately identified the pathogen from all samples tested. *Dematophora necatrix* multi-locus phylogeny and secondary meta-bolite profiling based on a combination of sequences from the internal transcribed spacer region (ITS), the large subunit (LSU) of nuclear rDNA, beta-tubulin (TUB2) and the second-largest subunit of RNA polymerase II (RPB2) were also reported (Wittstein et al. 2020). White root rot detection methods currently in use are based on microbial and molecular techniques, physiological parameters provided by imaging techniques, and machine learning methods. Data from the normalized difference vegetation index (NDVI) had the highest sensitivity and the lowest rate of false negatives. It has been reported as a quick and practical method for detecting trees affected by white root rot in avocado orchards (Pérez-Bueno et al. 2019).

3.4 Life Cycle

3.4.1 *Rosellinia necatrix*

When the mycelial inoculum of *R. necatrix* comes into contact with roots, mycelial masses and strands begin to propagate on the roots, penetrate from the lenticels of epidermal cells, and junctions between epidermal cells, and then invade the epidermis and xylem of the roots, forming fan-shaped mycelial strands. *R. necatrix* produces three types of spores: chlamydospores, which form under extreme environmental conditions from pyriform swellings in which protoplasm condenses and a wall forms, separating the spherical zone from the remaining hypha, Conidia are formed at the synnemata's end. Conidiogenous cells and ascospores form inside the perithecium and are expelled in a mucilaginous mass from the pore or papilla located on top of the perithecium when they reach maturity (Pérez-Jiménez 2006).

3.4.2 *Phytophthora cactorum*

Phytophthora inoculum that enters the apple plant through planting material, irrigation water, animals or implements may regenerate and produce long-term increases in soil inoculum content by colonizing live, abscissed apple organs such as leaves and fruit. Under field conditions, the population of *P. cactorum* is maximum during spring, declining in summer and autumn (Horner and Wilcox 1996). It survives by producing oospores and sporangium, which require near-saturated soil moisture, whereas zoospore release and motility are dependent on the presence of free water as the main infective structures. Mycelium and sporangia have intermediate levels of survival. Colonization of living tissues is associated with fungus growth and reproduction. The availability of water above-ground and in the soil, soil type, tree age, the height of the graft union, the interaction between scion and rootstock, and soil management techniques all influence disease incidence and development (Harris 1991).

3.5 Epidemiology

Phytophthora spp. symptoms are generally more severe in cool, moist climates, and they typically appear on the upper parts of stressed plants in late spring or early summer, especially in low-lying wet areas. Plants with severe root rot are frequently stunted and wilt in hot weather (Garrido et al. 2016). The Mediterranean maquis forest near agricultural lands, soil transferred to low-lying sections within orchards, infection sources carried by farmers from plots on the same farm, and infections via roots of adjacent trees within orchards are all potential root infection sources (Dafny-Yelin et al. 2008). *P. cactorum* population is relatively high at the bottom of the slope, with 50 and 70% propagule in 3 and 6 cm of soil, respectively, while it remains constant at 7 to 10 cm depth. After natural colonization in soil, fallen leaves, fruit and petals result in a significant increase in the soil population of *P. cactorum* (Horner and Wilcox 1996). Rosellinia grows well in soil with moisture holding capacities of 100 and 66% at temperatures ranging from 15 to 25°C, with the mycelial extension being more severe in infested soil (Mantell and Wheeler 1973). *D. necatrix* is most dangerous to apple plants between the age of 5 to 20 years; the disease spreads more quickly in clay soil with a pH of 6.1–6.5. Wet and waterlogged soil, as well as runoff, helps the disease spread quickly (Agarwal and Sharma 1966). Shiragane et al. (2019) identified the fungus on the roots of nine weed species from infested pear orchards using PCR assays with specific primers, revealing that weed roots aid in the spread of *R. necatrix*.

3.6 Integrated Disease Management

Control of white root rot is highly dependent on efforts to exclude the pathogen through the use of *R. necatrix* free propagating material and planting in non-infested soils, as it is primarily disseminated by propagating materials and can survive in the soil for many years (Schena et al. 2008). Rootstocks M9 and G11 are resistant to *P. cactorum*, whereas CG5087 and G935 rootstocks are less susceptible to *R. necatrix* than the M series (Choi et al. 2021). Consideration for decision-making regarding an integrated approach to prevent attack and minimize the effects of the white-rot agent (Carlucci et al. 2012) are (1) choosing suitable locations for new orchards, such as keeping new fruit tree orchards at a suitable distance from forests where *R. necatrix* is resident, given the occurrence of the aforementioned favorable conditions in forest soil; (2) adaptation of cropping practices to the specific soil features of fruit tree crops to reduce the risk of *Rosellinia* infection; for example, water stress or any other abiotic stress should be avoided as much as possible in the immediate post-transplant period; (3) adoption of measures to reduce the spread of the fungus in the event of previous attacks, such as removing wood residues from previous tree crops as much as possible; increasing soil aeration with deep ripping, or other tillage practices suitable for reducing the fraction of non-decomposed organic material in the soil.

3.6.1 Cultural Control

Disease suppressive soils protect plants from infection by soil-borne pathogens and serve as a rich resource for the discovery of beneficial microorganisms with novel antimicrobial and other plant protective properties (Exposito et al. 2017). Soil solarization with transparent polyethylene sheets (25 and 50 mm) raises soil temperatures by more than 10 and 6°C at depths of 10 and 30 cm, respectively, resulting in a reduction in the *Dematophora necatrix* population and also influences available nitrogen, potassium, organic carbon contents, electrical conductivity and pH, which rise after solarization (Sharma and Sharma 2002). The trees are in highly susceptible conditions during bud swell stage – injuries should be avoided at that time, graft union between rootstocks and scion should be kept 30 cm above the soil level to prevent the entry. Approach graft (inarch) is helpful in saving them from the pathogen in cases where the tree bark gets damaged (Gupta 1999). Bio-fumigation and the cultivation of non-host crops such as mustard and cabbage have successfully managed the disease to some extent, and when combined, solarization can reduce root rot disease by up to 98%, while amendment with oil cakes and plant leaves can also be beneficial in managing both soil-borne plant pathogens (Sharma and Negi 2013).

3.6.2 Biological Control

Endophytic fungi of the root *Crinipellistabtim* isolate colonizes the apple endosphere and rhizosphere and are highly effective against *Dematophora necatrix*; its soil application proved 93.55% disease control in glasshouse conditions and 84.95% disease control in field conditions (Pal et al. 2020). Under greenhouse conditions, the application of AMF and bacterial endophytes in combination with compost in infested soil suppressed *Phytophthora cactorum*, which causes strawberry crown rot, and can be a good biological control alternative for disease management (Hautsaloa et al. 2016). *In vitro*, the biocontrol fungus *Aureobasidium pullulans* (De Bary) G. Armaud inhibited 33–48% mycelial growth of *P. cactorum* and significantly reduced necrotic lesion size on leaves and disease severity on fruits (Iqbal et al. 2021). *Trichoderma* spp. applications combined with solarization using clear 50-m low-density polyethylene mulch reduced *P. cactorum* soil population by 88.9–99.0%, suggesting that they could be future alternatives to traditional chemicals for disease control (Porras et al. 2007). *Pseudomonas chlororaphis* PCL1606 (PcPCL1606), a model bacterium isolated from the avocado rhizosphere and applied in *Rosellinia*-infested soil, had a strong antifungal antagonism mediated by the production of 2-hexyl, 5-propil resorcinol (HPR), resulting in biological control of white root rot diseases in avocado (Tienda et al. 2020). *Bacillus amyloliquefaciens* strain HK34 effectively induced resistance in the leaves of Korean ginseng plants by up-regulating the genes *PgPR10*, *PgPR5* and *PgCAT*, and when applied as a soil drench once (99.1% disease control), as a priming treatment twice in the early stages (83.9% disease control) against *P. cactorum*, which causes foliar blight and root rot disease (Lee et al. 2015). Lists of fungal and bacterial bio-agents for biological control of collar and white root rot disease in apple has been represented in Table 3.1.

3.6.3 Chemical Control

Metalaxyl inhibited *Phytophthora cactorum* mycelial growth, sporangia production and zoospore germination *in vitro*, and its soil drenching resulted in a remission of typical collar rot symptoms under greenhouse conditions (Ellis et al. 1982). Metalaxyl, fosetyl-Al and metalaxyl alternated with *Enterobacter aerogenes* applied as a soil and trunk drench twice a year for the first 3 years and once a year for the next 7 years significantly reduced disease rating and percentage of root and crown infection, and increased fruit yield (Utkhede and Smith 1993; Sharma et al. 2014). Soil application of metalaxyl (as Ridomil5G) at 5 or 10 g ei: per tree and phosethyl AI (as Aliette 80 W) at 64 g ei per tree reduced the growth and viability of *P. cactorum* and can be used as curative treatments for *Phytophthora* root rot disease (Taylor and Washington 1984). Propamocarbchloridrate (Infinito® 687.5 CS) presented 14% greater efficacy than metalaxyl-m+mancozeb (Ridomil Gold® 68 WP) at 0.175% for the

TABLE 3.1

Lists of Fungal and Bacterial Bio-Agents for Biological Control of Collar and White Root Rot Disease in Apple

Bio-Agent	Pathogen	Mode of Action	References
Fungi			
Trichodermaviride	*Dematophora necatrix and Phytophthora cactorum*	Reduced *P. cactorum* soil population by 88.9–99.0%	Porras et al. 2007
Gliocladium virens	*D. necatrix*	*Gliocladium virens* showed 90.4% and 81.0 % white root rot disease control	Gupta and Sharma 2004
Aureobasidium pullulans (De Bary)	*P. cactorum*	Inhibited 33–48% mycelial growth of *P. cactorum* and significantly reduced necrotic lesion size on lea ves and disease severity on fruits	Iqbal et al. 2021
Crinipellistabtim	*D. necatrix*	Soil application proved 93.55% disease control in glasshouse conditions and 84.95 % disease control in field conditions	Pal et al. 2020
Bacteria			
Enterobacter aerogenes	*P. cactorum*	*Enterobacter aerogenes* applied as a soil and trunk drench twice a year reduced the disease rating and percentage of root and crown infection, and increased fruit yield	Utkhede and Smith 1993; Sharma et al. 2014
Bacillus amyloliquefaciens strain HK34	*P. cactorum*	Induced resistance in the leaves of Korean ginseng plants by upregulating the genes *PgPR10*, *PgPR5* and *PgCAT*	Lee et al. 2015
Pseudomonas chlororaphis PCL1606	*D. necatrix*	Strong antifungal antagonism mediated by the production of 2-hexyl, 5-propil resorcinol (HPR) against white root rot disease	Tienda et al. 2020

control of *Phytophthora* wilt in tomato after drenching in soil (Mendonça et al. 2015). Application of phosphorous acid at 200 and 215 g a.i./100 l, fosetyl-Al at 160 and 180 g a.i./100 l, and metalaxyl-M+mancozeb at 20+320 g a.i./100 l as soil drench significantly reduced the severity of root and crown rot disease of apricot and cherry caused by *Phytophthora palmivora* (Türkölmez and Dervis 2017). Bavistin (0.1 or 0.05%) in combination with *E. aerogenes* (bio-agents) demonstrated more than 92% white root rot disease control in pots after their simultaneous application, while *T. viride* and *Gliocladium virens* showed 90.4% and 81.0% disease control, respectively (Gupta and Sharma 2004). Fluazinam increases plant height and secondary root dry weight and effectively manages white root rot infested avocado plants after applications at 0.1% (w/v) and 1.0% (w/v) at 30-day intervals (Lopez-Herrera 2006). Microbiome analysis and characterization of endophytic bacteria will aid in the identification of disease resistance-conferring agents, which will aid in the search for novel and environmentally friendly disease control methods (Sawant et al. 2021).

3.6.4 Biotechnological Interventions for Disease Control

Root rot diseases are generally challenging to control and have a large influence on productivity. If a significant gene for resistance can be exploited, resistant cultivars may be developed by breeding (Tu 1992). *Rosellinia necatrix* is a fungus that causes white root rot disease in many plants, particularly apples. Since it can grow as a saprophyte and thrive in soil for a prolonged time on woody waste, it is difficult to completely eradicate (Pasini et al. 2016). *MYB TFs* are involved in crop defenses against a wide range of pathogens. However, *PnMYB2* is a very important transcription factor that regulates JA signaling, photosynthesis and disease-resistance genes in *P. notoginseng* against the defense responses for root rot pathogens (Qiu et al. 2022). Cai and coworkers in 2023 discovered that 46 DEGs were consistently up-regulated across *A. mellea* and *D. tabescens* treatments and these genes were shown to encode GSTs, hydrolases and ribonucleases, all of which may be involved in ARR resistance (Cai et al. 2023).

The RNAi approach was utilized to control root rot in tomatoes caused by *F. oxysporum* f. sp. *radicis-lycopersici* (*FORL*) by targeting three critical genes: *FoCYP51*, *FoChs1* and *FoEF2*. The intended naked dsRNA was supplied on hexagonal mono-dispersed and bio-degradable layered double hydroxide (LDH) nanosheets with 30–90 nm diameter, which provided exceptional protection and stability for dsRNA with long-term sustained release inside plant cells. The findings demonstrated that the three targeted genes played a functional role in lowering FORL's invasive growth on tomato fruits, confirming their involvement in FORL pathogenesis and suggesting that they could be effective targets for disease prevention via RNAi (Mosa and Youssef 2021). *Rosellinia necatrix megabirnavirus 1* (*RnMBV1*), which may belong to a new virus family and W779 virus is a unique bipartite dsRNA virus with virocontrol possibility (Chiba et al. 2009).

The expression levels of *RnDCL-2*, *RnAGL-2*, *RnRdRP-1* and *RnRdRP-2* in W97 strains inoculated with *RnMyRV3* or *RnMBV1* were 2–4 fold greater than in virus-free strains at 3 and 6 days of mycelial growth,. Furthermore, *RnAGL-1* expression increased after 6 days of mycelial development in the W97 strain infested with *RnMBV1* (Yaegashi et al. 2016). Jasmonic acid (JA) metabolism and disease resistance-related genes were activated in transgenic tobacco when *PnMYB2* protein were overexpressed and gave resistance to *F. solani* (Qiu et al. 2022). A total of 1876 (at 72 hours), 1612 (at 2 weeks) and 1047 (at 5 weeks) unique DEGs were detected for *A. mellea* fungal treatment. However, 325, 503 and 604 DEGs were commonly up-regulated between the two fungal treatments (*A. mellea* and *D. tabescens*) at 72 hours, 2 weeks and 5 weeks, respectively (Cai et al. 2023). Pattison et al. (2004) attempted to find red raspberry resistance gene analogs by employing degenerate primers constructed on conserved areas expressing known resistance genes. RAPD and AFLP markers were first utilized to create a molecular map in a BC1 red raspberry population of 70 individuals that segregated for resistance to *Phytophthora* root rot. RAPD markers associated with root rot resistance were discovered using bulk segregant analysis and QTL analysis. Both analyses discovered two common genetic areas that were expected to explain 50% of the phenotypic variation (Pattison et al. 2004).

3.7 Conclusion

Root rot, a major soil-borne disease in apple plantations, lives in the soil and spreads via free water. Collar rot is caused by *Phytophthora cactorum*, while white root rot is caused by *Dematophora necatrix*. A proper drainage system in the orchard can break the disease cycle by preventing free moisture. Breaking the disease triangle can help to reduce disease to some extent. It is critical to identify the pathogen at an early stage in order to apply the appropriate fungicide for drenching. If the pathogen is not correctly identified, the fungicide may remain ineffective due to differences in the cell walls of *Phytophthora* and *Dematophora* and their reaction with the cell wall. If collar rot is identified – Alliete, Ridomil, and Infinito are recommended fungicides, and if found white root rot – then bavistin and other systemic fungicides can be used for soil drenching 5–10 g/tree and are most effective when combined with cultural practices. Prior to infection, the use of bio-agent *Trichoderma, Bacillus* and *Pseudomonas* formulation at recommended level can help to reduce the actual population of the pathogen in infested soil, thereby reducing disease severity of root rot disease. Because fungicides have a harmful effect on humans and other animals due to their residual effect in soil and water, biological control has an advantage over chemical control through eco-friendly disease management with no toxic effect on humans and the environment. Because of the disease's complex soil-borne nature, it is difficult to manage at the field level; therefore, it is necessary to identify antagonistic fungi, bacteria and endophytes from root rot-infested apple plant root rhizosphere and use them in management in the future.

References

Agarwala RK, Sharma VC (1966) Epidemiology and control of root rot disease of apple tree caused by *Dematophora necatrix*. *Epidemiology Control* 37:285–290.

Arjona-López JM, Capote N, López-Herrera CJ (2019) Improved real-time PCR protocol for the accurate detection and quantification of *Rosellinianecatrix* in avocado orchards. *Plant Soil* 443(1):605–612.

Bock CH, Barbedo JG, Del Ponte EM, Bohnenkamp D, Mahlein AK (2020) From visual estimates to fully automated sensor-based measurements of plant disease severity: Status and challenges for improving accuracy. *Phytopathology Research* 2(1):1–30.

Bodah ET (2017) Root rot diseases in plants: A review of common causal agents and management strategies. *Agriculture Research & Technology: Open Access Journal* 5(3):555–661.

Cai L, Adelberg J, Naylor-Adelberg J, Schnabel G, Calle A, Li Z, ... and Saski CA (2023) Transcriptomics reveal the genetic coordination of early defense to Armillaria root rot (ARR) in Prunus spp. *Frontiers in Plant Science* 14:1181153.

Carlucci A, Manici LM, Colatruglio L, Caputo A, Frisullo S (2012) *Rosellinianecatrix* attack according to soil features in the Mediterranean environment. *Forest Pathology* 43(1):12–18.

Chen XR, Zhang BY, Xing YP, Li QY, Li YP, Tong YH, Xu JY (2014) Transcriptomic analysis of the phytopathogenic oomycete *Phytophthora cactorum* provides insights into infection-related effectors. *BMC Genomics* 15(1):1–24.

Chiba S, Salaipeth L, Lin YH, Sasaki A, Kanematsu S, Suzuki N (2009) A novel bipartite double-stranded RNA mycovirus from the white root rot fungus Rosellinia necatrix: Molecular and biological characterization, taxonomic considerations, and potential for biological control. *Journal of Virology* 83(24):12801–12812.

Choi BH, Kim CS, Jeong YJ, Park IH, Han SG, Yoon TM (2021). Resistance evaluation of G, CG, or M series apple rootstocks to soil-borne diseases (phytophthora root rot, white root rot, and southern blight) and Woolly Apple Aphid. *Horticulture Science Technology* 39(2):167–174.

Cline ET, Farr DF (2006) Synopsis of fungi listed as regulated plant pests by the USDA Animal and Plant Health Inspection Service: Notes on nomenclature, disease, plant hosts, and geographic distribution. *Plant Health Progress* 7(1):42.

Dafny-Yelin M, Mairesse O, Moy J, Dor S, Malkinson D (2008) Genetic diversity and infection sources of Rosellinianecatrix in northern Israel. *Phytopathologia Mediterranea* 57(1):37–47.

Ellis MA, Grove GG, Ferree DC (1982) Effects of metalaxyl on Phytophthora cactorum and collar rot of apple. *Phytopathology* 72(43):1431–1434.

Exposito RG, De Bruijn I, Postma J, Raaijmakers JM (2017) Current insights into the role of rhizosphere bacteria in disease suppressive soils. *Frontier Microbiology* 18(8):2529.

Garrido C, González-Rodríguez VE, Carbú M, Husaini AM, Cantoral JM (2016) 10 fungal diseases of strawberry and their diagnosis. *Strawberry: Growth, Development and Diseases* 17:157.

Gupta VK (1999) *Phytophthora cactorum* – A threatening pathogen of apple. *Indian Phytopathology* 52(2):105–113.

Gupta VK, Sharma K (2004) Integration of chemicals and biocontrol agents for managing white root rot of apple. *Acta Horticulturae 635:* 141–149.

Harris DC (1991) The Phytophthora diseases of apple. *Journal of Horticulture Science* 66(5):513–544.

Hautsalo J, Vestberg M, Parikka P, Kukkonen S, Karhu S, Tahvonen R (2016) Biological control of strawberry crown rot is substrate dependent phenomenon. *Journal of Berry Research* 6(1):65–79.

Horner IJ, Wilcox WF (1996) Spatial distribution of Phytophthora cactorum in New York apple orchard soils. *Population* 122:32.

Iqbal M, Jamshaid M, Zahid MA, Andreasson E, Vetukuri RR, Stenberg JA (2021) Biological control of strawberry crown rot, root rot and grey mould by the beneficial fungus *Aureobasidiumpullulans*. *BioControl* 66(4):535–545.

König S, Schwenkbier L, Pollok S, Riedel M, Wagner S, Popp J, Weber K, Werres S (2015) Potential of Ypt1 and ITS gene regions for the detection of *Phytophthora* species in a lab-on-a-chip DNA hybridization array. *Plant Pathology* 64(5):1176–1189.

Lee BD, Dutta S, Ryu H, Yoo SJ, Suh DS, Park K (2015) Induction of systemic resistance in Panax ginseng against *Phytophthora cactorum* by native *Bacillus amyloliquefaciens* HK34. *Journal of Ginseng Research* 39(3):213–220.

Lilja AT, Parikka PK, Pääskynkivi EA, Hantula JI, Vainio EJ, Vartiamäki HA, Lemmetty AH, Vestberg MV (2006) *Phytophthora cactorum* and *Colletotrichumacutatum*: Survival and Detection. *Agriculturae Conspectus Scientificus* 71(4):121–128.

López-Herrera CJ, Zea-Bonilla T (2006) Effects of benomyl, carbendazim, fluazinam and thiophanate methyl on white root rot of avocado. *Crop Protection* 26(8):1186–1192.

Mantell SH, Wheeler BE (1973) Rosellinia and white root rot of Narcissus in the Scilly isles. *Transactions of the British Mycological Society* 60(1):23-IN1.

Martin FN, Abad ZG, Balci Y, Ivors K (2012) Identification and detection of Phytophthora: Reviewing our progress, identifying our needs. *Plant Disease.* 96(8):1080–1103.

Mendonça LB, Coelho L, Stracieri J, Ferreira Junior JB, Tebaldi ND (2015) Chemical control of Phytophthora wilt in tomatoes. *Bioscience Journal* 31(4):1015–1023.

Mercado-Blanco J, Abrantes I, Barra Caracciolo A, Bevivino A, Ciancio A, Grenni P, Hrynkiewicz K, Kredics L, Proença DN (2018) Belowground microbiota and the health of tree crops. *Frontier Microbiology* 9:1006.

Mosa MA, Youssef K (2021) Topical delivery of host induced RNAi silencing by layered double hydroxide nanosheets: An efficient tool to decipher pathogenicity gene function of Fusarium crown and root rot in tomato. *Physiological and Molecular Plant Pathology* 115:101684.

Nellist CF, Armitage AD, Bates HJ, Sobczyk MK, Luberti M, Lewis LA, Harrison RJ (2021) Comparative analysis of host-associated variation in *Phytophthora cactorum*. *Frontier Microbiology* 12:1572.

Pal J, Sharma SK, Devi S, Sharma R, Raj H, Karn M, Verma S, Vedukola PR, Sharma A (2020) Screening, identification, and colonization of fungal root endophytes against *Dematophora necatrix:* A ubiquitous pathogen of fruit trees. *Egyptian Journal of Biological Pest Control* 30(1):1–4.

Pasini L, Prodorutti D, Pastorelli S, Pertot I (2016) Genetic diversity and biocontrol of Rosellinia necatrix infecting apple in northern Italy. *Plant Disease* 100(2):444–452.

Pattison JA, Samuelian SK, Weber CA (2004) Mapping Phytophthora root rot resistance in red raspberry using SCARS, CAPS, AFLPs, and RGAs. *Hortscience* 39(4):868B–868.

Pérez-Bueno ML, Pineda M, Vida C, Fernández-Ortuño D, Torés JA, De Vicente A, Cazorla FM, Barón M (2019) Detection of white root rot in avocado trees by remote sensing. *Plant Disease* 103(6):1119–1125.

Pérez-Jiménez RM (2006) A review of the biology and pathogenicity of *Roselliniane catrix*–the cause of white root rot disease of fruit trees and other plants. *Journal of Phytopathology* 154(5):257–266.

Pliego C, López-Herrera CA, Ramos C, Cazorla FM (2012) Developing tools to unravel the biological secrets of *Rosellinianecatrix*, an emergent threat to woody crops. *Molecular Plant Pathology* 13(3):226–239.

Porras M, Barrau C, Arroyo FT, Santos B, Blanco C, Romero F (2007) Reduction of *Phytophthora cactorum* in strawberry fields by Trichoderma spp. and soil solarization. *Plant Disease* 91(2):142–146.

Qiu B, Chen H, Zheng L, Su L, Cui X, Ge F, Liu D (2022) An MYB transcription factor modulates Panax notoginseng resistance against the Root Rot Pathogen Fusarium solani by regulating the Jasmonate acid signaling pathway and photosynthesis. *Phytopathology®* 112(6):1323–1334.

Sawant SS, Choi ED, Song J, Seo HJ (2021) Current status and future prospects of white root rot management in Pear Orchards: A review. *Research in Plant Disease* 27(3):91–98.

Schena L, Nigro F, Ippolito A (2008) Integrated management of Rosellinia necatrix root rot on fruit tree crops. In: Ciancio, A., Mukerji, K. (eds) *Integrated Management of Diseases Caused by Fungi, Phytoplasma and Bacteria*, pp. 137–158. Springer, Dordrecht.

Sewell GW, Wilson JF (1963) Branch, stem and collar rot of apple caused by Phytophthora species of the 'cactorum group'. *Nature* 200(4912):1229.

Sharma IM, Negi HS (2013) Soil Solarization, non-host crops, bio-fumigation and oil cakes in management of collar rot and white root rot diseases in apple. *Indian Journal of Plant Protection* 41(3):257–262.

Sharma IM, Negi HS, Sharma S (2014) Integrated management of collar rot in apple caused by Phytophthora cactorum. *Indian Phytopathology* 67(2):168–173.

Sharma M, Sharma SK (2002) Effect of soil solarization on soil microflora with special reference to *Dematophora necatrix* in apple nurseries. *Indian Phytopathology* 55(2):158–162.

Shiragane H, Usami T, Shishido M (2019) Weed roots facilitate the spread of *Rosellinianecatrix*, the causal agent of white root rot. *Microbes and Environments* 34(3):340–343.

Surina TA, Kopina MB, Smirnova AV, Shukhin DI (2020) Isolation and identification of Phytophthora species on nut-bearing plants. *Biorxiv*. https://doi.org/10.1101/2020.12.21.423714

Sztejnberg A, Madar Z (1980) Host range of *Dematophora necatrix*, the cause of white root rot disesase in fruit tree. *Plant Diseases* 64:662–664

Taylor PA, Washington WS (1984) Curative treatments for Phytophthora cactorum in peach trees using metalaxyl and phosethyl Al. *Australasian Plant Pathology* 13(3):31–33.

Tienda S, Vida C, Lagendijk E , De Weert S, Linares I, González-Fernández J, Guirado E, de Vicente A, Cazorla FM (2020) Soil application of a formulated biocontrolrhizobacterium, Pseudomonas chlororaphis PCL1606, induces soil suppressiveness by impacting specific microbial communities. *Frontier Microbiology* 11:1874.

Tsao PH (1990) Why many phytophthora root rots and crown rots of tree and horticultural crops remain undetected 1. *EPPO Bulletin* 20(1):11–17.

Tu JC (1992) Management of root rot diseases of peas, beans, and tomatoes. *Canadian Journal of Plant Pathology* 14(1):92–99.

Türkölmez Ş, Dervis S (2017) Activity of metalaxyl-m+ mancozeb, fosetyl-al, and phosphorous acid against Phytophthora crown and root rot of apricot and cherry caused by Phytophthorapalmivora. *Plant Protection Science* 53(4):216–225.

Utkhede RS, Smith EM (1993) Long-term effects of chemical and biological treatments on crown and root rot of apple trees caused by *Phytophthora cactorum*. *Soil Biology and Biochemistry* 25(3):383–386.

Watanabe T (1992) Sporulation of Dematophora necatrix in vitro and its pathogenicity. *Japanese Journal of Phytopathology* 58(1):65–71.

Wittstein K, Cordsmeier A, Lambert C, Wendt L, Weber J, Wurzler N, Petrini LE, Stadler M (2020) Identification of Rosellinia species as producers of cyclodepsipeptide PF1022 A and resurrection of the genus Dematophora as inferred from polythetic taxonomy. *Studies in Mycology* 96(1):1–6.

Yaegashi H, Shimizu T, Ito T, Kanematsu S (2016) Differential inductions of RNA silencing among encapsidated double-stranded RNA mycoviruses in the white root rot fungus Rosellinia necatrix. *Journal of Virology* 90(12):5677–5692.

Yang M, Duan S, Mei X, Huang H, Chen W, Liu Y, Guo C, Yang T, Wei W, Liu X, He X (2018). The *Phytophthora cactorum* genome provides insights into the adaptation to host defense compounds and fungicides. *Scientific Reports* 8(1):1–1.

Zumaquero A, Kanematsu S, Nakayashiki H, Matas A, Martínez-Ferri E, Barceló-Muñóz A, Pliego-Alfaro F, Lopez-Herrera C, Cazorla FM, Pliego C (2019) Transcriptome analysis of the fungal pathogen *Rosellinianecatrix* during infection of a susceptible avocado rootstock identifies potential mechanisms of pathogenesis. *BMC Genomics* 20(1):1–4.

4

Biotechnology for Grape Improvement

Anuradha Upadhyay, Supriya Aglawe, Sharmistha Naik, Roshni R. Samarth and Komal D. Thorat

4.1 Introduction

Grapevine (*Vitis vinifera* L.) is an important fruit crop, with significant economic value. Grapes are grown worldwide over a varied range of climate and soil conditions. Grapes are grown for diverse purposes like fresh consumption, dried or processed into wine, juice or jelly. Being a perennial crop, grapevine experiences several biotic and abiotic stresses during its lifecycle. Changing climate scenarios pose additional risks to the sustainable production of quality grapes. Climate changes, the emergence of new pests and diseases, and changing consumer preferences require continuous efforts for grape improvement. The grape improvement involves the development of new grape varieties or improving existing ones to meet the challenges and to better meet the needs of growers, processors and consumers. In the past, conventional breeding has been used for grape improvement. However, the perennial nature of grape, long gestation period, heterozygosity, linkage drag, etc., hamper the progress of grape breeding programs. To overcome these limitations, several biotechnological approaches have been developed to complement conventional breeding.

4.2 Molecular Markers for Characterization of Grape Germplasm

Grapes have huge genetic variability. In order to exploit it, characterization of available grape germplasm is mandatory. Grape cultivars have been identified and characterized by ampelographic descriptors, which are based on visual observations, however, ampelographic descriptors have many limitations. Later, biochemical-like isoenzymes, phenolics and aromatic

 DOI: 10.1201/9781003625674-4

compound-based methods of grapevine characterization were introduced and used as a supplementary method for ampelographic characterization. Problems associated with enzyme extraction, lack of reproducibility and scarcity of discriminating enzymes hinder the widespread use of biochemical methods in grape characterization. Introduction of molecular markers overcome difficulties associated with biochemical methods and offer the next level of simplicity and ease. Molecular markers are DNA-based markers that are not influenced by environmental factors and give accurate results. Molecular marker systems *viz.*, RFLP, RAPD, AFLP, SSR, SNP, etc., have been successfully used for grape characterization.

4.2.1 Restriction Fragment Length Polymorphism (RFLP)

RFLP is the first molecular marker developed and is the only hybridization-based marker system. Variation due to insertion, deletion and point mutation can result in an altered restriction site and finally give the altered length of DNA fragments. RFLP is a co-dominant marker and has reproducibility. It is mainly used for cultivar and rootstock identification, to differentiate between cultivars, and for identification of kin relationships in a set of germplasm (Bourquin et al. 1993; Bourquin et al. 1995). Requirements of large amounts of pure DNA and skill requirements restricted the use of RFLP marker application in germplasm characterization.

4.2.2 Random Amplified Polymorphic DNA (RAPD)

The RAPD marker system uses a short random primer in PCR, which offers technical simplicity and ease of use. Moreover, sequence information is also not required. RAPD can also be converted into another PCR-based marker system like SCAR, SNP, etc., by cloning the RAPD band of interest and further its sequencing. RAPD markers have been used for grape germplasm characterization by many researchers (Collins and Symons 1993; Jeanjaques et al. 1993). The major disadvantage of RAPD is no reproducibility.

4.2.3 Amplified Fragment Length Polymorphism (AFLP)

AFLP is the combination of the RFLP and RAPD marker systems. AFLP can be visualized by selective amplification of restriction fragmented DNA. AFLP gives reproducibility, requires less DNA compared to RFLP, and does not require DNA sequence information. Like RAPD, AFLP is a dominant marker, so it is not suitable for parentage analysis. There are several reports on the use of AFLP in grapes mainly for assessing genetic similarity among different varieties and to study genetic relationships in germplasm (Cervera et al. 2000, 2002; Fanizza et al. 2003). AFLP markers are especially useful for differentiating clones of a variety, which are generally not differentiated by SSR markers (Scott et al. 2000; Fanizza et al. 2005b; Blaich et al. 2007; Moncada and Hinrichsen 2007; Stajner et al. 2009b; Upadhyay et al. 2011; Shinde et al. 2013). AFLP markers have been used along with other marker systems like SSR to study genetic diversity, and relatedness between grape cultivars and to characterize grape rootstocks (Fossati et al. 2001; Labra et al. 2001; Upadhyay et al. 2007; Baneh et al. 2009; Lopez et al. 2009).

4.2.4 Simple Sequence Repeat (SSR)

Simple sequence repeats (SSRs) also known as microsatellites are simple tandem repeats of mono-, di-, tri- and tetra-nucleotide motifs. SSR markers have many advantages *viz.*, co-dominant nature, reproducibility, transferability between labs, high polymorphism and good coverage throughout the genome. Its technical simplicity including many advantages and ease of multiplexing made it the marker of choice among plant scientists. A large number of *Vitis* SSR markers have been developed and used in various laboratories (Thomas and Scott 1993; Thomas et al. 1994; Bowers et al. 1996; Sefc et al. 1999; Bowers et al. 1999b; Di Gaspero et al. 2005). *Vitis* SSR markers developed from these studies are now in the public domain and play an important role in grape germplasm characterization. This et al. (2004) developed a standard set of microsatellite reference alleles for the identification of grape cultivars. As SSR markers are highly polymorphic and multiallelic, it develops a unique profile of a grape cultivar, which has a great practical implication in viticulture such as identification of different cultivars, their parental lineage, etc.

Among developed *Vitis* SSRs markers, few were chosen as a standard set of markers for varietal identification, including loci VVS2, VVMD27, VVMD7, VVMD5, VrZAG79, VrZAG62 (This et al. 2004). Markers VVMD25, VVMD28 and VVMD32 identified by Bowers et al. (1999b) also have been used along with the standard set of SSR markers for varietal identification. *Vitis* cultivars from various regions and countries have been characterized by SSR markers (Lopes et al. 1999; Ladoukakis et al. 2005; González-Andrés et al. 2007; Laiadi et al. 2009; Stajner et al. 2009a; Štajner et al. 2011; Tomić et al. 2012; Upadhyay et al. 2013; Baránková et al. 2020; Bibi et al. 2020; Jiménez-Cantizano et al. 2020; Li et al. 2021). Identification of parentage, phylogeny and kinship analysis is important in grape breeding as it helps in selecting breeding parents, identifying the parents of a cultivar and establishing the origin of a cultivar. Based on an analysis of 30 microsatellite loci, Bowers and Meredith (1997) determined Cabernet Franc and Sauvignon Blanc as the parents of Cabernet Sauvignon. Similarly, Bowers et al. (1999a) analyzed 300 grape cultivars with 32 markers and established common parentage for Chardonnay, Gamay noir, Aligoté and Melon. Subsequently, SSR markers have been widely used to establish the parentage of grape genotypes and rootstocks (Piljac et al. 2002;

Crespan 2003; Akkak et al. 2007; Upadhyay et al. 2007; Vouillamoz et al. 2007; Laucou et al. 2008; Cipriani et al. 2010; Cao et al. 2020).

4.2.5 Single Nucleotide Polymorphism (SNP)

SNPs occur due to changes in single nucleotides by natural or artificial mutations. SNP can be bi-, tri- and tetra-allelic, but most of the SNPs are bi-allelic in nature. Based on the position of SNP in the genome, there are two types of SNPs, non-genic SNPs and genic SNPs. Genic SNPs are of two types, exonic and intronic SNPs. Based on the effect of exonic SNP on the coding region, it is further divided into two types, synonymous SNP, which does not cause a change in amino acid sequence, and non-synonymous SNP, which causes a change in amino acid sequence and ultimately changes function of the gene. With advancements in sequencing technologies, the detection of SNPs has become easy and economical. Abundant in number, good coverage, availability of a lot of genomic data, easy for multiplexing and automation has made SNP as a popular marker amongst plant researchers. Identification of SNPs from the already available databases and the experimental approach are the two approaches mainly used for SNP identification. Both methods have been exploited for SNP identification in grapes. Salmaso et al. (2005) identified 247 SNPs based on SSCP and resequencing of 50 gene fragments in nine grape genotypes. Lijavetzky et al. (2007) combined a re-sequencing approach and SNPlex technology for high-throughput detection of SNPs and marker development. The developed SNP markers were used for cultivar identification, linkage mapping and genetic diversity studies using SNPlex™ genotyping. Troggio et al. (2008) compared the efficiency of SSCP and minisequencing for SNP detection and found a combination of multiplexing with minisequencing as a flexible and reliable technique for SNPs detection in grapes. Subsequently, SNPlex genotyping was successfully used to investigate a large number of SNPs in a large number of samples (Pindo et al. 2008). Advancement in next-generation sequencing has allowed the identification of a large number of SNPs in grape. To facilitate SNP genotyping of a large pool of grape germplasm, SNP arrays Vitis9kSNP containing 9000 SNPs (Myles et al. 2010) and Vitis18kSNP array containing 18775 SNPs (Illumina Inc., San Diego, California) have been developed. SNP array-based large-scale analysis of cultivated and wild accessions from different geographical regions has thrown light on the evolutionary, domestication and phylogenetic aspects of grapes (Myles et al. 2011; De Lorenzis et al. 2015; Sunseri et al. 2018; De Lorenzis et al. 2019) and assessed the genetic diversity (Mercati et al. 2016; Laucou et al. 2018; Bianchi et al. 2020). As against SSR markers, SNP markers have lower polymorphic information content and therefore identification of the set of polymorphic SNPs suitable of varietal identification is crucial. Cabezas et al. (2011) identified a standard set of 48 SNPs to identify grape cultivars. These SNPs have good coverage all over the genome and have a similar cultivar identification power as 16 SSR markers. More recently, Wang et al. (2022) established a SNP molecular identification system based on 25 KASP markers for grape cultivars. These markers were found to be highly efficient in cultivar identification with high accuracy and efficiency at a low cost.

4.3 Marker-Assisted Selection

Marker-assisted selection (MAS) is a method to identify plants that carry specific genetic markers associated with desirable traits. MAS allows breeders to quickly and accurately identify plants with the desired trait, thus reducing the time and resources needed for development of varieties by traditional breeding methods. Grape is a perennial crop and varietal development through conventional breeding takes a long time. Thus, MAS is a viable alternative for progeny selection at an early stage. In grape, MAS has been successfully used for the early selection of progenies for seedlessness and disease resistance.

4.3.1 Seedlessness

Seedless varieties are preferred by the consumers and therefore, developing seedless varieties is a major research focus for grape breeders. Early exclusion of seeded progenies gives a big boost to these efforts. Therefore, emphasis has been on developing markers for marker-assisted selection of seedless progenies. Striem et al. (1996) first reported that RAPD markers significantly correlated with seedlessness-related traits. Lahogue et al. (1998) developed a SCAR marker SCC8 closely linked to seedlessness locus *sd*I. The usefulness of this marker was demonstrated in seedless × seedless and seeded × seeded populations as well as a set of seeded and seedless varieties (Adam-Blondon et al. 2001). Another SCAR marker SCF27 developed by Mejía and Hinrichsen (2003) showed co-segregation with seedlessness. Detailed analyses identified microsatellite marker VMC7F2 (Cabezas et al. 2006; Mejía et al. 2007), closely linked to seedless QTL, and SDI, and thus useful for marker-assisted selection of seedless segregant. Stenospermocarpic bred cultivars contain 198-bp alleles at VMC7F2 and share a main haplotype around this locus not found in seeded cultivars (Karaagac et al. 2012). Mejía et al. (2011) proposed VvAGL11 as the most promising candidate gene for seedlessness and an intragenic STS marker p3_vvAGL11 tightly linked to the traits. This marker has been validated in several segregating populations (Bergamini et al. 2013).

4.3.2 Disease Resistance

MAS is specifically useful for breeding for disease resistance as MAS reduces the dependence of trait expression on the environment and allows the selection of desired plants without exposing them to pathogens. Grape production worldwide is mostly affected by mildew diseases and therefore research has focused on the development of markers closely linked to resistance loci for mildew diseases. Based on bulk segregant analysis, Dalbo (1998) identified one RAPD and one AFLP marker closely linked to major QTL for powdery mildew resistance in Illinois 547. These markers were converted to CAPS markers and showed strong association with powdery mildew in four different crosses (Dalbó et al. 2001). The resistance QTL in Illinois547 was later designated as REN2 and is located on chromosome 14 (www.vivc.de/). *M. rotundifolia* exhibits resistance against powdery mildew, which is determined by single dominant gene *RUN1*. Three AFLP markers closely linked to powdery mildew resistance gene *RUN1* were identified by Pauquet et al. (2001) in *M. rotundifolia*-derived BC5 full-sib family. These markers were absent in all the susceptible plants but present in all resistant individuals. In another study, Donald et al. (2002) identified three tightly linked resistance gene analog (RGA) markers closely linked to *RUN*1 locus. Two markers (GLP1–12 and MHD145) co-segregated with the resistance phenotype in 167 tested progenies, while the third marker (MHD98) was positioned 2.4 cM from the *Run1* locus. The detailed genetic and physical mapping of *RUN1* locus (Barker et al. 2005) allowed the identification of 20 new closely linked genetic markers including SSR marker VMC4f3.1. Powdery mildew resistance in *Vitis vinifera* variety Kishmish Vatkana is controlled by a single dominant gene *Ren1*. Katula-Debreceni et al. (2010) developed a multiplex PCR and agarose gel electrophoresis-based method for the simultaneous detection of both *Run1* and *Ren1* in the BC5 population and observed that the presence of at least one resistant allele imparted resistance against powdery mildew. Grape variety Regent is resistant to powdery mildew and is extensively used as a resistance source in breeding programs. Two SCAR markers linked to resistance loci in Regent were developed (Akkurt et al. 2007) and successfully used for the selection of resistant seedlings. This variety has *REN*3 and *REN*9 loci for powdery mildew, both located on chromosome 15 (Zendler et al. 2017). Subsequently, more powdery mildew resistance loci were mapped and associated markers have been identified from different grape genotypes. Table 4.1 contains details of known powdery mildew resistance loci and associated markers.

Downy mildew is another devastating disease in grapes and worldwide attempts are being made to understand the genetics of resistance against downy mildew. Several downy mildew resistance QTLs have been identified from different sources. Merdinoglu et al. (2003) were the first to identify molecular markers linked to downy mildew resistance genes in muscadine grapes using bulk segregant analysis. They identified QTL *RPV*1 and

TABLE 4.1

Summary of GWAS (Genome-Wide Association Mapping) in Grapevine

Sr. No	Traits	Populations Size	No. of Markers	No. of Marker-Trait Association	References
1	Berry color	289	5,110	Several	Myles et al. (2011)
2	Berry shape	304	4,241,729	10	Zhang et al. (2017b)
3	Muscat flavor	96	187	51	Yang et al. (2017)
4	Seedlessness	199	4,14,223	82	Zhang et al. (2017a)
5	Resistance to *Coniella diplodiella*	81	160	5	Zhang et al. (2017c)
6	Single berry weight (SBW), No. of bunches per plant (NBPP)	86	26,893	SBW-2, NBPP – 1	Marrano et al. (2018)
7	Flower sex (FS), Berry skin color (BSC), seedlessness (SL), Flavor (Fl), Cluster Weight (CW), wine acidity (WA), Maturity period (MP)	783	10,207	FS-2, BSC - 40, SL-8, Fl – 5, CW – 1, WA – 2, MP – 1	Laucou et al. (2018)
8	Berry color (BC), Berry development period (BDP), cluster size (CS), Berry weight (BW), Berry flesh texture (BFT), Berry flavor (BF)	179	32,311	BC -5, BDP – 2, CS – 2, BW – 4, BFT – 1, BF – 1	Guo et al. (2019)
9	Berry shape, Seeds, Cluster compactness, Berry composition, Aromatic composition, Berry weight, Brix	472	8.73-9.07 M	Several	Liang et al. (2019)
10	Resistance to *Plamopara viticola* in *Vitis vinifera*	132	12,828	3	Sargolzaei et al. (2020)
11	Resitance to *Colletotrichum* spp.	350	77,126	2	Jang et al. (2020)
12	Resistance to *Coniella diplodiella*	386	88,877	6	Zhang et al. (2020b)
13	Mite-related traits	399	4,523	1	LaPlante et al. (2021)
14	Seed-to-berry ratio	88	6.86 M	1	Magris et al. (2021)
15	Drought resistance	100	7,133	24	Trenti et al. (2021)
16	Cold tolerance	118	1.04M	2	Wang et al. (2021b)
17	Multiple traits	279	63,000	Several	Flutre et al. (2022)
18	Berry-related traits	356	1,599	52	Park et al. (2022)
19	Berry shape	279	5,66,129	122	Zhang et al. (2022a)
20	Berry cracking	287	6,01,261	5	Zhang et al. (2022b)

12 markers were found to be closely linked to this QTL. A major QTL for downy mildew resistance was identified by Welter et al. (2007) in a population derived from Regent and Lemberger. This QTL is located on chromosome 18 and a marker UDV112 explaining 37% variation was identified. Subsequently, several markers linked to various alleles of RPV3 detected in different grape genotypes have been identified (Bellin et al. 2009; Di Gaspero et al. 2012; van Heerden et al. 2014; Zyprian et al. 2016; Vezzulli et al. 2019). Welter et al. (2007) also identified a minor QTL Rpv4 on chromosome 4 and two associated markers were reported. In subsequent years, several workers reported the identification of downy mildew resistance QTLs in different genotypes and at present 35 QTLs for downy mildew resistance and their associated markers are known (www.vivc.de/ of Loci for Traits in Grapevine.pdf). These markers have been successfully used for marker-assisted selection of resistance progenies (Hvarleva et al. 2009; Saifert et al. 2018; Ruiz-García et al. 2021; Deore et al. 2021; Akkurt et al. 2022). Pierce's disease (PD) is economically significant in the southern United States and South America. Bacterium *Xylella fastidiosa* is the causal agent of PD. PdR1, a major gene for resistance against PD was mapped on LG 14, and two markers VMCNg3h8 and VVIN64 were closely linked to this gene (Riaz et al. 2006). Refined mapping of this region identified two alleles of PdR1 and closely linked markers for each allele (Riaz et al. 2008). These markers have been successfully used for MAS and introgression of PD resistance into wine, table and raisin grapes (Riaz et al. 2009).

4.4 Linkage and QTL Mapping

Linkage maps play a crucial role in plant genetics as a tool for identifying genomic regions for desired traits, genes and closely linked markers, thus improving breeding programs through marker-assisted selection. Early efforts using interspecific populations and the limited number of markers resulted in the development of several linkage maps for grapes (Lodhi et al. 1995; Grando et al. 2003). Riaz et al. (2004) developed a framework linkage map of *V. vinifera* using microsatellite markers. The mapping population of 153 plants was derived from a cross of grape cultivars Riesling and Cabernet Sauvignon. In total, 152 microsatellites and one EST marker were mapped. Adam-Blondon et al. (2004) mapped 245 SSR markers on the *V. vinifera* genome using two different mapping populations. Subsequently, several saturated maps have been reported both for *vinifera* as well as interspecific hybrids using SSR (Lowe and Walker 2006; Doligez et al. 2006a; Di Gaspero et al. 2007; Salmaso et al. 2008; Vezzulli et al. 2008; Riaz et al. 2012) and AFLP (Doucleff et al. 2004; Troggio et al. 2007) markers. With the availability of next-generation sequencing techniques, several sequencing-based marker systems were developed, which helped in the construction of high-density linkage maps for several grape genotypes. Linkage maps using SNP (Troggio et al. 2007; Salmaso et al. 2008; Vezzulli et al. 2008; Wang et al. 2012; Zhu et al. 2018), SRAP (Guo et al. 2014), SLAF-seq (Guo et al. 2015; Wang et al. 2017) markers have been reported. Most of these maps were developed for the identification of QTL regions for different traits like disease resistance, and morphological and quality traits.

Quantitative trait locus (QTL) mapping is a powerful tool to identify the genetic basis of complex traits. Mapping of genetic loci linked with these complex traits allows breeders to develop strategies for the selection of progenies, thus improving the efficacy and efficiency of breeding work. Identification of candidate genes in QTL regions for different traits helps better understand complex traits and develop strategies for the improvement of these traits through biotechnological approaches like transgenesis and genome editing. In grapes, QTL mapping has been used to study the genetic basis of many important horticultural traits like disease resistance, fruit quality, bunch traits and abiotic stress tolerance. The first report on QTL mapping for disease resistance was by Dalbó et al. (2001) who identified a major QTL for powdery mildew using an interspecific mapping population. Subsequently, several other QTLs have been identified for powdery mildew resistance in different grape genotypes (Fischer et al. 2004; Welter et al. 2007; Riaz et al. 2011; Barba et al. 2014). The first QTL for downy mildew resistance in Muscadine grapes, Rpv1 was mapped using BC2 segregating population (Merdinoglu et al. 2003). Subsequently, QTL mapping has been used to identify the genomic regions for downy mildew resistance in different *Vitis* species. The details of these QTLs and references can be accessed at www.vivc.de. The consumer acceptability for table grapes depends on the grape quality. Quality traits like seedlessness, berry size, berry weight, etc., are important and therefore, focus has been to identify the genetic determinants for these traits. QTL mapping using F1 progenies of two partially seedless genotypes identified major and minor QTLs for seedlessness and berry weight (Doligez et al. 2002). Later, Cabezas et al. (2006) and Mejía et al. (2007) confirmed major QTLs for seedlessness on LG18 and minor QTLs on different linkage groups. QTL analysis for berry size (Mejía et al. 2007; Richter et al. 2019), berry weight (Doligez et al. 2002; Fanizza et al. 2005a; Doligez et al. 2013; Wang et al. 2020) and other cluster architecture traits like rachis length, pedicle length (Correa et al. 2014; Richter et al. 2019) revealed presence for several major and minor alleles and influence of seasonal variation on these traits. Several common QTL regions were reported for these traits, indicating pleiotropic influence of these genomic regions. In the case of wine grapes, parameters like aroma, sugar, and acids determine the wine quality. The QTLs related to these quality parameters like aroma (Sevini et al. 2004; Doligez et al. 2006b; Battilana et al. 2009; Wang et al. 2020; Koyama et al. 2022), sugars and acidity (Chen et al. 2015; Bayo-Canha et al. 2019; Reshef et al. 2022) have been identified in different mapping populations. Besides these quality traits, QTLs for other parameters like berry firmness (Correa et al. 2016; Crespan et al. 2021), fruitfulness (Doligez et al. 2010), maturity period (Fechter et al. 2014; Zhao et al. 2016) have been identified. These analyses have not only helped identify the genomic regions controlling these complex traits but also have potential as useful resources for the identification of candidate genes and closely linked markers for their use in grape breeding.

4.5 Genome-Wide Association Analysis

Although linkage and QTL mapping have helped dissect the genetics of several traits in the grapevine, the information generated so far is limited. The requirement of segregating population, pose severe limitations for determining the genetic determinant of these traits in different genetic backgrounds. The genome-wide association mapping (GWAS) approach has opened up wide opportunities to fill these gaps. GWAS has the potential to identify genetic variants with high resolution without the need for mapping populations as GWAS can be performed on natural populations. With the advancement of NGS-based marker detection technologies, high-density SNPs are used for GWAS thus allowing the high resolution in detecting genomic regions associated with a trait. Due to these reasons, in recent years, grape researchers have focused on the use of GWAS for mapping complex traits especially commercially useful bunch and berry traits. Table 4.1 summarizes the reports of GWAS for the identification of loci for different traits.

4.6 Allele Mining

Any trait-specific grape improvement program is undertaken without compromising its commercial value, such as bunch and berry characters for table purposes; skin thickness and pulp content for raisin; while flavor and aroma components for wine purposes. But most of the commercial traits are very complex. Therefore, allele mining for specific genes of interest offers in-depth analysis for the identification of superior genotypes. In grape breeding, floral biology has a significant impact. Most of the commercial varieties belong to *V. vinifera* and consist of hermaphrodite flowers; while the wild relatives are dioecious in nature. The hermaphrodite nature is derived from the male wild genotypes (Picq et al. 2014). In the *Vitis*, *SEX-DETERMINING REGION (SDR)* is located on chromosome 2 (Riaz et al. 2006; Battilana et al. 2013). Genes *VviINP1* (mutation in gene lead to male sterility), *VviYABBY3* (mutation in gene lead to female sterility), *VviAPRT3* (pistil abortion), *VviFSEX* (associated with dioecy) and *VviPLATZ1* (controls female flower morphology) have been reported to be involved in the sex determination of the vines (Coito et al. 2017; Massonnet et al. 2020; Iocco-Corena et al. 2021). Environment plays a critical role in the phenological processes of the vine from bud burst to harvest, which are complex in nature and expressed from the gene and environment interaction. The *TERMINAL FLOWER 1 (TFL1)* gene is reported to be involved in flowering delay and ramose inflorescence (Fernandez et al. 2010). The higher frequency of allelic variation was reported due to changes in promoter and intron regions of the gene and were associated with berry size, bunch width and time of flowering (Fernandez et al. 2014). Differential response in flowering time genes reported includes *VvHUA2a* and *VvGAI1* for late flowering; while *VvTM6* and *VvWNK6* for early flowering (Boss and Thomas 2002; Kamal et al. 2019).

Being the influencing trait for the economic value of the produce, bunch architecture has acquired much attention. Using the SNP approach, several allelic variations are detected for bunch architecture-related traits (Tello et al. 2016; Grimplet et al. 2019). Allelic variation in *VvGRF4* transcription factor has been reported to be associated with bunch compactness (Richter et al. 2020; Rossmann et al. 2020). The sequence analysis of cell wall protein-encoding gene "*VviUCC1*" has detected many SNPs involved in variation in cluster compactness-related traits (Tello et al. 2016, 2020). Most of the selections derived from mutations and hybridizations are based on the morphological appearance of the berry and bunch. The gene *VvNAC26* is reported to be associated with berry development. Variations in this gene is mainly originated from SNPs and INDELs (Tello et al. 2015; Muñoz-Espinoza et al. 2020). This gene has also shown its interaction with MADS-box gene "*VvMADS9*" associated with berry and seed development (Zhang et al. 2021a). Concerning berry firmness, potential genes associated were linked to cation exchange, cellulose synthase, invertase, abscisic-aldehyde oxidase-like, endoglucanase 3 genes, and transcription factors (Correa et al. 2016; Jiang et al. 2020; Ma et al. 2020). The berry ripening is a complex process and several signaling pathways are involved in it (Fortes et al. 2015; Balic et al. 2018; Li et al. 2022). Ethylene (1-aminocyclopropane-1-carboxylic acid oxidase (ACO) in ethylene synthesis) has played an important role in berry ripening. A NAC transcription factor "*VviCARPO*" (Controlled Adjustment of Ripening and maturation of Plant Organs) is identified as a key berry maturation factor. It upregulates genes (*VviSGR1* and *VviMYBA1*) involved in chlorophyll degradation and anthocyanin accumulation.

Berry's skin color has a significant impact on the quality and value of the product. A wide range of color variations from green-yellow to dark blue are documented, originating from natural or artificial selections expressed due to differences in intensity and composition of polyphenol content. Studies have confirmed the role of Myb transcription factors and variations in *VvMybA* genes responsible for skin color (Kobayashi et al. 2002; Kobayashi et al. 2004;

Walker et al. 2006; Azuma et al. 2007; This et al. 2007; Fournier-Level et al. 2010; Carrasco et al. 2015; Ferreira et al. 2019). These variations in *VvMYBA* genes resulted from mutations, insertions (*Gret1* in the promoter region) or deletions. Also, the colored mesocarp variations are the outcome of insertions of variable repeats in the *VvMybA1* promoter region expressed as differences in anthocyanin content (Röckel et al. 2020). Among the selection criteria for the table grape variety, seedlessness lies in the top priority. Therefore, decoding the genetics of this trait plays a crucial role in grape improvement. In *V. vinifera*, seed abortion due to stenospermocarpy is linked with *SEED DEVELOPMENT INHIBITOR (SDI)* locus. The detrimental point mutations in the two genes accommodated in the *sdi* locus resulted in stenospermocarpy due to the substitution of arginine to leucine in the *VviAGL11* gene (Royo et al. 2018). Further, B3 genes showed differential expression in seeded and seedless cultivars (Ahmad et al. 2019). Also, the role of homeobox transcription factor and DNA methylation of *VvHB58* gene have been suggested to play a role in ovule abortion in seedless cultivars (Li et al. 2017). Several recessive genes are documented to be involved in seedlessness. Three candidate genes *viz. Vitvi01g00455, Vitvi08g01528* and *Vitvi18g01237* have exhibited a recessive effect on seedlessness (Kim et al. 2020). The sensory component has significant relevance to wines. Aromas are the results of several complex metabolic pathways (major aroma-contributing components: mono- and sesquiterpenes, methoxypyrazines, furan derivatives, lipoxygenase pathway products and phenylpropanoid pathway products) (Li et al. 2019). Several aromatic varieties comprise a compound α-terpineol (VviTer, α-terpineol synthase) (Martin and Bohlmann 2004; Yang et al. 2017). Also, sesquiterpene synthases (terpene synthase (*VviTPS*) genes) and their variants (VviTPS04 and VviTPS10, VvTPS24) are well documented for flavor and aroma (Drew et al. 2015; Smit et al. 2021). A gene responsible for muscat flavor (admired largely in wine industry) is 1-deoxy-D-xylulose 5-phosphate synthase (*VviDXS*), which is located on chromosome 5 (Battilana et al. 2009; Dunchene et al. 2017). Variants of these alleles are reported in different cultivars (Emanuelli et al. 2010; Dunchene et al. 2017; Morcia et al. 2021). The presence of green bell pepper aroma is known to be due to 3-isobutyl-2-methoxypyrazine. Alleles of *VviOMT3* were found responsible for methoxypyrazine in Cabernet Sauvignon (Dunlevy et al. 2013).

Grapes are being cultivated in wider environmental regimes and hence adapted several mechanisms for survival under various stresses. Abscisic acid (ABA) is known to be a principal component associated with stress tolerance under water deficit, salinity, chilling, etc. ABA receptor genes (*VvPYL1, VvPYL4, VvPYL7* and *VvPYL8*) and peroxiredoxin *PrxII C* gene (*VvPrxII* C) respond under abiotic stress (Haddad and Japelaghi 2015; Yanxia 2020). Further, some secondary metabolites such as resveratrol synthase (*STS1*), flavonoid, and stilbene synthesis (*STS11, STS12* and *STS13*) genes also contribute to stress tolerance. Grapevine is prone to several fungal diseases with the downy mildew and powdery mildew as the major ones. Downy mildew is mainly observed in humid regions. Using GWAS approach, three loci (*Rpv29*, *Rpv30* and *Rpv31*) against *Plasmopara viticola* were identified, which will be useful for the development of homozygous resistant lines similar to that of *Rpv1* and *Rpv3* loci (Sánchez-Mora et al. 2017; Sargolzaei et al. 2020). Low sporulation level was noted in the genotypes harbouring *Rpv29*, *Rpv30* and *Rpv31* loci. Mapping of resistance interval underlying TIR-NB-LRR (TNL) gene pair in wild grape haplotype (Rpv3-1) was performed by Foria et al. (2020). The imbalance of allelic expression in resistant heterozygotes was observed indicating the role of cis-acting regulatory variation in the expression of wild haplotypes. The diversity in haplotypes mainly originated from the movement of transposable elements. Deep sequencing of four putative genes for downy mildew susceptibility (*VvDMR6.1*, *VvDMR6.2*, *VvDLO1* and *VvDLO2*) has led to the identification of new source recessive inherited resistance (Pirrello et al. 2021). Two haplotypes identified at *VvDMR6.2* were notably presented in resistant genotypes. The information will be helpful in the molecular-assisted breeding program. Stilbenes synthase (STS) is a principal enzyme for phytoalexin resveratrol production and disease resistance. A resistant allele (*VpSTS*) from *Vitis pseudoreticulata* shares around 99% amino acid identity for the coding region with the susceptible *VvSTS* allele from *V. vinifera*; while the significant difference was mainly observed for the promoter region (Jiao et al. 2016). Elevated salicylic acid and the pathogen-related response of the promoter at the molecular level was reported in wild allele. The higher accumulation of resveratrol is also reported in some *Vitis sylvestris* clones (Duan et al. 2016). Differences in the transcripts of *MYB14* promotors have been observed in Hoe29 and of Ke83 genotypes. Only the *MYB14* promotors of Hoe29 responded against *Plasmopara viticola* infection.

4.7 Somatic Embryogenesis

Somatic embryogenesis is a remarkable biological phenomenon in which non-zygotic somatic cells differentiate into embryos to regenerate fertile plants (Rose et al. 2010). This is an asexual process that utilizes the intrinsic totipotency of somatic cells to de-differentiate and re-differentiate into fully functional plants (Sahrawat et al. 2003). The first somatic embryogenesis was reported by Steward et al. (1958) in *Daucus carota* cell suspensions. During the past few decades, somatic embryogenesis has been described in a large number of plants (George et al. 2008). There are two pathways of somatic embryogenesis viz., indirect and direct based on the presence or absence of callus, respectively. Most of the somatic embryogenesis protocols consist of the initiation of embryogenic cultures, multiplication/proliferation of embryogenic calli, maturation of somatic embryos and plant regeneration from these embryos. For the initiation of embryogenesis, somatic plant cells contain all the genetic information necessary to create a complete plant. Plant growth hormones and stresses induced during cutting and preparing explants play an important role in mediating signal transduction cascade leading to DNA methylation and down regulation of gene expression. This results in a series of cell divisions producing unorganized callus that finally leads to somatic embryogenesis (Dudits et al. 1995). Explant source is important for the success of somatic embryogenesis. The average percentage of somatic embryogenesis can vary between different tissues of the same genotype. The percentage of somatic embryogenesis obtained from ovaries was reported to be about 2-fold higher than that of anthers (Vidal et al. 2009). The culture medium provides critical nutrients for the growth of explants at different developmental stages, and its composition is a major factor determining the probability of successful plant regeneration. Although several tissue culture media are available, MS (Murashige and Skoog 1962) and NN (Nitsch and Nitsch 1969) are widely used in grapes. The hormones play different roles at different stages, such as embryogenic culture initiation, embryo development, embryo maturation and embryo germination. Somatic embryogenesis was first developed in grapes in 1970 in order to recover dihaploid plants for improvement programs in grapes (Martinelli and Gribaudo 2001). Since then, many attempts have been made for the establishment of haploid plants through anther culture. Recently, somatic embryogenesis has been used in many improvement programs for grapes and is one of the most important tools for any kind of *in vitro* improvement of a crop. Types of grape explants, media and hormonal compositions used in some of the successful reports on grape somatic embryogenesis are summarised in Table 4.2. In addition to hormones, some of the media given in Table 4.2 also contain other additives such as activated charcoal, polyvinylpyrrolidone, casein hydrolysate, etc.

The developmental stages of somatic embryogenesis resemble zygotic embryogenesis in terms of developmental and regulatory mechanisms (Dodeman et al. 1997). Therefore, somatic embryogenesis provides an excellent model for studying zygotic embryo development in higher plants. In addition to the developmental studies, somatic embryos have been used for cryopreservation employing encapsulation-dehydration or encapsulation-vitrification in grape cultivars and rootstocks (Wang et al. 2004). Encapsulation of mature embryos has been proposed as an alternate technology for storage in seed repositories. No viruses were detected in plants raised from somatic embryos from mother plants infected with phloem limited viruses (Gambino et al. 2009). Grapevine Fanleaf Virus (GFLV) could be eliminated through somatic embryogenesis alone (Gambino et al. 2009) or in combination with thermotherapy. Somatic embryogenesis also has the potential to produce a large number of plants through *in vitro* propagation and has now become a routine protocol for micropropagation in many plants. More recently, more emphasis is being given on production of full-grown transgenic plants from genetically transformed cells. All these aspects have been reviewed by Guan et al. (2016).

TABLE 4.2

Explants, Type of Media, and Hormonal Combinations Used in Some of the Somatic Embryogenesis Protocols in Grapes

	Culture Medium Used. Hormonal Concentrations are Given in Brackets as µM			
Explant	**Induction of Embryogenic Callus**	**Induction of Somatic Embryos**	**Somatic Embryo Germination and/or Plant Regeneration**	**References**
Anther filaments	½ MS + 2,4-D (1) + TDZ (1) or ½ MS + 2,4-D (10) + TDZ (10)	½ MS	Not available	Nakajima and Matsuta (2003)
Flowers, anthers & ovaries	NN major elements + MS minor elements including Fe-EDTA + B-5 vitamins + 2,4-D (4.5) + BAP (8.9)	NN major elements + MS minor elements including Fe-EDTA + B-5 vitamins + NOA (10) + BAP (1) + IAA (20)	Not available	Gambino et al. (2007)
Stigma/Style	NN + BAP (9) + NOA (5)	NN	NN	Morgana et al. (2004)
Ovules, stigma, style and anthers	NN + BAP (4.5 or 9) + NOA (5 or 9.9)	NN	NN	Carimi et al. (2005)
Anthers and ovaries	NN + 2,4-D (2.5) + NOA (2.5) + 4-CPPU (5)	½ MS	½ MS	Kikkert et al. (2005)
Anthers and ovaries	MS + 2,4-D (4) + BA (1 or 1.3)	½ MS	½ MS	López-Pérez et al. (2008)
Nodal and leaf explants	½ MS + 2,4-D (9) + BAP (4.5)	½ MS + IAA (20) + NOA (10) + BAP (1)	½ MS + BAP (0.5)	Maillot et al. (2006)
Gynoecia and anthers	NN + 2,4-D (4.5) + BAP (8.9)	NN + BAP (13.3)	MS + BAP (4.4)	Pinto-Sintra (2007)
Buds	NB + 2,4-D (5) + BA (2.5) + NAA (5)	Modified MS	C2D + BA (0.4) + IBA (0.25)	Araya et al. (2008)
Anthers and ovaries	NN major elements + MS minor elements including Fe-EDTA + B-5 vitamins + 2,4-D (4.5) + BAP (8.9)	½ MS	½ MS + IAA (10) + GA (1)	Cadavid Labrada et al. (2008)
Leaves, pistils and stamen	MS + 2,4-D (5) + BAP (4.4)	Modified MS	Modified MS + BA (1)	Dhekney et al. (2009)
Anthers and ovaries	NN salts + MS vitamins + 2,4-D (2.5) + NOA (2.5) + 4-CPPU (5)	NN salts + MS vitamins + 2,4-D (2.5) + NOA (2.5) + 4-CPPU (5)	½ MS	Vidal et al. (2009)

TABLE 4.2 (Continued)

Explants, Type of Media, and Hormonal Combinations Used in Some of the Somatic Embryogenesis Protocols in Grapes

	Culture Medium Used. Hormonal Concentrations are Given in Brackets as µM			
Explant	**Induction of Embryogenic Callus**	**Induction of Somatic Embryos**	**Somatic Embryo Germination and/or Plant Regeneration**	**References**
Anthers and ovaries	NN salts + MS vitamins + 2,4-D (4.5) + BA (9)	NN salts + MS vitamins + NOA (10) + BA (1) + IAA (20)	NN salts + MS vitamins + IAA (10) + GA (1)	Prado et al. (2010)
Leaves and petioles	NN + 2,4-D (9) + BA (4.4)	Modified MS	MS + BA (1)	Dhekney et al. (2011a)
Stamen filaments	NN salts + MS vitamins + 2,4-D (4.5) + TDZ (4.5)	NN salts + MS vitamins	½ MS + BA (0.5) + IBA (0.05)	Acanda et al. (2013)
Leaves	½ MS + BA (4.5) + NOA (5)	½ MS	WPM + BA (.05) + IBA (0.05)	Nookaraju and Agrawal (2013)
Nodal explants	½ MS + 2,4-D (18) + BAP (9)	½ MS + IAA (20) + NOA (10) + BAP (1)	½ MS + BAP (0.5)	Maillot et al. (2016)
Flower bud filaments from dormant cane	½ MS + 2,4-D (1) + TDZ (1)	MS	MS + NAA (0.1) + BAP (2.5)	Nakajima et al. (2020)

B-5 = B-5 Medium; C2D = C2D Medium; MS = Murashige and Skoog Medium; NN = Nitsch and Nitsch Medium; WPM = Woody Plant Medium; NB = Modified medium NB® (PhytoTechnology Labs., Shawnee Mission, KS); 2,4-D = 2,4–dichlorophenoxyacetic acid; 4-CPPU = N-(2-chloro-4-pyridyl)-N-phenoxylurea; ABA = Abscisic acid; BA = 6-Benzyladenine; BAP = 6-Benzylaminopurine; GA = Gibberellic acid; IAA = Indol acetic acid; IASP = Indol-3-acetyl-L-aspartic acid; IBA = Indol butyric acid; NAA = Napthalene acetic acid; NOA = 2-napthoxyacetic acid; TDZ = Thidiazuron.

Source: Modified from Zhang et al. (2021b).

4.8 Genetic Transformation

Genetic improvement of cultivated plants is one of the major strategies to improve food production. In crop plants, genetic make-up, agro-techniques and biotic/abiotic stress-related factors can affect yield contribution to the extent of about 50–60%, 25–30% and 20–35%, respectively. Thus, the development of improved cultivars having superior genetic make-up with an in-built ability to fight biotic/abiotic stresses would be one of the major inputs to accomplish increased production. Although, classical breeding led to the development of high-yielding varieties of almost all crops and significantly improved food production, it generally remains a "hit or miss" technique. Genetic engineering or genetic transformation has two distinct advantages over classical breeding. The first is selectivity. In classical breeding, the entire genome of parent plants is recombined, thereby, needing several cycles of backcrossing and selection to eliminate undesirable traits. In contrast, genetic transformation can introduce a single gene for a desired trait without disturbing a plant's genetic make-up. Secondly, classical breeding is confined to only cross-fertile plants and thus uses limited variability. Recent developments in molecular biology and genetic transformation have made it possible to identify, isolate and transfer any desirable gene from any living organism to plants. This has brought the whole genetic diversity of the living world together, which was unthinkable some 40 years ago. From the information provided by Zhang et al. (2021b), some of the important transgenic attempts made during the last two decades are summarized in Table 4.3.

To date, grape improvement through genetic transformation has mainly been focused on biotic stress resistance (fungi and viruses). *Agrobacterium*-mediated transformation is the method of choice for stable transformation of grapevine, but most of the genetic transformation is concentrated on only a few grape varieties. The development of efficient somatic embryogenesis protocols has significantly accelerated the regeneration of transgenic plants from transformed cells. Biosafety issues and strict regulations are likely to restrict genetic transformation activities to rootstock improvement only. However, recent developments in gene-editing techniques open safer frontiers in grape improvement using molecular tools. Gene-editing techniques (mainly CRISPR/Cas9) coupled with efficient plant regeneration through somatic embryogenesis are likely to facilitate the production of a large number of superior plants for diverse economically important traits in several grapevine cultivars and rootstocks.

4.9 Gene/Genome Editing

Gene or genome editing is the sequence-specific modifications of the genomes. Genome-editing tools in combination with the available grape genome sequence has the potential to revolutionize grape improvement. Among many genome-editing tools developed during the last few years, clustered regularly interspaced short palindromic repeats/CRISPR-associated protein 9 system (CRISPR/Cas9) has been successfully demonstrated in grapes. Five types of CRISPR/Cas9 target sites have been identified and characterized in grapes (Wang et al. 2016). Ren et al. (2016) used suspension cells of the grape

TABLE 4.3

Some of the Attempts That Have Obtained Stable Grapevine Transgenics Through Somatic Embryogenesis

Explants	Transgene(s)	Transformation Method	Objective	References
Somatic embryos	*GFLV-CP* (Grapevine fanleaf virus coat protein) and 4 antifreeze protein genes (*Atf11, Atf62, Atf78* and *B5*)	*Agrobacterium tumefaciens*	Resistance to GFLV	Gutoranov et al. (2001)
Embryonic calli	*PGIP* (pear polygalacturonase inhibiting protein)	*A. tumefaciens*	Resistance to Gray mold	Agüero et al. (2005)
Leaf discs	*Chitinase* and *RIP* (ribosome-inactivating protein)	*A. tumefaciens*	Resistance to fungal diseases	Bornhoff et al. (2005)
Somatic embryos	*egfp* and *codA* (cytosine deaminase)	*A. tumefaciens*	Production of marker-free transgenics	Dutt et al. (2008)
Embryogenic callus	*IR MPc GFLV* (inverted repeat silencing movement protein from GFLV)	*A. tumefaciens*	Resistance to GFLV	Jardak-Jamoussi et al. (2009)
Embryogenic callus	*nptII* gene and XVE-*Cre/loxP* sequence	*A. tumefaciens*	To develop protocol to remove marker gene	Dalla Costa et al. (2010)
Somatic embryos	Overexpression of *vvtl-1* (a thaumatin like protein)	*A. tumefaciens*	Broad spectrum fungal disease resistance.	Dhekney et al. (2011b)
Somatic embryos	Overexpression of *chitinase* and *β-1,3-glucanase* genes.	*A. tumefaciens*	Tolerance to downy mildew.	Nookaraju and Agrawal (2012)
Somatic embryos	Overexpression of *LIMA-A* (Synthetic gene encoding lytic peptide)	*A. tumefaciens*	Resistance to Pierces's disease	Li et al. (2015)
Embryogenic callus	Overexpression of *VaTLP* (thaumatin like protein)	*A. tumefaciens*	Downy mildew resistance	He et al. (2017)
Embryogenic calli	Overexpression of *VaPUB* (stress responsive U-box protein gene)	*A. tumefaciens*	Disease resistance	Jiao et al. (2017)
Somatic embryos	Overexpression of *VpPR10.1*	*A. tumefaciens*	Downy mildew resistance	Ma et al. (2018)
Pro-embryonic mass	CRISPR/Cas9 *VvWRKY52* (WRKY transcription factor)	*A. tumefaciens*	CRISPR/Cas-9 protocol development	Wang et al. (2018)

(continued)

TABLE 4.3 (Continued)

Some of the Attempts That Have Obtained Stable Grapevine Transgenics Through Somatic Embryogenesis

Explants	Transgene(s)	Transformation Method	Objective	References
Leaf discs	Overexpression of *AgNHX1* (Na^+/H^+ antiporter gene)	*A. tumefaciens*	Salt tolerance	Soliman (2018)
Proembryonic mass	Overexpression of *VpSTS29/STS-2*	*A. tumefaciens*	Powdery mildew resistance	Xu et al. (2019)
Proembryonic mass	Overexpression of *VlbZIP30* (leucine zipper transcription factor)	*A. tumefaciens*	Drought tolerance	Tu et al. (2020)
Proembryonic mass	CRISPR/Cas9 *VvMLO3* and *VvMLO4*	*A. tumefaciens*	Powdery mildew resistance	Wan et al. (2020)

variety Chardonnay to edit the L-idonate dehydrogenase gene (*IdnDH*) and observed 100% mutation frequency and no off-target events were detected. In another study, Nakajima et al. (2017) transformed the embryonic calli of Neo Muscat with a synthetic sgRNA expression construct targeting the phytoene desaturase (*VvPDS*) gene. The regenerated plants were albino; however, the ratio of mutated cells was higher in lower, older leaves compared to that in newly appearing upper leaves. CRISPR/Cas9 editing efficiency depends on various factors. While optimizing editing efficiency in grape, Ren et al. (2019) observed that single guide RNA (sgRNA) with 65% GC content yielded the highest editing efficiency. They also observed genotype's effect on editing efficiency and efficiency was more in grape variety "41B" than in "Chardonnay". Among the different promoters studied, *VvU3/U6* and *UBQ2* promoters could significantly increase the editing efficiency in grape (Ren et al. 2021) and were useful for efficient multiplex genome editing of tonoplastic monosaccharide transporters TMT1 and TMT2. Using whole genome sequence analysis of Cas9 edited and wild-type grape plants, Wang et al. (2021a) established high specificity of CRISPR/Cas9 for genome editing of grapevine. In grapes, targeted gene editing has been used to improve resistance against powdery mildew by modifying *MLO* genes (Malnoy et al. 2016; Giacomelli et al. 2019; Wan et al. 2020) and *VviDEL1* gene (Olivares et al. 2021), resistance against downy mildew by editing *DMR6* (Giacomelli et al. 2022) and *VvPR4* gene (Jiang et al. 2020), whereas CRISPR/cas9 assisted knockout of *VvWRKY52* increased the resistance to *Botrytis* (Wang et al. 2018). CRISPR/Cas9 editing also demonstrated the role of SL biosynthesis genes *VvCCD8* in the control of grapevine shoot branching (Ren et al. 2020). Sunitha and Rock (2020) reported the usefulness of this technology for the functional characterization of non-coding RNA and MYB transcription factors.

4.10 Crop Biotechnology and Biosafety Regulatory Framework

The application of biotechnology in crop improvement is multifaceted. Molecular breeding, genetic engineering and genome editing are the tools of biotechnology that can be exploited for grape improvement at a commercial scale. As grape molecular breeding uses marker-assisted selection and does not use any transgene directly, products developed through molecular breeding or MAS need not undergo any biosafety regulation before being released as a variety. However, varietal development through transgenic approaches requires biosafety regulation clearance. Many transgenic grapes have been developed and tested in the field in countries like the USA, Australia and some European countries (Campos et al. 2021). However, there is no commercially released transgenic grape cultivar. Transgenic plants developed by genetic modification contains transgenes from a diverse source. To overcome the risk associated with genetically modified crops and products derived from them, transgenic crops are tested for their biosafety. Biosafety assessment of transgenic plants is extensive and mainly includes molecular characterization, allergenicity and toxicity assessment and environmental impact. Many countries have their own biosafety regulatory framework. Moreover, there are institutes and protocols such as the Cartagena Protocol, Codex Alimentarius Commission (CAC), etc., which give guidelines and control the movement of genetically modified organisms between the countries [Regulation (EC) No 1946/2003; Cartagena Protocol on Biosafety, 2000]. The European Union has very stringent biosafety regulations and a complex legal framework for GMOs. The European legislation on GMOs consists of many legal instruments such as Directive 90/219/EEC, Directive 90/220/EEC, Directive 2009/41/EC, etc. These instruments are supported by a number of guidelines, rules and recommendations. The USA Government has its own biosafety regulations, policies and guidelines for GMOs. Different agencies are extensively involved in the biosafety framework in the USA such as USDA's Animal and Plant Health Inspection Service (USDA-APHIS), Food and Drug Administration (FDA), Environmental Protection Agency (EPA), Department of Health and Human Services, etc.

The regulation of genome editing and other new breeding techniques varies from country to country (El-Mounadi et al. 2020). The existing biosafety framework for genome-edited crop plants is mainly process-based and product-based. The process-based approach mainly focuses on the technique of origin involved in the development of new crop plant variety; whereas, the product-based approach mainly focusses on the risk caused by the food product derived from genome-edited crops. Countries that follow a product-based approach do not consider genome-edited crops as GMOs rather they consider them as the mutagenesis-derived products (Zhang et al. 2020a). However, countries that follow a process-based approach consider genome-edited crop plants as a GMO. Countries like America, Australia, China Brazil, Chile, etc., follow a product-based approach and European Union follows a process-based approach (Wolt et al. 2016; Chen et al. 2019). Other than this, genome genome-edited crops can be categorized into three types namely site-directed nuclease-1 (SDN-1), site-directed nuclease-2 (SDN-2) and site-directed nuclease-3 (SDN-3). Genome-edited plants having small insertions or deletions and no foreign DNA comes under SDN-1. In the SDN-2 category, up to 20 nucleotide changes can be allowed for gene correction or modification at one or more places using the homologous template. Genome-edited plants having foreign gene comes under the SDN-3 category (Wolt 2017). Based on the above classification, different countries have adapted diverse regulatory frameworks for genome-edited plants.

4.11 Conclusion and Future Prospects

So far, several biotechnological tools especially molecular marker-based techniques have enhanced our knowledge of the genetic architecture of grapes and have been used for grape improvement. The future of biotechnology for grape improvement looks promising. Further advancement in gene editing, high-throughput phenotyping, and omics techniques have the potential to revolutionize the ways biotechnology could be used to improve the grapes and develop varieties that meet the needs of growers, consumers and the environment.

References

Acanda Y, Prado MJ, González MV, Rey M. 2013. Somatic embryogenesis from stamen filaments in grapevine (*Vitis vinifera* L. cv. Mencía): Changes in ploidy level and nuclear DNA content. *In Vitro Cell Dev Biol Plant* **49**: 276–284, doi: 10.1007/s11627-013-9499-7.

Adam-Blondon A-F, Lahogue-Esnault F, Bouquet A, Boursiquot JM, This P. 2001. Usefulness of two SCAR markers for marker-assisted selection of seedless grapevine cultivars. *Vitis* **40**: 147–155, doi: 10.5073/vitis.2001.40.147-155.

Adam-Blondon AF, Roux C, Claux D, Butterlin G, Merdinoglu D, This P. 2004. Mapping 245 SSR markers on the *Vitis vinifera* genome: A tool for grape genetics. *Theor Appl Genet* **109**: 1017–1027, doi: 10.1007/s00122-004-1704-y.

Agüero C, Uratsu S, Greve C, Powell A, Labavitch J, Meredith C, Dandekar A. 2005. Evaluation of tolerance to Pierce's disease and Botrytis in transgenic plants of *Vitis vinifera* L. expressing the pear PGIP gene. *Mol Plant Pathol* **6**: 43–51, doi: 10.1111/j.1364-3703.2004.00262.x.

Ahmad B, Zhang S, Yao J, Rahman MU, Hanif M, Zhu Y, Wang X. 2019. Genomic organization of the B3-domain transcription factor family in grapevine (*Vitis vinifera* L.) and expression during seed development in seedless and seeded cultivars. *Int J Mol Sci* **20**: 4553, doi: 10.3390/ijms20184553.

Akkak A, Boccacci P, Botta R. 2007. Cardinal grape parentage: A case of a breeding mistake. *Genome* **50**: 325–328, doi: 10.1139/g06-145.

Akkurt M, Şenses I, Aktürk B, Tozlu I, Özer N, Uzun HI. 2022. Marker assisted selection (MAS) for downy mildew resistance in grapevines using *Rpv3.1* associated markers. *Not Bot Horti Agrobo* **50**: 12708, doi: 10.15835/nbha50112708.

Akkurt M, Welter L, Maul E, Töpfer R, Zyprian E. 2007. Development of SCAR markers linked to powdery mildew (*Uncinula necator*) resistance in grapevine (*Vitis vinifera* L. and Vitis sp.). *Mol Breed* **19**: 103–111, doi: 10.1007/s11032-006-9047-9.

Araya S, Prieto H, Hinrichsen P. 2008. An efficient buds culture method for the regeneration via somatic embryogenesis of table grapes 'Red Globe' and 'Flame Seedless'. *Vitis* **47**: 251, doi: 10.5073/vitis.2008.47.251-252.

Azuma A, Kobayashi S, Yakushiji H, Yamada M, Mitani N, Sato A. 2007. *VvmybA1* genotype determines grape skin color. *Vitis* **46**: 154–155, doi: 10.5073/vitis.2007.46.154–155.

Balic I, Vizoso P, Nilo-Poyanco R, Sanhueza D, Olmedo P, Sepúlveda P, Arriagada C, Defilippi BG, Meneses C, Campos-Vargas R. 2018. Transcriptome analysis during ripening of table grape berry cv. Thompson Seedless. *PLoS One* **13**: e0190087, doi: 10.1371/journal.pone.0190087.

Baneh HD, Mohammadi SA, Mahmoudzadeh H, Mattia Fd, Labra M. 2009. Analysis of SSR and AFLP markers to detect genetic diversity among selected clones of grapevine (*Vitis vinifera* L.) cv. Keshmeshi. *South African J Enol Vitic* **30**: 38–42, doi: 10.21548/30-1-1422.

Baránková K, Sotolář R, Baránek M. 2020. Identification of rare traditional grapevine cultivars using SSR markers and their geographical location within the Czech Republic. *Czech J Genet Plant Breed* **56**: 71–78, doi: 10.17221/61/2019-CJGPB.

Barba P, Cadle-Davidson L, Harriman J, Glaubitz JC, Brooks S, Hyma K, Reisch B. 2014. Grapevine powdery mildew resistance and susceptibility loci identified on a high-resolution SNP map. *Theor Appl Genet* **127**: 73–84, doi: 10.1007/s00122-013-2202-x.

Barker CL, Donald T, Pauquet J, Ratnaparkhe MB, Bouquet A, Adam-Blondon AF, Thomas MR, Dry I. 2005. Genetic and physical mapping of the grapevine powdery mildew resistance gene, *Run*1, using a bacterial artificial chromosome library. *Theor Appl Genet* **111**: 370–377, doi: 10.1007/s00122-005-2030-8.

Battilana J, Costantini L, Emanuelli F, Sevini F, Segala C, Moser S, Velasco R, Versini G, Grando MS. 2009. The 1-deoxy-d-xylulose 5-phosphate synthase gene co-localizes with a major QTL affecting monoterpene content in grapevine. *Theor Appl Genet* **118**: 653–669, doi: 10.1007/s00122-008-0927-8.

Battilana J, Lorenzi S, Moreira FM, Moreno-Sanz P, Failla O, Emanuelli F, Grando MS. 2013. Linkage mapping and molecular diversity at the flower sex locus in wild and cultivated grapevine reveal a prominent SSR haplotype in hermaphrodite plants. *Mol Biotechnol* **54**: 1031–1037, doi: 10.1007/s12033-013-9657-5.

Bayo-Canha A, Costantini L, Fernández-Fernández JI, Martínez-Cutillas A, Ruiz-García L. 2019. QTLs related to berry acidity identified in a wine grapevine population grown in warm weather. *Plant Mol Biol Rep* **37**: 157–169, doi: 10.1007/s11105-019-01145-6.

Bellin D, Peressotti E, Merdinoglu D, Wiedemann-Merdinoglu S, Adam-Blondon AF, Cipriani G, Morgante M, Testolin R, Di Gaspero G. 2009. Resistance to *Plasmopara viticola* in grapevine 'Bianca' is controlled by a major dominant gene causing localised necrosis at the infection site. *Theor Appl Genet* **120**: 163–176, doi: 10.1007/s00122-009-1167-2.

Bergamini C, Cardone MF, Anaclerio A, Perniola R, Pichierri A, Genghi R, Alba V, Forleo LR, Caputo AR, Montemurro C et al. 2013. Validation assay of p3_VvAGL11 marker in a wide range of genetic background for early selection of stenospermocarpy in *Vitis vinifera* L. *Mol Biotechnol* **54**: 1021–1030, doi: 10.1007/s12033-013-9654-8.

Bianchi D, Brancadoro L, De Lorenzis G. 2020. Genetic diversity and population structure in a *Vitis* spp. core collection investigated by SNP markers. *Diversity* **12**: 103, doi: 10.3390/d12030103.

Bibi AC, Gonias ED, Doulis AG. 2020. Genetic diversity and structure analysis assessed by SSR markers in a large collection of *Vitis* cultivars from the Island of Crete, Greece. *Biochem Genet* **58**: 294–321, doi: 10.1007/s10528-019-09943-z.

Blaich R, Konradi J, Rühl E, Forneck A. 2007. Assessing genetic variation among Pinot noir (*Vitis vinifera* L.) clones with AFLP markers. *Am J Enol Vitic* **58**: 526–529.

Bornhoff BA, Harst M, Zyprian E, Töpfer R. 2005. Transgenic plants of *Vitis vinifera* cv. Seyval blanc. *Plant Cell Rep* **24**: 433–438.

Boss PK, Thomas MR. 2002. Association of dwarfism and floral induction with a grape 'green revolution' mutation. *Nature* **416**: 847–850, doi: 10.1038/416847a.

Bourquin JC, Otten L, Walter B. 1995. PCR-RFLP analysis of *Vitis, Ampelopsis* and *Parthenocissus* and its application to the identification of rootstocks. *Vitis* **34**: 103–108, doi: 10.5073/vitis.1995.34.103-108.

Bourquin JC, Sonko A, Otten L, Walter B. 1993. Restriction fragment length polymorphism and molecular taxonomy in *Vitis vinifera* L. *Theor Appl Genet* **87**: 431–438, doi: 10.1007/BF00215088.

Bowers J, Boursiquot JM, This P, Chu K, Johansson H, Meredith C. 1999a. Historical genetics: The parentage of Chardonnay, Gamay, and other wine grapes of northeastern France. *Science* **285**: 1562–1565.

Bowers JE, Dangl GS, Meredith CP. 1999b. Development and characterization of additional microsatellite DNA markers for grape. *Am J Enol Vitic* **50**: 243–246.

Bowers JE, Dangl GS, Vignani R, Meredith CP. 1996. Isolation and characterization of new polymorphic simple sequence repeat loci in grape, *Vitis vinifera* L. *Genome* **39**: 628–633, doi: 10.1139/g96-080.

Bowers JE, Meredith CP. 1997. The parentage of a classic wine grape, Cabernet Sauvignon. *Nat Genet* **16**: 84–87, doi: 10.1038/ng0597-84.

Cabezas J, Ibanez J, Lijavetzky D, Velez D, Bravo G, Rodriguez V, Carreno I, Jermakow A, Carreno J, Ruiz-Garcia L et al. 2011. A 48 SNP set for grapevine cultivar identification. *BMC Plant Biol* **11**: 153, doi: 10.1186/1471-2229-11-153.

Cabezas JA, Cervera MT, Ruiz-García L, Carreño J, Martínez-Zapater JM. 2006. A genetic analysis of seed and berry weight in grapevine. *Genome* **49**: 1572–1585, doi: 10.1139/g06-122.

Cadavid Labrada A, Medina C, Martinelli L, Arce-johnson P. 2008. Somatic embryogenesis and efficient regeneration of *Vitis vinifera* L. 'Carménère' plants. *Vitis* **47**: 73–74, doi: 10.5073/vitis.2008.47.73-74.

Campos G, Chialva C, Miras S, Lijavetzky D. 2021. New technologies and strategies for grapevine breeding through genetic transformation. *Front Plant Sci* **12**: 767522, doi: 10.3389/fpls.2021.767522.

Cao S, Stringer S, Gunawan G, McGregor C, Conner PJ. 2020. Genetic diversity and pedigree analysis of muscadine grape using SSR Markers. *J Am Soc Hort Sci* **145**: 143–151, doi: 10.21273/jashs04856-20.

Carimi F, Barizza E, Gardiman M, Schiavo Fl. 2005. Somatic embryogenesis from stigmas and styles of grapevine. *In Vitro Cell Dev Biol Plant* **41**: 249–252, doi: 10.1079/IVP2004617.

Carrasco D, De Lorenzis G, Maghradze D, Revilla E, Bellido A, Failla O, Arroyo-García R. 2015. Allelic variation in the *VvMYBA1* and *VvMYBA2* domestication genes in natural grapevine populations (*Vitis vinifera* subsp. sylvestris). *Plant Syst Evol* **301**: 1613–1624, doi: 10.1007/s00606-014-1181-y.

Cervera MT, Cabezas JA, Rodríguez-Torres I, Chávez J, Cabello F, Martínez-Zapater J. 2002. Varietal diversity within grapevine accessions of cv. Tempranillo. *Vitis* **41**: 33–36, doi: 10.5073/vitis.2002.41.33-36.

Cervera MT, Cabezas JA, Sánchez-Escribano E, Cenis J, Martínez-Zapater J. 2000. Characterization of genetic variation within table grape varieties (*Vitis vinifera* L.) based on AFLP markers. *Vitis* **39**: 109–114, doi: 10.5073/vitis.2000.39.109-114.

Chen J, Wang N, Fang L-C, Liang Z-C, Li S-H, Wu B-H. 2015. Construction of a high-density genetic map and QTLs mapping for sugars and acids in grape berries. *BMC Plant Biol* **15**: 28, doi: 10.1186/s12870-015-0428-2.

Chen K, Wang Y, Zhang R, Zhang H, Gao C. 2019. CRISPR/Cas genome editing and precision plant breeding in agriculture. *Annu Rev Plant Biol* **70**: 667–697, doi: 10.1146/annurev-arplant-050718-100049.

Cipriani G, Spadotto A, Jurman I, Di Gaspero G, Crespan M, Meneghetti S, Frare E, Vignani R, Cresti M, Morgante M et al. 2010. The SSR-based molecular profile of 1005 grapevine (*Vitis vinifera* L.) accessions uncovers new synonymy and parentages, and reveals a large admixture amongst varieties of different geographic origin. *Theor Appl Genet* 121: 1569–1585, doi: 10.1007/s00122-010-1411-9.

Coito JL, Ramos MJN, Cunha J, Silva HG, Amâncio S, Costa MMR, Rocheta M. 2017. *VviAPRT3* and *VviFSEX*: Two genes involved in sex specification able to distinguish different flower types in *Vitis*. *Front Plant Sci* **8**: 98, doi: 10.3389/fpls.2017.00098.

Collins GG, Symons RH. 1993. Polymorphisms in grapevine DNA detected by the RAPD PCR technique. *Plant Mol Biol Rep* **11**: 105–112, doi: 10.1007/BF02670467.

Correa J, Mamani M, Muñoz-Espinoza C, González-Agüero M, Defilippi BG, Campos-Vargas R, Pinto M, Hinrichsen P. 2016. New stable QTLs for berry firmness in table grapes. *Am J Enol Vitic* **67**: 212–217, doi: 10.5344/ajev.2015.15049.

Correa J, Mamani M, Muñoz-Espinoza C, Laborie D, Muñoz C, Pinto M, Hinrichsen P. 2014. Heritability and identification of QTLs and underlying candidate genes associated with the architecture of the grapevine cluster (*Vitis vinifera* L.). *Theor Appl Genet* **127**: 1143–1162, doi: 10.1007/s00122-014-2286-y.

Crespan M. 2003. The parentage of Muscat of Hamburg. *Vitis* **42**: 193–197, doi: 10.5073/vitis.2003.42.193-197.

Crespan M, Migliaro D, Vezzulli S, Zenoni S, Tornielli GB, Giacosa S, Paissoni MA, Río Segade S, Rolle L. 2021. A major QTL is associated with berry grape texture characteristics. *OENO One* **55**: 183–206, doi: 10.20870/oeno-one.2021.55.1.3994.

Dalbo MA. 1998. Genetic mapping, QTL analysis, and marker-assisted selection for disease resistance loci in grapes. Vol Ph. D. Cornell University, Ithaca, New York.

Dalbó MA, Ye GN, Weeden NF, Wilcox WF, Reisch BI. 2001. Marker-assisted selection for powdery mildew resistance in grapes. *J Am Soc Hort Sci* **126**: 83–89, doi: 10.21273/jashs.126.1.83.

Dalla Costa L, Mandolini M, Poletti V, Martinelli L. 2010. Comparing 17-β-estradiol supply strategies for applying the XVE-Cre/loxP system in grape gene transfer (*Vitis vinifera* L.). *Vitis* **49**: 201–208, doi: 10.5073/vitis.2010.49.201-208.

De Lorenzis G, Chipashvili R, Failla O, Maghradze D. 2015. Study of genetic variability in *Vitis vinifera* L. germplasm by high-throughput Vitis18kSNP array: The case of Georgian genetic resources. *BMC Plant Biol* **15**: 154, doi: 10.1186/s12870-015-0510-9.

De Lorenzis G, Mercati F, Bergamini C, Cardone MF, Lupini A, Mauceri A, Caputo AR, Abbate L, Barbagallo MG, Antonacci D et al. 2019. SNP genotyping elucidates the genetic diversity of Magna Graecia grapevine germplasm and its historical origin and dissemination. *BMC Plant Biol* **19**: 7, doi: 10.1186/s12870-018-1576-y.

Deore P, Samarth RR, Mane V, Upadhyay A, Sawant IS. 2021. Marker assisted selection for downy mildew resistance in table grapes. *Indian J Hortic* **78**: 239–243, doi: 10.5958/0974-0112.2021.00034.7.

Dhekney SA, Li ZJT, Gray DJ. 2011a. Factors influencing induction and maintenance of *Vitis rotundifolia* Michx. embryogenic cultures. *Plant Cell Tissue Organ Cult* **105**: 175–180, doi: 10.1007/s11240-010-9849-7.

Dhekney SA, Li ZT, Compton ME, Gray DJ. 2009. Optimizing initiation and maintenance of *Vitis* embryogenic cultures. *Hortscience* **44**: 1400–1406, doi: 10.21273/HORTSCI.44.5.1400.

Dhekney SA, Li ZT, Gray DJ. 2011b. Grapevines engineered to express cisgenic *Vitis vinifera* thaumatin-like protein exhibit fungal disease resistance. *In Vitro Cell Dev Biol Plant* **47**: 458–466, doi: 10.1007/s11627-011-9358-3.

Di Gaspero G, Cipriani G, Adam-Blondon AF, Testolin R. 2007. Linkage maps of grapevine displaying the chromosomal locations of 420 microsatellite markers and 82 markers for R-gene candidates. *Theor Appl Genet* **114**: 1249–1263, doi: 10.1007/s00122-007-0516-2.

Di Gaspero G, Cipriani G, Marrazzo MT, Andreetta D, Castro MJP, Peterlunger E, Testolin R. 2005. Isolation of (AC)n-microsatellites in *Vitis vinifera* L. and analysis of genetic background in grapevines under marker assisted selection. *Mol Breed* **15**: 11–20, doi: 10.1007/s11032-004-1362-4.

Di Gaspero G, Copetti D, Coleman C, Castellarin SD, Eibach R, Kozma P, Lacombe T, Gambetta G, Zvyagin A, Cindrić P et al. 2012. Selective sweep at the *Rpv3* locus during grapevine breeding for downy mildew resistance. *Theor Appl Genet* **124**: 277–286, doi: 10.1007/s00122-011-1703-8.

Dodeman VL, Ducreux G, Kreis M. 1997. Zygotic embryogenesis versus somatic embryogenesis. *J Exp Bot* **48**: 1493–1509.

Doligez A, Adam-Blondon AF, Cipriani G, Di Gaspero G, Laucou V, Merdinoglu D, Meredith CP, Riaz S, Roux C, This P. 2006a. An integrated SSR map of grapevine based on five mapping populations. *Theor Appl Genet* **113**: 369–382, doi: 10.1007/s00122-006-0295-1.

Doligez A, Audiot E, Baumes R, This P. 2006b. QTLs for muscat flavor and monoterpenic odorant content in grapevine (*Vitis vinifera* L.). *Mol Breed* **18**: 109–125, doi: 10.1007/s11032-006-9016-3.

Doligez A, Bertrand Y, Dias S, Grolier M, Ballester J-F, Bouquet A, This P. 2010. QTLs for fertility in table grape (*Vitis vinifera* L.). *Tree Genet Genomes* **6**: 413–422, doi: 10.1007/s11295-009-0259-0.

Doligez A, Bertrand Y, Farnos M, Grolier M, Romieu C, Esnault F, Dias S, Berger G, François P, Pons T et al. 2013. New stable QTLs for berry weight do not colocalize with QTLs for seed traits in cultivated grapevine (*Vitis vinifera* L.). *BMC Plant Biol* **13**: 217, doi: 10.1186/1471-2229-13-217.

Doligez A, Bouquet A, Danglot Y, Lahogue F, Riaz S, Meredith C, Edwards K, This P. 2002. Genetic mapping of grapevine (*Vitis vinifera* L.) applied to the detection of QTLs for seedlessness and berry weight. *Theor Appl Genet* **105**: 780–795, doi: 10.1007/s00122-002-0951-z.

Donald TM, Pellerone F, Adam-Blondon AF, Bouquet A, Thomas MR, Dry IB. 2002. Identification of resistance gene analogs linked to a powdery mildew resistance locus in grapevine. *Theor Appl Genet* **104**: 610–618, doi: 10.1007/s00122-001-0768-1.

Doucleff M, Jin Y, Gao F, Riaz S, Krivanek AF, Walker MA. 2004. A genetic linkage map of grape, utilizing *Vitis rupestris* and *Vitis arizonica*. *Theor Appl Genet* **109**: 1178–1187, doi: 10.1007/s00122-004-1728-3.

Drew DP, Andersen TB, Sweetman C, Møller BL, Ford C, Simonsen HT. 2015. Two key polymorphisms in a newly discovered allele of the *Vitis vinifera TPS*24 gene are responsible for the production of the rotundone precursor α-guaiene. *J Exp Bot* **67**: 799–808, doi: 10.1093/jxb/erv491.

Duan D, Fischer S, Merz P, Bogs J, Riemann M, Nick P. 2016. An ancestral allele of grapevine transcription factor MYB14 promotes plant defence. *J Exp Bot* **67**: 1795–1804, doi: 10.1093/jxb/erv569.

Dudits D, Györgyey J, Bögre L, Bakó L. 1995. Molecular Biology of Somatic Embryogenesis. In *In Vitro Embryogenesis in Plants*, (ed. TA Thorpe), pp. 267–308. Springer Netherlands, Dordrecht, doi: 10.1007/978-94-011-0485-2_8.

Dunchene E, Butterlin G, Claudel P, Dumas V, Jaegli N, Hugueney P, Arnold G, Merdinoglu D. 2017. Genetic determinism of the 'Muscat' flavour in grapevine (*Vitis vinifera* L.) cultivars. *Acta Hortic* **1157**: 87–92, doi: 10.17660/ActaHortic.2017.1157.14.

Dunlevy JD, Dennis EG, Soole KL, Perkins MV, Davies C, Boss PK. 2013. A methyltransferase essential for the methoxypyrazine-derived flavour of wine. *Plant J* **75**: 606–617, doi: 10.1111/tpj.12224.

Dutt M, Li ZT, Dhekney SA, Gray DJ. 2008. A co-transformation system to produce transgenic grapevines free of marker genes. *Plant Sci* **175**: 423–430, doi: 10.1016/j.plantsci.2008.06.014.

El-Mounadi K, Morales-Floriano ML, Garcia-Ruiz H. 2020. Principles, applications, and biosafety of plant genome editing using CRISPR-Cas9. *Front Plant Sci* **11**: 56, doi: 10.3389/fpls.2020.00056.

Emanuelli F, Battilana J, Costantini L, Le Cunff L, Boursiquot J-M, This P, Grando MS. 2010. A candidate gene association study on muscat flavor in grapevine (*Vitis vinifera* L.). *BMC Plant Biol* **10**: 241, doi: 10.1186/1471-2229-10-241.

Fanizza G, Chaabane R, Lamaj F, Ricciardi L, Resta P. 2003. AFLP analysis of genetic relationships among aromatic grapevines (*Vitis vinifera*). *Theor Appl Genet* **107**: 1043–1047, doi: 10.1007/s00122-003-1354-5.

Fanizza G, Lamaj F, Costantini L, Chaabane R, Grando MS. 2005a. QTL analysis for fruit yield components in table grapes (*Vitis vinifera*). *Theor Appl Genet* **111**: 658–664, doi: 10.1007/s00122-005-2016-6.

Fanizza G, Lamaj F, Resta P, Ricciardi L, Savino V. 2005b. Grapevine cvs Primitivo, Zinfandel and Crljenak kastelanski: Molecular analysis by AFLP. *Vitis* **44**: 147–148, doi: 10.5073/vitis.2005.44.147–148.

Fechter I, Hausmann L, Zyprian E, Daum M, Holtgräwe D, Weisshaar B, Töpfer R. 2014. QTL analysis of flowering time and ripening traits suggests an impact of a genomic region on linkage group 1 in *Vitis*. *Theor Appl Genet* **127**: 1857–1872, doi: 10.1007/s00122-014-2310-2.

Fernandez L, Le Cunff L, Tello J, Lacombe T, Boursiquot JM, Fournier-Level A, Bravo G, Lalet S, Torregrosa L, This P et al. 2014. Haplotype diversity of *VvTFL1A* gene and association with cluster traits in grapevine (*V. vinifera*). *BMC Plant Biol* **14**: 209–209, doi: 10.1186/s12870-014-0209-3.

Fernandez L, Torregrosa L, Segura V, Bouquet A, Martinez-Zapater JM. 2010. Transposon-induced gene activation as a mechanism generating cluster shape somatic variation in grapevine. *Plant J* **61**: 545–557, doi: 10.1111/j.1365-313X.2009.04090.x.

Ferreira V, Matus JT, Pinto-Carnide O, Carrasco D, Arroyo-García R, Castro I. 2019. Genetic analysis of a white-to-red berry skin color reversion and its transcriptomic and metabolic consequences in grapevine (*Vitis vinifera* cv. 'Moscatel Galego'). *BMC Genom* **20**: 952, doi: 10.1186/s12864-019-6237-5.

Fischer BM, Salakhutdinov I, Akkurt M, Eibach R, Edwards KJ, Töpfer R, Zyprian EM. 2004. Quantitative trait locus analysis of fungal disease resistance factors on a molecular map of grapevine. *Theor Appl Genet* **108**: 501–515, doi: 10.1007/s00122-003-1445-3.

Flutre T, Le Cunff L, Fodor A, Launay A, Romieu C, Berger G, Bertrand Y, Terrier N, Beccavin I, Bouckenooghe V et al. 2022. A genome-wide association and prediction study in grapevine deciphers the genetic architecture of multiple traits and identifies genes under many new QTLs. *G3: Genes Genomes Genet* **12**: jkac103, doi: 10.1093/g3journal/jkac103.

Foria S, Copetti D, Eisenmann B, Magris G, Vidotto M, Scalabrin S, Testolin R, Cipriani G, Wiedemann-Merdinoglu S, Bogs J et al. 2020. Gene duplication and transposition of mobile elements drive evolution of the *Rpv3* resistance locus in grapevine. *Plant J* **101**: 529–542, doi: 10.1111/tpj.14551.

Fortes AM, Teixeira RT, Agudelo-Romero P. 2015. Complex interplay of hormonal signals during grape berry ripening. *Molecules* **20**: 9326–9343, doi: 10.3390/molecules20059326.

Fossati T, Labra M, Castiglione S, Failla O, Scienza A, Sala F. 2001. The use of AFLP and SSR molecular markers to decipher homonyms and synonyms in grapevine cultivars: The case of the varietal group known as "Schiave". *Theor Appl Genet* **102**: 200–205, doi: 10.1007/s001220051636.

Fournier-Level A, Lacombe T, Le Cunff L, Boursiquot JM, This P. 2010. Evolution of the *VvMybA* gene family, the major determinant of berry colour in cultivated grapevine (*Vitis vinifera* L.). *Heredity* **104**: 351–362, doi: 10.1038/hdy.2009.148.

Gambino G, Di Matteo D, Gribaudo I. 2009. Elimination of Grapevine fanleaf virus from three Vitis vinifera cultivars by somatic embryogenesis. *Eur J Plant Pathol* **123**: 57–60, doi: 10.1007/s10658-008-9342-x.

Gambino G, Ruffa P, Vallania R, Gribaudo I. 2007. Somatic embryogenesis from whole flowers, anthers and ovaries of grapevine (*Vitis* spp.). *Plant Cell Tissue Organ Cult* **90**: 79–83, doi: 10.1007/s11240-007-9256-x.

George EF, Hall MA, Klerk G-JD. 2008. Somatic embryogenesis. In *Plant Propagation by Tissue Culture: Volume 1 The Background*, (ed. EF George, et al.), pp. 335–354. Springer Netherlands, Dordrecht, doi: 10.1007/978-1-4020-5005-3_9.

Giacomelli L, Zeilmaker T, Malnoy M, Rouppe van der Voort J, Moser C. 2019. Generation of mildew-resistant grapevine clones via genome editing. *Acta Hortic* **1249**: 195–200, doi: 10.17660/ActaHortic.2019.1248.28.

Giacomelli L, Zeilmaker T, Scintilla S, Salvagnin U, van der Voort JR, Moser C. 2022. Vitis vinifera plants edited in DMR6 genes show improved resistance to downy mildew. *bioRxiv*: 488768.

González-Andrés F, Martín J, Yuste J, Rubio J, Arranz C, Ortiz J. 2007. Identification and molecular biodiversity of autochthonous grapevine cultivars in the 'Comarca del Bierzo', León, Spain. *Vitis* **46**: 71–76, doi: 10.5073/vitis.2007.46.71-76.

Grando MS, Bellin D, Edwards KJ, Pozzi C, Stefanini M, Velasco R. 2003. Molecular linkage maps of *Vitis vinifera* L. and *Vitis riparia* Mchx. *Theor Appl Genet* **106**: 1213–1224, doi: 10.1007/s00122-002-1170-3.

Grimplet J, Ibáñez S, Baroja E, Tello J, Ibáñez J. 2019. Phenotypic, hormonal, and genomic variation among *Vitis vinifera* clones with different cluster compactness and reproductive performance. *Front Plant Sci* **9**: 1917, doi: 10.3389/fpls.2018.01917.

Guan Y, Li S-G, Fan X-F, Su Z-H. 2016. Application of somatic embryogenesis in woody plants. *Front Plant Sci* **7**, doi: 10.3389/fpls.2016.00938.

Guo D-L, Zhao H-L, Li Q, Zhang G-H, Jiang J-F, Liu C-H, Yu Y-H. 2019. Genome-wide association study of berry-related traits in grape (*Vitis vinifera* L.) based on genotyping-by-sequencing markers. *Hortic Res* **6**: 11, doi: 10.1038/s41438-018-0089-z.

Guo Y, Lin H, Liu Z, Zhao Y, Guo X, Li K. 2014. SSR and SRAP marker-based linkage map of *Vitis vinifera* L. *Biotechnol Biotechnol Equip* **28**: 221–229, doi: 10.1080/13102818.2014.907996.

Guo Y, Shi G, Liu Z, Zhao Y, Yang X, Zhu J, Li K, Guo X. 2015. Using specific length amplified fragment sequencing to construct the high-density genetic map for *Vitis* (Vitis vinifera L. × Vitis amurensis Rupr.). *Front Plant Sci* **6**: 393, doi: 10.3389/fpls.2015.00393.

Gutoranov G, Tsvetkov I, Tsolova V, Atanassov A. 2001. Genetically Engineered grapevines carrying GFLV coat protein and antifreeze genes. *Agric Conspec Sci* **66**: 69–74.

Haddad R, Japelaghi RH. 2015. Isolation of grape peroxiredoxin gene responding to abiotic stresses. *Russ J Plant Physiol* **62**: 856–865, doi: 10.1134/S1021443715050088.

He R, Wu J, Zhang Y, Agüero CB, Li X, Liu S, Wang C, Walker MA, Lu J. 2017. Overexpression of a thaumatin-like protein gene from *Vitis amurensis* improves downy mildew resistance in *Vitis vinifera* grapevine. *Protoplasma* **254**: 1579–1589, doi: 10.1007/s00709-016-1047-y.

Hvarleva T, Bakalova A, Rusanov K, Diakova G, Ilieva I, Atanassov A, Atanassov I. 2009. Toward marker assisted selection for fungal disease resistance in grapevine. *Biotechnol Biotechnol Equip* **23**: 1431–1435, doi: 10.2478/V10133-009-0008-4.

Iocco-Corena P, Chaïb J, Torregrosa L, Mackenzie D, Thomas MR, Smith HM. 2021. *VviPLATZ1* is a major factor that controls female flower morphology determination in grapevine. *Nat Commun* **12**: 6995, doi: 10.1038/s41467-021-27259-8.

Jang HA, Oo MM, Kim D-G, Yoon H-Y, Kim M-R, Lee K-S, Thinn KSZ, Arif S, Geng JG, Min J. 2020. CC-NBS-LRR, a set of *VvCRP* markers, can distinguish cultivars with ripe rot resistance to *Colletotrichum* pathogens in grapevine. *Hortic Environ Biotechnol* **61**: 915–927, doi: 10.1007/s13580-020-00290-2.

Jardak-Jamoussi R, Winterhagen P, Bouamama B, Dubois C, Mliki A, Wetzel T, Ghorbel A, Reustle G. 2009. Development and evaluation of a GFLV inverted repeat construct for genetic transformation of grapevine. *Plant Cell Tissue Organ Cult* **97**: 187–196, doi: 10.1007/s11240-009-9514-1.

Jeanjaques I, Defontaine A, Hallet JN. 1993. Characterization of *Vitis vinifera* cultivars by random amplified polymorphic DNA markers. *Vitis* **32**: 189–190, doi: 10.5073/vitis.1993.32.189-190.

Jiang J, Fan X, Zhang Y, Tang X, Li X, Liu C, Zhang Z. 2020. Construction of a high-density genetic map and mapping of firmness in grapes (*Vitis vinifera* L.) based on whole-genome resequencing. *Int J Mol Sci* **21**: 797, doi: 10.3390/ijms21030797.

Jiao L, Zhang Y, Lu J. 2017. Overexpression of a stress-responsive U-box protein gene VaPUB affects the accumulation of resistance related proteins in *Vitis vinifera* 'Thompson Seedless'. *Plant Physiol Biochem* **112**: 53–63, doi: 10.1016/j.plaphy.2016.12.019.

Jiao Y, Xu W, Duan D, Wang Y, Nick P. 2016. A stilbene synthase allele from a Chinese wild grapevine confers resistance to powdery mildew by recruiting salicylic acid signalling for efficient defence. *J Exp Bot* **67**: 5841–5856, doi: 10.1093/jxb/erw351.

Jiménez-Cantizano A, Muñoz-Martín A, Amores-Arrocha A, Sancho-Galán P, Palacios V. 2020. Identification of red grapevine cultivars (*Vitis vinifera* L.) preserved in ancient vineyards in Axarquia (Andalusia, Spain). *Plants* **9**: 1572, doi: 10.3390/plants9111572.

Kamal N, Ochßner I, Schwandner A, Viehoever P, Hausmann L, Toepfer R, Weisshaar B, Holtgräwe D. 2019. Characterization of genes and alleles involved in the control of flowering time in grapevine. *PLoS One* **14**: e0214703, doi: 10.1371/journal.pone.0214703.

Karaagac E, Vargas AM, de Andrés MT, Carreño I, Ibáñez J, Carreño J, Martínez-Zapater JM, Cabezas JA. 2012. Marker assisted selection for seedlessness in table grape breeding. *Tree Genet Genomes* **8**: 1003–1015, doi: 10.1007/s11295-012-0480-0.

Katula-Debreceni D, Lencsés AK, Szőke A, Veres A, Hoffmann S, Kozma P, Kovács LG, Heszky L, Kiss E. 2010. Marker-assisted selection for two dominant powdery mildew resistance genes introgressed into a hybrid grape population. *Sci Hortic* **126**: 448–453, doi: 10.1016/j.scienta.2010.08.012.

Kikkert J, Striem M, Vidal J, Wallace P, Barnard J, Reisch B. 2005. Long-term study of somatic embryogenesis from anthers and ovaries of 12 grapevine (*Vitis* sp.) genotypes. *In Vitro Cell Dev Biol Plant* **41**: 232–239, doi: 10.1079/IVP2004609.

Kim MS, Hur YY, Kim JH, Jeong SC. 2020. Genome resequencing, improvement of variant calling, and population genomic analyses provide insights into the seedlessness in the genus *Vitis*. *G3: Genes Genomes Genet* **10**: 3365–3377, doi: 10.1534/g3.120.401521.

Kobayashi S, Goto-Yamamoto N, Hirochika H. 2004. Retrotransposon-induced mutations in grape skin color. *Science* **304**: 982, doi: 10.1126/science.1095011.

Kobayashi S, Ishimaru M, Hiraoka K, Honda C. 2002. *Myb*-related genes of the Kyoho grape (*Vitis labruscana*) regulate anthocyanin biosynthesis. *Planta* **215**: 924–933, doi: 10.1007/s00425-002-0830-5.

Koyama K, Kono A, Ban Y, Bahena-Garrido SM, Ohama T, Iwashita K, Fukuda H, Goto-Yamamoto N. 2022. Genetic architecture of berry aroma compounds in a QTL (quantitative trait loci) mapping population of interspecific hybrid grapes (*Vitis labruscana* × *Vitis vinifera*). *BMC Plant Biol* **22**: 458, doi: 10.1186/s12870-022-03842-z.

Labra M, Winfield M, Ghiani A, Grassi F, Sala F, Scienza A, Failla O. 2001. Genetic studies on Trebbiano and morphologically related varieties by SSR and AFLP markers. *Vitis* **40**: 187–190, doi: 10.5073/vitis.2001.40.187-190.

Ladoukakis ED, Lefort F, Sotiri P, Bacu A, Kongjika E, Roubelakis-Angelakis KA. 2005. Genetic characterization of Albanian grapevine cultivars by microsatellite markers. *OENO One* **39**: 109–119, doi: 10.20870/oeno-one.2005.39.3.895.

Lahogue F, This P, Bouquet A. 1998. Identification of a codominant SCAR marker linked to the seedlessness character in grapevine. *Theor Appl Genet* **97**: 950–959, doi: 10.1007/s001220050976.

Laiadi Z, Bentchikou MM, Bravo G, Cabello F, Martinez-Zapater JM. 2009. Molecular identification and genetic relationships of Algerian grapevine cultivars maintained at the germplasm collection of Skikda (Algeria). *Vitis* **48**: 25–32, doi: 10.5073/vitis.2009.48.25-32.

LaPlante ER, Fleming MB, Migicovsky Z, Weber MG. 2021. Genome-wide association study reveals a genomic region associated with mite-recruitment phenotypes in the domesticated grapevine (*Vitis vinifera*). *Genes* **12**: 1013, doi: 10.3390/genes12071013.

Laucou V, Boursiquot JM, Lacombe T, Bordenave L, Decroocq S, Ollat N. 2008. Parentage of grapevine rootstock Fercal finally elucidated. *Vitis* **47**: 163–167, doi: 10.5073/vitis.2008.47.163-167.

Laucou V, Launay A, Bacilieri R, Lacombe T, Adam-Blondon AF, Bérard A, Chauveau A, de Andrés MT, Hausmann L, Ibáñez J et al. 2018. Extended diversity analysis of cultivated grapevine *Vitis vinifera* with 10K genome-wide SNPs. *PLoS One* **13**: e0192540, doi: 10.1371/journal.pone.0192540.

Li B, Fan X, Zhang Y, Liu C, Jiang J. 2021. Genetic diversity and population structure analysis of Chinese wild grape using simple sequence repeat markers. *J Am Soc Hort Sci* **146**: 158–168, doi: 10.21273/jashs05016-20.

Li Y, Zhang S, Dong R, Wang L, Yao J, van Nocker S, Wang X. 2019. The grapevine homeobox gene *VvHB58* influences seed and fruit development through multiple hormonal signaling pathways. *BMC Plant Biol* **19**: 523, doi: 10.1186/s12870-019-2144-9.

Li Y, Zhu Y, Yao J, Zhang S, Wang L, Guo C, van Nocker S, Wang X. 2017. Genome-wide identification and expression analyses of the homeobox transcription factor family during ovule development in seedless and seeded grapes. *Sci Rep* **7**: 12638, doi: 10.1038/s41598-017-12988-y.

Li Y-M, Sun X-R, You J-L, Li X-J, Li B, Xie Z-S. 2022. Transcriptional analysis of the early ripening of 'Summer Black' grape in response to abscisic acid treatment. *Sci Hortic* **299**: 111054, doi: 10.1016/j.scienta.2022.111054.

Li ZT, Hopkins DL, Gray DJ. 2015. Overexpression of antimicrobial lytic peptides protects grapevine from Pierce's disease under greenhouse but not field conditions. *Transgenic Res* **24**: 821–836, doi: 10.1007/s11248-015-9876-6.

Liang Z, Duan S, Sheng J, Zhu S, Ni X, Shao J, Liu C, Nick P, Du F, Fan P. 2019. Whole-genome resequencing of 472 *Vitis* accessions for grapevine diversity and demographic history analyses. *Nat Commun* **10**: 1190, doi: 10.1038/s41467-019-09135-8.

Lijavetzky D, Cabezas JA, Ibáñez A, Rodríguez V, Martínez-Zapater JM. 2007. High throughput SNP discovery and genotyping in grapevine (*Vitis vinifera* L.) by combining a re-sequencing approach and SNPlex technology. *BMC Genom* **8**: 424, doi: 10.1186/1471-2164-8-424.

Lodhi MA, Daly MJ, Ye GN, Weeden NF, Reisch BI. 1995. A molecular marker based linkage map of *Vitis*. *Genome* **38**: 786–794, doi: 10.1139/g95-100.

Lopes MS, Sefc KM, Eiras Dias E, Steinkellner H, Laimer da Camara Machado M, da Camara Machado A. 1999. The use of microsatellites for germplasm management in a Portuguese grapevine collection. *Theor Appl Genet* **99**: 733–739, doi: 10.1007/s001220051291.

Lopez M, Cid N, Gonzalez MV, Cuenca B, Prado MJ, Rey M. 2009. Microsatellite and AFLP analysis of autochthonous grapevine cutlivars from Galicia (Spain). *Am J Enol Vitic* **60**: 215–222, doi: 10.5344/ajev.2009.60.2.215.

López-Pérez AJ, Velasco L, Pazos-Navarro M, Dabauza M. 2008. Development of highly efficient genetic transformation protocols for table grape Sugraone and Crimson Seedless at low *Agrobacterium* density. *Plant Cell Tissue Organ Cult* **94**: 189–199, doi: 10.1007/s11240-008-9404-y.

Lowe KM, Walker MA. 2006. Genetic linkage map of the interspecific grape rootstock cross Ramsey (*Vitis champinii*) x Riparia Gloire (*Vitis riparia*). *Theor Appl Genet* **112**: 1582–1592, doi: 10.1007/s00122-006-0264-8.

Ma H, Xiang G, Li Z, Wang Y, Dou M, Su L, Yin X, Liu R, Wang Y, Xu Y. 2018. Grapevine *VpPR10.1* functions in resistance to *Plasmopara viticola* through triggering a cell death-like defence response by interacting with *VpVDAC3*. *Plant Biotechnol J* **16**: 1488–1501, doi: 10.1111/pbi.12891.

Ma L, Sun L, Guo Y, Lin H, Liu Z, Li K, Guo X. 2020. Transcriptomc analysis of table grapes (*Vitis vinifera* L.) identified a gene network module associated with berry firmness. *PLoS One* **15**: e0237526, doi: 10.1371/journal.pone.0237526.

Magris G, Jurman I, Fornasiero A, Paparelli E, Schwope R, Marroni F, Di Gaspero G, Morgante M. 2021. The genomes of 204 *Vitis vinifera* accessions reveal the origin of European wine grapes. *Nat Commun* **12**: 7240, doi: 10.1038/s41467-021-27487-y.

Maillot P, Deglène-Benbrahim L, Walter B. 2016. Efficient somatic embryogenesis from meristematic explants in grapevine (*Vitis vinifera* L.) cv. Chardonnay: An improved protocol. *Trees* **30**: 1377–1387, doi: 10.1007/s00468-016-1374-9.

Maillot P, Kieffer F, Walter B. 2006. Somatic embryogenesis from stem nodal sections of grapevine. *Vitis* **45**: 185–189, doi: 10.5073/vitis.2006.45.185-189.

Malnoy M, Viola R, Jung M-H, Koo O-J, Kim S, Kim J-S, Velasco R, Nagamangala Kanchiswamy C. 2016. DNA-free genetically edited grapevine and apple protoplast using CRISPR/Cas9 ribonucleoproteins. *Front Plant Sci* **7**: 1904, doi: 10.3389/fpls.2016.01904.

Marrano A, Micheletti D, Lorenzi S, Neale D, Grando MS. 2018. Genomic signatures of different adaptations to environmental stimuli between wild and cultivated *Vitis vinifera* L. *Hortic Res* **5**: 34, doi: 10.1038/s41438-018-0041-2.

Martin DM, Bohlmann J. 2004. Identification of *Vitis vinifera* (-)-alpha-terpineol synthase by *in silico* screening of full-length cDNA ESTs and functional characterization of recombinant terpene synthase. *Phytochem* **65**: 1223–1229, doi: 10.1016/j.phytochem.2004.03.018.

Martinelli L, Gribaudo I. 2001. Somatic Embryogenesis in Grapevine. In *Molecular Biology & Biotechnology of the Grapevine*, (ed. KA Roubelakis-Angelakis), pp. 327–351. Springer Netherlands, Dordrecht, doi:10.1007/978-94-017-2308-4_13.

Massonnet M, Cochetel N, Minio A, Vondras AM, Lin J, Muyle A, Garcia JF, Zhou Y, Delledonne M, Riaz S et al. 2020. The genetic basis of sex determination in grapes. *Nat Commun* **11**: 2902, doi: 10.1038/s41467-020-16700-z.

Mejía N, Gebauer M, Muñoz L, Hewstone N, Muñoz C, Hinrichsen P. 2007. Identification of QTLs for seedlessness, berry size, and ripening date in a Seedless x Seedless table grape progeny. *Am J Enol Vitic* **58**: 499–507, doi: 10.5344/ajev.2007.58.4.499.

Mejía N, Hinrichsen P. 2003. A new, highly assertive SCAR marker potentially useful to assist selection for seedlessness in table grape breeding. *Acta Hortic* **603**: 559–564, doi: 10.17660/ActaHortic.2003.603.74.

Mejía N, Soto B, Guerrero M, Casanueva X, Houel C, de los Ángeles Miccono M, Ramos R, Le Cunff L, Boursiquot J-M, Hinrichsen P et al. 2011. Molecular, genetic and transcriptional evidence for a role of VvAGL11 in stenospermocarpic seedlessness in grapevine. *BMC Plant Biol* **11**: 57, doi: 10.1186/1471-2229-11-57.

Mercati F, De Lorenzis G, Brancadoro L, Lupini A, Abenavoli MR, Barbagallo MG, Di Lorenzo R, Scienza A, Sunseri F. 2016. High-throughput 18K SNP array to assess genetic variability of the main grapevine cultivars from Sicily. *Tree Genet Genomes* **12**: 59, doi: 10.1007/s11295-016-1021-z.

Merdinoglu D, Wiedeman-Merdinoglu S, Coste P, Dumas V, Haetty S, Butterlin G, Greif C. 2003. Genetic analysis of downy mildew resistance derived from *Muscadinia rotundifolia*. *Acta Hortic* **603**: 451–456, doi: 10.17660/ActaHortic.2003.603.57.

Moncada X, Hinrichsen P. 2007. Limited genetic diversity among clones of red wine cultivar Carmenere as revealed by microsatellite and AFLP markers. *Vitis* **46**: 174–180, doi: 10.5073/vitis.2007.46.174-180.

Morcia C, Tumino G, Raimondi S, Schneider A, Terzi V. 2021. Muscat flavor in grapevine: A digital PCR assay to track allelic variation in VvDXS gene. *Genes* **12**: 747, doi: 10.3390/genes12050747.

Morgana C, De Lorenzo R, Carimi F. 2004. Somatic embryogenesis of *Vitis vinifera* L. (cv. Sugraone) from stigma and style culture. *Vitis* **43**: 169–173, doi: 10.5073/vitis.2004.43.169-173.

Muñoz-Espinoza C, Di Genova A, Sánchez A, Correa J, Espinoza A, Meneses C, Maass A, Orellana A, Hinrichsen P. 2020. Identification of SNPs and InDels associated with berry size in table grapes integrating genetic and transcriptomic approaches. *BMC Plant Biol* **20**: 365, doi: 10.1186/s12870-020-02564-4.

Murashige T, Skoog FK. 1962. A revised medium for rapid growth and bio assays with tobacco tissue cultures. *Physiol Plant* **15**: 473–497, doi: 10.1111/j.1399-3054.1962.tb08052.x.

Myles S, Boyko AR, Owens CL, Brown PJ, Grassi F, Aradhya MK, Prins B, Reynolds A, Chia J-M, Ware D et al. 2011. Genetic structure and domestication history of the grape. *Proc Natl Acad Sci USA* **108**: 3530–3535, doi: 10.1073/pnas.1009363108.

Myles S, Chia J-M, Hurwitz B, Simon C, Zhong GY, Buckler E, Ware D. 2010. Rapid genomic characterization of the genus *Vitis*. *PLoS One* **5**: e8219, doi: 10.1371/journal.pone.0008219.

Nakajima I, Ban Y, Azuma A, Onoue N, Moriguchi T, Yamamoto T, Toki S, Endo M. 2017. CRISPR/Cas9-mediated targeted mutagenesis in grape. *PLoS One* **12**: e0177966, doi: 10.1371/journal.pone.0177966.

Nakajima I, Endo M, Haji T, Moriguchi T, Yamamoto T. 2020. Embryogenic callus induction and Agrobacterium-mediated genetic transformation of 'Shine Muscat' grape. *Plant Biotechnol* **37**: 185–194, doi: 10.5511/plantbiotechnology.20.0527a.

Nakajima I, Matsuta N. 2003. Somatic embryogenesis from filaments of *Vitis vinifera* L. and *Vitis labruscana* Bailey. *Vitis* **42**: 53–54.

Nitsch JP, Nitsch C. 1969. Haploid plants from pollen grains. *Science* **163**: 85–87, doi: 10.1126/science.163.3862.85.

Nookaraju A, Agrawal D. 2013. Use of amino acids for a highly efficient somatic embryogenesis in grapevine 'Crimson Seedless'. *Vitis* **52**: 137–140.

Nookaraju A, Agrawal DC. 2012. Enhanced tolerance of transgenic grapevines expressing chitinase and β-1,3-glucanase genes to downy mildew. *Plant Cell Tissue Organ Cult* **111**: 15–28, doi: 10.1007/s11240-012-0166-1.

Olivares F, Loyola R, Olmedo B, Miccono MdlÁ, Aguirre C, Vergara R, Riquelme D, Madrid G, Plantat P, Mora R et al. 2021. CRISPR/Cas9 Targeted editing of genes associated with fungal susceptibility in *Vitis vinifera* L. cv. Thompson Seedless using geminivirus-derived replicons. *Front Plant Sci* **12**: 791030, doi: 10.3389/fpls.2021.791030.

Park M, Vera D, Kambrianda D, Gajjar P, Cadle-Davidson L, Tsolova V, El-Sharkawy I. 2022. Chromosome-level genome sequence assembly and genome-wide association study of *Muscadinia rotundifolia* reveal the genetics of 12 berry-related traits. *Hortic Res* **9**: uhab011, doi: 10.1093/hr/uhab011.

Pauquet J, Bouquet A, This P, Adam-Blondon AF. 2001. Establishment of a local map of AFLP markers around the powdery mildew resistance gene *Run*1 in grapevine and assessment of their usefulness for marker assisted selection. *Theor Appl Genet* **103**: 1201–1210, doi: 10.1007/s001220100664.

Picq S, Santoni S, Lacombe T, Latreille M, Weber A, Ardisson M, Ivorra S, Maghradze D, Arroyo-Garcia R, Chatelet P et al. 2014. A small XY chromosomal region explains sex determination in wild dioecious *V. vinifera* and the reversal to hermaphroditism in domesticated grapevines. *BMC Plant Biol* **14**: 229, doi: 10.1186/s12870-014-0229-z.

Piljac J, Maletic E, Kontic JK, Dangl GS, Pejic I, Mirosevic N, Meredith CP. 2002. The parentage of Posip bijeli, a major white wine cultivar of Croatia. *Vitis* **41**: 83–87, doi: 10.5073/vitis.2002.41.83-87.

Pindo M, Vezzulli S, Coppola G, Cartwright DA, Zharkikh A, Velasco R, Troggio M. 2008. SNP high-throughput screening in grapevine using the SNPlex (TM) genotyping system. *BMC Plant Biol* **8**: 12, doi: 10.1186/1471-2229-8-12.

Pinto-Sintra AL. 2007. Establishment of embryogenic cultures and plant regeneration in the Portuguese cultivar 'Touriga Nacional' of *Vitis vinifera* L. *Plant Cell Tissue Organ Cult* **88**: 253–265, doi: 10.1007/s11240-006-9199-7.

Pirrello C, Zeilmaker T, Bianco L, Giacomelli L, Moser C, Vezzulli S. 2021. Mining grapevine downy mildew susceptibility genes: A resource for genomics-based breeding and tailored gene editing. *Biomolecules* **11**: 181, doi: 10.3390/biom11020181.

Prado MJ, Grueiro MP, González MV, Testillano PS, Domínguez C, López M, Rey M. 2010. Efficient plant regeneration through somatic embryogenesis from anthers and ovaries of six autochthonous grapevine cultivars from Galicia (Spain). *Sci Hortic* **125**: 342–352, doi: 10.1016/J.SCIENTA.2010.04.019.

Ren C, Guo Y, Kong J, Lecourieux F, Dai Z, Li S, Liang Z. 2020. Knockout of VvCCD8 gene in grapevine affects shoot branching. *BMC Plant Biol* **20**: 47, doi: 10.1186/s12870-020-2263-3.

Ren C, Liu X, Zhang Z, Wang Y, Duan W, Li S, Liang Z. 2016. CRISPR/Cas9-mediated efficient targeted mutagenesis in Chardonnay (*Vitis vinifera* L.). *Sci Rep* **6**: 32289, doi: 10.1038/srep32289.

Ren C, Liu Y, Guo Y, Duan W, Fan P, Li S, Liang Z. 2021. Optimizing the CRISPR/Cas9 system for genome editing in grape by using grape promoters. *Hortic Res* **8**: 52, doi: 10.1038/s41438-021-00489-z.

Ren F, Ren C, Zhang Z, Duan W, Lecourieux D, Li S, Liang Z. 2019. Efficiency optimization of CRISPR/Cas9-mediated targeted mutagenesis in grape. *Front Plant Sci* **10**: 612, doi: 10.3389/fpls.2019.00612.

Reshef N, Karn A, Manns DC, Mansfield AK, Cadle-Davidson L, Reisch B, Sacks GL. 2022. Stable QTL for malate levels in ripe fruit and their transferability across Vitis species. *Hortic Res* **9**: uhac009, doi: 10.1093/hr/uhac009.

Riaz S, Dangl GS, Edwards KJ, Meredith CP. 2004. A microsatellite marker based framework linkage map of *Vitis vinifera* L. *Theor Appl Genet* **108**: 864–872, doi: 10.1007/s00122-003-1488-5.

Riaz S, Hu R, Walker MA. 2012. A framework genetic map of *Muscadinia rotundifolia*. *Theor Appl Genet* **125**: 1195–1210, doi: 10.1007/s00122-012-1906-7.

Riaz S, Krivanek AF, Xu K, Walker MA. 2006. Refined mapping of the Pierce's disease resistance locus, PdR1, and Sex on an extended genetic map of *Vitis rupestris x V. arizonica*. *Theor Appl Genet* **113**: 1317–1329, doi: 10.1007/s00122-006-0385-0.

Riaz S, Tenscher AC, Graziani R, Krivanek AF, Ramming DW, Walker MA. 2009. Using marker-assisted selection to breed Pierce's disease-resistant grapes. *Am J Enol Vitic* **60**: 199–207.

Riaz S, Tenscher AC, Ramming DW, Walker MA. 2011. Using a limited mapping strategy to identify major QTLs for resistance to grapevine powdery mildew (*Erysiphe necator*) and their use in marker-assisted breeding. *Theor Appl Genet* **122**: 1059–1073, doi: 10.1007/s00122-010-1511-6.

Riaz S, Tenscher AC, Rubin J, Graziani R, Pao SS, Walker MA. 2008. Fine-scale genetic mapping of two Pierce's disease resistance loci and a major segregation distortion region on chromosome 14 of grape. *Theor Appl Genet* **117**: 671–681, doi: 10.1007/s00122-008-0802-7.

Richter R, Gabriel D, Rist F, Töpfer R, Zyprian E. 2019. Identification of co-located QTLs and genomic regions affecting grapevine cluster architecture. *Theor Appl Genet* **132**: 1159–1177, doi: 10.1007/s00122-018-3269-1.

Richter R, Rossmann S, Gabriel D, Töpfer R, Theres K, Zyprian E. 2020. Differential expression of transcription factor- and further growth-related genes correlates with contrasting cluster architecture in *Vitis vinifera* 'Pinot Noir' and *Vitis* spp. genotypes. *Theor Appl Genet* **133**: 3249–3272, doi: 10.1007/s00122-020-03667-0.

Röckel F, Moock C, Braun U, Schwander F, Cousins P, Maul E, Töpfer R, Hausmann L. 2020. Color Intensity of the red-fleshed berry phenotype of *Vitis vinifera* teinturier grapes varies due to a 408 bp duplication in the promoter of *VvmybA1*. *Genes* **11**: 891, doi: 10.3390/genes11080891.

Rose RJ, Mantiri FR, Kurdyukov S, Chen S-K, Wang X-D, Nolan KE, Sheahan MB. 2010. Developmental Biology of Somatic Embryogenesis. In *Plant Developmental Biology – Biotechnological Perspectives: Volume 2,* (eds. EC Pua, MR Davey), pp. 3–26. Springer Berlin Heidelberg, Berlin, Heidelberg, doi:10.1007/978-3-642-04670-4_1.

Rossmann S, Richter R, Sun H, Schneeberger K, Töpfer R, Zyprian E, Theres K. 2020. Mutations in the miR396 binding site of the growth-regulating factor gene *VvGRF4* modulate inflorescence architecture in grapevine. *Plant J* **101**: 1234–1248, doi: 10.1111/tpj.14588.

Royo C, Torres-Pérez R, Mauri N, Diestro N, Cabezas JA, Marchal C, Lacombe T, Ibáñez J, Tornel M, Carreño J et al. 2018. The major origin of seedless grapes is associated with a missense mutation in the *MADS*-Box gene *VviAGL11*. *Plant Physiol* **177**: 1234–1253, doi: 10.1104/pp.18.00259.

Ruiz-García L, Gago P, Martínez-Mora C, Santiago JL, Fernádez-López DJ, Martínez MdC, Boso S. 2021. Evaluation and pre-selection of new grapevine genotypes resistant to downy and powdery mildew, obtained by cross-breeding programs in spain. *Front Plant Sci* **12**: 674510, doi: 10.3389/fpls.2021.674510.

Sahrawat AK, Becker D, Lütticke S, Lörz H. 2003. Genetic improvement of wheat via alien gene transfer, an assessment. *Plant Sci* **165**: 1147–1168, doi: https://doi.org/10.1016/S0168-9452(03)00323-6.

Saifert L, Sánchez-Mora FD, Assumpção WT, Zanghelini JA, Giacometti R, Novak EI, Vesco LLD, Nodari RO, Eibach R, Welter LJ. 2018. Marker-assisted pyramiding of resistance loci to grape downy mildew. *Pesqui Agropecu Bras* **53**: 602–610, doi: 10.1590/S0100-204X2018000500009.

Salmaso M, Faes G, Segala C, Stefanini M, Salakhutdinov I, Zyprian E, Toepfer R, Grando MS, Velasco R. 2005. Genome diversity and gene haplotypes in the grapevine (*Vitis vinifera* L.), as revealed by single nucleotide polymorphisms. *Mol Breed* **14**: 385–395, doi: 10.1007/s11032-005-0261-7.

Salmaso M, Malacarne G, Troggio M, Faes G, Stefanini M, Grando MS, Velasco R. 2008. A grapevine (*Vitis vinifera* L.) genetic map integrating the position of 139 expressed genes. *Theor Appl Genet* **116**: 1129–1143, doi: 10.1007/s00122-008-0741-3.

Sánchez-Mora FD, Saifert L, Zanghelini J, Assumpção WT, Guginski-Piva CA, Giacometti R, Novak EI, Klabunde GH, Eibach R, Dal Vesco L et al. 2017. Behavior of grape breeding lines with distinct resistance alleles to downy mildew (*Plasmopara viticola*). *Crop Breed Appl Biotechnol* **17**: 141–149, doi: 10.1590/1984-70332017v17n2a21

Sargolzaei M, Maddalena G, Bitsadze N, Maghradze D, Bianco PA, Failla O, Toffolatti SL, De Lorenzis G. 2020. *Rpv29, Rpv30 and Rpv31*: Three novel genomic loci associated with resistance to *Plasmopara viticola* in *Vitis vinifera*. *Front Plant Sci* **11**: 562432, doi: 10.3389/fpls.2020.562432.

Scott KD, Ablett EM, Lee LS, Henry RJ. 2000. AFLP markers distinguishing an early mutant of Flame Seedless grape. *Euphytica* **113**: 243–247, doi: 10.1023/A:1003977408214.

Sefc KM, Regner F, Turetschek E, Glössl J, Steinkellner H. 1999. Identification of microsatellite sequences in *Vitis riparia* and their applicability for genotyping of different *Vitis* species. *Genome* **42**: 367–373, doi: 10.1139/g98-168.

Sevini F, Marino R, Grando MS, Moser S, Versini G. 2004. Mapping candidate genes and QTLs for aroma content in grape. *Acta Hortic* **652**: 439–446, doi: 10.17660/ActaHortic.2004.652.58.

Shinde MP, Upadhyay A, Aher LB, Karibasappa GS. 2013. Molecular marker analysis to differentiate a clonal selection of Centennial Seedless grapevine. *Afr J Biotechnol* **12**: 1594–1597.

Smit SJ, Vivier MA, Young PR. 2021. Seeing the forest through the (Phylogenetic) trees: Functional characterisation of grapevine terpene synthase (*VviTPS*) paralogues and orthologues. *Plants* **10**: 1520, doi: 10.3390/plants10081520.

Soliman H. 2018. Production of genetically modified grape (*Vitis vinifera* L.) plants. *Int J Hortic Food Sci* **2**: 111–120, doi: 10.22161/ijhaf.2.4.1.

Stajner N, Angelova E, Bozinovic Z, Petkov M, Javornik B. 2009a. Microsatellite marker analysis of Macedonian grapevines (*Vitis vinifera* L.) compared to Bulgarian and Greek cultivars. *J Int Sci Vigne Vin* **43**: 29–34, doi: 10.20870/oeno-one.2009.43.1.805.

Stajner N, Jakse J, Javornik B, Masuelli RW, Martinez LE. 2009b. Highly variable AFLP and S-SAP markers for the identification of Malbec and Syrah clones. *Vitis* **48**: 145–150, doi: 10.5073/vitis.2009.48.145-150.

Štajner N, Rusjan D, Korošec-Koruza Z, Javornik B. 2011. Genetic characterization of old Slovenian grapevine varieties of *Vitis vinifera* L. by microsatellite genotyping. *Am J Enol Vitic* **62**: 250–255, doi: 10.5344/ajev.2011.10011.

Steward FC, Mapes MO, Mears K. 1958. Growth and organized development of cultured cells. II. Organization in cultures grown from freely suspended cells. *Am J Bot* **45**: 705–708, doi: 10.2307/2439728.

Striem MJ, Ben-Hayyim G, Spiegel-Roy P. 1996. Identifying molecular genetic markers associated with seedlessness in grape. *J Am Soc Hort Sci* **121**: 758–763, doi: 10.21273/jashs.121.5.758.

Sunitha S, Rock CD. 2020. CRISPR/Cas9-mediated targeted mutagenesis of TAS4 and MYBA7 loci in grapevine rootstock 101-14. *Transgenic Res* **29**: 355–367, doi: 10.1007/s11248-020-00196-w.

Sunseri F, Lupini A, Mauceri A, De Lorenzis G, Araniti F, Brancadoro L, Dattola A, Gullo G, Zappia R, Mercati F. 2018. Single nucleotide polymorphism profiles reveal an admixture genetic structure of grapevine germplasm from Calabria, Italy, uncovering its key role for the diversification of cultivars in the Mediterranean Basin. *Aust J Grape Wine Res* **24**: 345–359, doi: doi.org/10.1111/ajgw.12339.

Tello J, Torres-Pérez R, Flutre T, Grimplet J, Ibáñez J. 2020. *VviUCC1* nucleotide diversity, linkage disequilibrium and association with rachis architecture traits in grapevine. *Genes* **11**: 598, doi: 10.3390/genes11060598.

Tello J, Torres-Pérez R, Grimplet J, Carbonell-Bejerano P, Martínez-Zapater JM, Ibáñez J. 2015. Polymorphisms and minihaplotypes in the *VvNAC26* gene associate with berry size variation in grapevine. *BMC Plant Biol* **15**: 253, doi: 10.1186/s12870-015-0622-2.

Tello J, Torres-Perez R, Grimplet J, Ibáñez J. 2016. Association analysis of grapevine bunch traits using a comprehensive approach. *Theor Appl Genet* **129**: 227–242, doi: 10.1007/s00122-015-2623-9.

This P, Jung A, Boccacci P, Borrego J, Botta R, Costantini L, Crespan M, Dangl GS, Eisenheld C, Ferreira-Monteiro F et al. 2004. Development of a standard set of microsatellite reference allelles for identification of grape cultivars. *Theor Appl Genet* **109**: 1448–1458, doi: 10.1007/s00122-004-1760-3.

This P, Lacombe T, Cadle-Davidson M, Owens CL. 2007. Wine grape (*Vitis vinifera* L.) color associates with allelic variation in the domestication gene *VvmybA1*. *Theor Appl Genet* **114**: 723–730, doi: 10.1007/s00122-006-0472-2.

Thomas MR, Cain P, Scott NS. 1994. DNA typing of grapevines: A universal methodology and database for describing cultivars and evaluating genetic relatedness. *Plant Mol Biol* **25**: 939–949, doi: 10.1007/BF00014668.

Thomas MR, Scott NS. 1993. Microsatellite repeats in grapevine reveal DNA polymorphisms when analysed as sequence-tagged sites, STSs. *Theor Appl Genet* **86**: 985–990, doi: 10.1007/BF00211051.

Tomić L, Štajner N, Jovanović-Cvetković T, Cvetković M, Javornik B. 2012. Identity and genetic relatedness of Bosnia and Herzegovina grapevine germplasm. *Sci Hortic* **143**: 122–126, doi: 10.1016/j.scienta.2012.05.023.

Trenti M, Lorenzi S, Bianchedi PL, Grossi D, Failla O, Grando MS, Emanuelli F. 2021. Candidate genes and SNPs associated with stomatal conductance under drought stress in *Vitis*. *BMC Plant Biol* **21**: 7, doi: 10.1186/s12870-020-02739-z.

Troggio M, Malacarne G, Coppola G, Segala C, Cartwright DA, Pindo M, Stefanini M, Mank R, Moroldo M, Morgante M et al. 2007. A dense single-nucleotide polymorphism-based genetic linkage map of grapevine (Vitis vinifera L.) anchoring Pinot Noir bacterial artificial chromosome contigs. *Genetics* **176**: 2637–2650, doi: 10.1534/genetics.106.067462.

Troggio M, Malacarne G, Vezzulli S, Faes G, Salmaso M, Velasco R. 2008. Comparison of different methods for SNP detection in grapevine. *Vitis* **47**: 21–30, doi: 10.5073/vitis.2008.47.21-30.

Tu M, Wang X, Yin W, Wang Y, Li Y, Zhang G, Li Z, Song J, Wang X. 2020. Grapevine VlbZIP30 improves drought resistance by directly activating VvNAC17 and promoting lignin biosynthesis through the regulation of three peroxidase genes. *Hortic Res* **7**: 150, doi: 10.1038/s41438-020-00372-3.

Upadhyay A, Aher LB, Shinde MP, Mundankar KY, Datre A, Karibasappa GS. 2013. Microsatellite analysis to rationalize grape germplasm in India and development of a molecular database. *Plant Genet Res* **11**: 225–233, doi: 10.1017/S1479262113000117.

Upadhyay A, Lalitkumar BA, Gaudar Shivshankar K. 2011. Detection of variation among clonal selections of grapevine (*Vitis vinifera* L.) 'Kishmish Chernyi' by AFLP analysis. *J Hortic Sci Biotechnol* **86**: 230–234, doi: 10.1080/14620316.2011.11512753.

Upadhyay A, Saboji MD, Reddy S, Deokar K, Karibasappa GS. 2007. AFLP and SSR marker analysis of grape rootstocks in Indian grape germplasm. *Sci Hortic* **112**: 176–183, doi: 10.1016/j.scienta.2006.12.011.

van Heerden CJ, Burger P, Vermeulen A, Prins R. 2014. Detection of downy and powdery mildew resistance QTL in a 'Regent'×'RedGlobe' population. *Euphytica* **200**: 281–295, doi: 10.1007/s10681-014-1167-4.

Vezzulli S, Malacarne G, Masuero D, Vecchione A, Dolzani C, Goremykin V, Mehari ZH, Banchi E, Velasco R, Stefanini M et al. 2019. The Rpv3-3 haplotype and stilbenoid induction mediate downy mildew resistance in a grapevine interspecific population. *Front Plant Sci* **10**: 234, doi: 10.3389/fpls.2019.00234.

Vezzulli S, Troggio M, Coppola G, Jermakow A, Cartwright D, Zharkikh A, Stefanini M, Grando MS, Viola R, Adam-Blondon AF et al. 2008. A reference integrated map for cultivated grapevine (*Vitis vinifera* L.) from three crosses, based on 283 SSR and 501 SNP-based markers. *Theor Appl Genet* **117**: 499–511, doi: 10.1007/s00122-008-0794-3.

Vidal JR, Rama J, Taboada L, Martin C, Ibañez M, Segura A, González-Benito ME. 2009. Improved somatic embryogenesis of grapevine (*Vitis vinifera*) with focus on induction parameters and efficient plant regeneration. *Plant Cell Tissue Organ Cult* **96**: 85–94, doi: 10.1007/s11240-008-9464-z.

Vouillamoz JF, Monaco A, Costantini L, Stefanini M, Scienza A, Grando MS. 2007. The parentage of 'Sangiovese', the most important Italian wine grape. *Vitis* **46**: 19–22, doi: 10.5073/vitis.2007.46.19-22.

Walker AR, Lee E, Robinson SP. 2006. Two new grape cultivars, bud sports of Cabernet Sauvignon bearing pale-coloured berries, are the result of deletion of two regulatory genes of the berry colour locus. *Plant Mol Biol* **62**: 623–635, doi: 10.1007/s11103-006-9043-9.

Wan D-Y, Guo Y, Cheng Y, Hu Y, Xiao S, Wang Y, Wen Y-Q. 2020. CRISPR/Cas9-mediated mutagenesis of VvMLO3 results in enhanced resistance to powdery mildew in grapevine (*Vitis vinifera*). *Hortic Res* **7**: 116, doi: 10.1038/s41438-020-0339-8.

Wang F-Q, Fan X-C, Zhang Y, Sun L, Liu C-H, Jiang J-F. 2022. Establishment and application of an SNP molecular identification system for grape cultivars. *J Integr Agric* **21**: 1044–1057, doi: 10.1016/S2095-3119(21)63654-7.

Wang H, Yan A, Sun L, Zhang G, Wang X, Ren J, Xu H. 2020. Novel stable QTLs identification for berry quality traits based on high-density genetic linkage map construction in table grape. *BMC Plant Biol* **20**: 411, doi: 10.1186/s12870-020-02630-x.

Wang J, Su K, Guo Y, Xing H, Zhao Y, Liu Z, Li K, Guo X. 2017. Construction of a high-density genetic map for grape using specific length amplified fragment (SLAF) sequencing. *PLoS One* **12**: e0181728, doi: 10.1371/journal.pone.0181728.

Wang N, Fang L, Xin H, Wang L, Li S. 2012. Construction of a high-density genetic map for grape using next generation restriction-site associated DNA sequencing. *BMC Plant Biol* **12**: 148, doi: 10.1186/1471-2229-12-148.

Wang Q, Mawassi M, Sahar N, Li P, Violeta C-T, Gafny R, Sela I, Tanne E, Perl A. 2004. Cryopreservation of Grapevine (*Vitis* spp.) Embryogenic cell suspensions by Encapsulation–Vitrification. *Plant Cell Tissue Organ Cult* **77**: 267–275, doi: 10.1023/B:TICU.0000018393.58928.b1.

Wang X, Tu M, Wang D, Liu J, Li Y, Li Z, Wang Y, Wang X. 2018. CRISPR/Cas9-mediated efficient targeted mutagenesis in grape in the first generation. *Plant Biotechnol J* **16**: 844–855, doi: 10.1111/pbi.12832.

Wang X, Tu M, Wang Y, Yin W, Zhang Y, Wu H, Gu Y, Li Z, Xi Z, Wang X. 2021a. Whole-genome sequencing reveals rare off-target mutations in CRISPR/Cas9-edited grapevine. *Hortic Res* **8**: 114, doi: 10.1038/s41438-021-00549-4.

Wang Y, Liu X, Ren C, Zhong G-Y, Yang L, Li S, Liang Z. 2016. Identification of genomic sites for CRISPR/Cas9-based genome editing in the Vitis vinifera genome. *BMC Plant Biol* **16**: 96, doi: 10.1186/s12870-016-0787-3.

Wang Y, Xin H, Fan P, Zhang J, Liu Y, Dong Y, Wang Z, Yang Y, Zhang Q, Ming R. 2021b. The genome of Shanputao (*Vitis amurensis*) provides a new insight into cold tolerance of grapevine. *Plant J* **105**: 1495–1506, doi: 10.1111/tpj.15127.

Welter LJ, Göktürk-Baydar N, Akkurt M, Maul E, Eibach R, Töpfer R, Zyprian EM. 2007. Genetic mapping and localization of quantitative trait loci affecting fungal disease resistance and leaf morphology in grapevine (*Vitis vinifera* L.). *Mol Breed* **20**: 359–374, doi: 10.1007/s11032-007-9097-7.

Wolt JD. 2017. Safety, security, and policy considerations for plant genome editing. *Prog Mol Biol Transl Sci* **149**: 215–241, doi: 10.1016/bs.pmbts.2017.03.005.

Wolt JD, Wang K, Yang B. 2016. The regulatory status of genome-edited crops. *Plant Biotechnol J* **14**: 510–518, doi: 10.1111/pbi.12444.

Xu W, Ma F, Li R, Zhou Q, Yao W, Jiao Y, Zhang C, Zhang J, Wang X, Xu Y et al. 2019. VpSTS29/STS2 enhances fungal tolerance in grapevine through a positive feedback loop. *Plant Cell Environ* **42**: 2979–2998, doi: 10.1111/pce.13600.

Yang X, Guo Y, Zhu J, Niu Z, Shi G, Liu Z, Li K, Guo X. 2017. Genetic diversity and association study of aromatics in grapevine. *J Am Soc Hort Sci* **142**: 225–231.

Yanxia Z. 2020. Analysis of *PYL* genes and their potential relevance to stress tolerance and berry ripening in grape. *J Am Soc Hort Sci* **145**: 308–317, doi: 10.21273/JASHS04942-20.

Zendler D, Schneider P, Töpfer R, Zyprian E. 2017. Fine mapping of *Ren3* reveals two loci mediating hypersensitive response against Erysiphe necator in grapevine. *Euphytica* **213**: 68, doi: 10.1007/s10681-017-1857-9.

Zhang C, Cui L, Fang J. 2022a. Genome-wide association study of the candidate genes for grape berry shape-related traits. *BMC Plant Biol* **22**: 42, doi: 10.1186/s12870-022-03434-x.

Zhang C, Wu J-Y, Cui L-W, Fang J-G. 2022b. Mining of candidate genes for grape berry cracking using a genome-wide association study. *J Integr Agric* **21**: 2291–2304, doi: 10.1016/S2095-3119(21)63881-9.

Zhang D, Hussain A, Manghwar H, Xie K, Xie S, Zhao S, Larkin RM, Qing P, Jin S, Ding F. 2020a. Genome editing with the CRISPR-Cas system: An art, ethics and global regulatory perspective. *Plant Biotechnol J* **18**: 1651–1669, doi: 10.1111/pbi.13383.

Zhang H, Fan X, Zhang Y, Jiang J, Liu C. 2017a. Identification of favorable SNP alleles and candidate genes for seedlessness in *Vitis vinifera* L. using genome-wide association mapping. *Euphytica* **213**: 1–13, doi: 10.1007/s10681-017-1919-z.

Zhang H, Liu Z, Fan X, Zhang C, Liwen1 C, Liu C, Fang J. 2017b. Genome-wide association mapping of berry shape traits via the reduced representation sequencing in grape. *Acta Hortic Sin* **44**: 1959–1968, doi: 10.16420/j.issn.0513-353x.2016-0845.

Zhang S, Dong R, Wang Y, Li X, Ji M, Wang X. 2021a. NAC domain gene *VvNAC26* interacts with *VvMADS9* and influences seed and fruit development. *Plant Physiol Biochem* **164**: 63–72, doi: 10.1016/j.plaphy.2021.04.031.

Zhang X, Wu Y, Li Z, Song C-B, Wang X. 2021b. Advancements in plant regeneration and genetic transformation of grapevine (*Vitis* spp.). *J Integr Agric* **20**: 1407–1434, doi: 10.1016/S2095-3119(20)63586-9.

Zhang Y, Fan X, Li Y, Sun H, Jiang J, Liu C. 2020b. Restriction site-associated DNA sequencing reveals the molecular genetic diversity of grapevine and genes related to white rot disease. *Sci Hortic* **261**: 108907, doi: 10.1016/j.scienta.2019.108907.

Zhang Y, Feng L, Fan X, Jiang J, Zheng X-B, Sun H, Chonghuai L. 2017c. Genome-wide assessment of population structure, linkage disequilibrium and resistant QTLs in Chinese wild grapevine. *Sci Hortic* **215**: 59–64, doi: 10.1016/j.scienta.2016.12.014.

Zhao YH, Su K, Guo YH, Ma HF, Guo XW. 2016. Molecular genetic map construction and QTL analysis of fruit maturation period in grapevine. *Genet Mol Res* **15**, doi:10.4238/gmr.15028040.

Zhu J, Guo Y, Su K, Liu Z, Ren Z, Li K, Guo X. 2018. Construction of a highly saturated genetic map for *Vitis* by next-generation Restriction Site-associated DNA Sequencing. *BMC Plant Biol* **18**: 347, doi: 10.1186/s12870-018-1575-z.

Zyprian E, Ochßner I, Schwander F, Šimon S, Hausmann L, Bonow-Rex M, Moreno-Sanz P, Grando MS, Wiedemann-Merdinoglu S, Merdinoglu D et al. 2016. Quantitative trait loci affecting pathogen resistance and ripening of grapevines. *Mol Genet Genom* **291**: 1573–1594, doi: 10.1007/s00438-016-1200-5.

5

Genetic Diversity Studies in Potato (Solanum tuberosum L.) Using Various Molecular Markers

Namo Dubey, Swati Verma, Jagesh Kumar Tiwari, Dalamu, Vikas Mangal, Salej Sood, Rasna Zinta, Jyoti Chaudhary, S.K. Luthra and Ajay Kumar Thakur

5.1 Introduction

Potato (*Solanum tuberosum* L.) is an important tuber crop contributing to global food and nutritional security after rice, wheat and maize (Liu et al. 2021). Potatoes serve as a major source of calories and fulfills the food security demands of Asia and South America (Scott and Suarez 2012). Thus, this tuber crop has a significant role in meeting the global nutrition and sustainable development goals of zero hunger and food security of the United Nation (www.fao.org/3/cc2211en/cc2211en.pdf). Potatoes serve as a very relevant and stable component of the daily human diet, having high nutritional value and yield potential (Koch et al. 2019). However, limited genetic variation makes potato crop more vulnerable to disease and insect epidemics. As witnessed by Ireland in 1845, genetic vulnerability had devastating effects on food security, when the late blight disease caused by *Phytophthora infestans* destroyed the potato crop. The Irish famine resulted in the death of millions of Irish people due to starvations and the exodus to other countries of another million people (Majeed et al. 2017). Some biotic or abiotic stresses in potato cannot be managed by chemical sprays. Major abiotic stresses (high temperature, drought, soil salinity and nutrient) and biotic stresses (late blight, bacterial wilt, root-knot nematode and viruses) adversely affect plant development, tuberization, tuber bulking, thus influencing both tuber yield and quality (Tiwari et al. 2022). Global food demand is expected to increase by 59%–98% between 2005 and 2050, the aforesaid factors, along with a rapidly growing global population are the major challenges in attaining global food security. The growing concerns over food security demand

DOI: 10.1201/9781003625674-5

crop improvement using diverse approaches for sustainable agriculture (McKenzie and Williams 2015).

Genetic variation present in the germplasm of a crop can be used for genetic improvement in that crop. Progress in the improvement of a crop through crop breeding is driven by the availability of genetic diversity or variability in the germplasm or genetic repository of that crop (Swarup et al. 2021). With increasing population and more food demand compounded with disease pests' pressure and climate change, enhancing potato production is necessary through development of new varieties using diverse genetic resources like wild/semi and cultivated *Solanum* species in potato. Genetic diversity offers the opportunity for the development of potato cultivars with biotic and abiotic stress tolerance and improved quality. Conventionally, potato cultivars are differentiated on the basis of morphological and physiological traits, including plant architecture, flower type, leaf color and shape, flesh color, skin, tuber, tuber color and shape, the depth of the eyes, flower color, sprouting, etc. (Van Eck et al. 2007). However, these traits are highly influenced by environmental factors and thus give misleading information (Govindaraj et al. 2015). Further, these traditional methods are expensive, time-consuming and skill and labor demanding. The correct identification of potato cultivars is an essential prerequisite for their registration to protect plant breeders' rights, their use as donors in breeding programs, seed certification, etc., which necessitates the use of molecular markers to establish their uniqueness, identity and genetic diversity evaluation for identification of genetically diverse genotypes for use in breeding. In plant biology and genetics, molecular markers are commonly employed to explore genetic diversity within plant populations, assess genetic relatedness among accessions, and investigate the evolutionary links between various plant species (Mohammadi and Prasanna 2003). The release of the first potato reference genome in 2011 proved to be a turning point in potato molecular breeding. Developments made in marker-assisted selection have made the potato breeding programs more efficient (Marhadour and Prodhomme 2023). Molecular markers are free from environmental influences and developmental stages of the plants, and show Mendelian inheritance, thus they are the best descriptors rather than morphological and biochemical markers (Govindaraj et al. 2015).

Various types of molecular markers including random amplified polymorphic DNA (RAPD), restriction fragment length polymorphism (RFLP), amplified fragment length polymorphism (AFLP), inter simple sequence repeat (ISSR), simple sequence repeats (SSRs) and simple nucleotide polymorphism (SNPs) have been frequently used in potato to estimate the genetic diversity, establish the genetic inter-relationships, perform taxonomic studies and to develop efficient fingerprinting tools (Table 5.1). We summarize and discuss the use of various molecular markers in genetic diversity studies in potato germplasm and varieties.

TABLE 5.1

A Catalog of DNA Markers Used in Various Studies to Assess the Genetic Diversity of Potato

Marker Type	Study Purpose	References
RFLP	First potato map	Bonierbale et al. (1988)
RFLP	High heterozygosity observed in diploid potato genetics lines	Gebhardt et al. (1989)
RFLP	Highly saturated map for comparison with tomato	Tanksley et al. (1992)
RFLP	Integration of known loci and new markers	Jacobs et al. (1995)
AFLP	Genetic diversity and interrelationship among 16 commercial potato cultivars	Milbourne et al. (1997)
AFLP	Genetic diversity among three potato species cultivars	Bamberg and del Rio (2016)
AFLP	Genetic diversity among 288 potato germplasm accessions	Jian et al. (2017)
SSR	Fingerprinting	Milbourne et al. (1998)
SSR	Genetic relationships among potato (*S. tuberosum*) cultivars	Provan et al. (1996); Schneider and Douches (1997)
SSR	Genetic diversity among 16 accessions of potato	Milbourne et al. (1997)
SSR	Studying phylogenetic relationships among wild and domesticated potato species	Ashkenazi et al. (2001)
SSR	SSR markers to distinguish *S. tuberosum* subsp. *tuberosum* from *S. tuberosum* subsp. *andigena*	Raker and Spooner (2002)
SSR	Genetic diversity among 900 native potato accessions	Ghislain et al. (2004)
SSR	Genetic diversity among 41 potato accessions	Barandalla et al. (2006)
SSR	Genetic diversity among 105 potato cultivars	Galarreta et al. (2011)
SSR	Genetic diversity among 47 potato cultivars	Salimi et al. (2016)
SSR	Genetic diversity among 138 potato cultivars	Anoumaa et al. (2017)
SSR	Genetic diversity among 48 potato cultivars	Tiwari et al. (2018)
SSR	Genetic diversity among 20 potato cultivars	Tillault and Yevtushenko (2019)
SSR	Genetic diversity among 82 potato cultivars	Tiwari et al. (2019)
SSR	Genetic diversity among 292 potato cultivars	Wang et al. (2019)
SSR	Genetic diversity among 47 potato cultivars	Patil et al. (2020)
SSR	Genetic diversity among 25 potato cultivars	Muhammad et al. (2022)
SSR	Genetic diversity among 12 potato cultivars	Rahman et al. (2022)
ISSR	Validation for fingerprinting	Prevost and Wilkinson (1999)
RAPD	Genetic diversity in a set of nine potato somaclonal variants and three check varieties	Shamsuzzaman et al. (2021)
RAPD	Genetic diversity evaluation among 10 local potato genotypes	Islam et al. (2022)

5.2 Origin and Evolutionary History of Potato

Potato was first cultivated by indigenous people living in the high-altitude regions of South America's Andes Mountains. This region includes modern-day Peru, Bolivia and Ecuador (Sandweiss and Richardson 2008). These early growers most likely collected wild potato species and carefully propagated those with desired qualities like larger tuber size, better flavor and adaptability to different altitudes and climates. In these tribes, potatoes became a main food source, providing much-needed nutrition in the harsh Andean terrain (McNeill 1999). During their journeys to South America in the 16th century, Spanish explorers, including Francisco Pizarro, came across the potato. The potato was brought back to Europe as part of the Columbian Exchange. However, because of its unfamiliarity and relationship with other likely toxic relatives of the *Solanaceae* family, potato was initially received with mistrust and opposition in Europe (de Santayana et al. 2010). Over the past, potato gained popularity in Europe, particularly in areas where other main crops struggled to thrive. Agronomists such as Antoine-Augustin Parmentier in France promoted potato eating and cultivation in the 18th century (Earle 2017). Despite great damages like the Irish Potato Famine, potato cultivation kept spreading worldwide. Different regions developed their own potato cultivation and growing methods, and worldwide development of many potato varieties was observed. Potato is currently a vital food crop farmed on practically every continent, with China, India and Russia being major producers (Birch et al. 2012). In the early 17th century, Portuguese sailors introduced potato to India. Later, the British expanded its cultivation to North India. Uttar Pradesh, West Bengal, Bihar, Punjab and Gujarat contributing significantly to total potato production in India. India has had an extraordinary increase in potato production over the last four to five decades, outpacing other major food crops in terms of yearly compound growth rate for area, production and productivity (Chand et al. 2011). From its humble beginnings in the Andes to its current prominence as an important global food crop, potato's journey highlights the connectivity of cultures, landscapes and agriculture around the world.

5.3 Initiation of Molecular Marker Studies for Decoding the Genetic Diversity in Potato

Utilization of DNA markers to discern genetic variations, diversity and evolutionary connections proves advantageous in pinpointing genetic and population structures, as well as unveiling diversity among cultivars originating from diverse geographical regions (Parita et al. 2018). The origin of molecular marker studies in decoding genetic diversity in potatoes can be traced back to the development of molecular biology techniques that allowed scientists to analyze and characterize DNA at the molecular level (Spooner 2005). The first potato genetic maps were created in the late 1980s, based upon sexual recombination frequencies and chromosomal identification (Bonierbale et al. 1988). Soon after, the use of PCR-based markers for the selection of desired features emerged (Meksem et al. 1995). Because of high levels of genetic diversity within potato, it became imperative to screen more individuals than usual to accurately assess a marker's association with a specific phenotypic trait (Bonierbale et al. 1988; Torres-Rendon et al. 2009). Finding relevant markers in diploid variants is easier than in tetraploid types (Diwan et al. 2000). Furthermore, discovery of new genetic markers can be influenced by the age and type of polyploidy (Huang et al. 2016). Within the realm of potato, PCR-based techniques have been extensively employed for DNA fingerprinting and genotyping. For instance, various studies have utilized RAPD, SSR, ISSR and AFLP for decoding the genetic diversity in potatoes, which have been summarized in further sections of this book chapter.

5.4 Potato Genome Sequencing Consortium

The DNA sequence information retrieved from potato genome sequencing consortium has immensely contributed to speeding up the molecular marker development in potato. These were the collaborative efforts of the Potato Genome Sequencing Consortium (PGSC), comprising 26 research institutes from 13 nations (India, the United States, Ireland, China, Chile, Argentina, Brazil, the United Kingdom, Netherlands, New Zealand, Peru, Poland and Russia), which led to the successful sequencing of the potato genome (Visser et al. 2009). Through a combination of cutting-edge sequencing technologies including Pyrosequencing and Illumina GAIIx, as well as traditional Sanger sequencing, a vast amount of raw data comprising 96.6 gigabases (Gb) was generated. This data was then meticulously processed, resulting in the assembly of the 844 megabase (Mb) potato genome into ungapped super-scaffolds (Pham et al. 2020). The PGSC sequenced two diploid potato genotypes: the heterozygous diploid line RH89-039-16 (RH) from the *S. tuberosum* Group *Tuberosum* genetics, and the doubled monoploid clone DM1-3 (DM) from the *S. tuberosum* Group Phureja (Watanabe 2015).

These super-scaffolds covered 93.9% of the overall genome size, which stands at 840 Mb. The aligned sequence, spanning 99% of the genome, was organized into 443 super-scaffolds. The potato's genetic makeup encompasses 12 chromosomes, with a haploid length spanning about 840 million base pairs

(Visser et al. 2009). The initial efforts of the PGSC focused on sequencing "RH". The subsequent phase of the project involved leveraging a collection of diploid potato genomic library of 78,000 bacterial artificial chromosome (BAC) clones. This assortment has undergone fingerprinting and alignment, resulting in approximately 7,000 contigs within a physical map (Visser et al. 2009). Furthermore, the sequences of the BAC ends have been decoded and are accessible to the public. Roughly 30,000 BACs have been affixed to the Ultra High Density genetic map of potatoes, which is constructed from 10,000 distinct AFLP markers. Within this integrated genetic-physical map, a range of 50 to 150 selected BACs have presently been pinpointed for each individual chromosome (Patil et al. 2017). The PGSC database provides information on maps and markers, genes, genomes and sequences, breeding tools, pathways and phenotypes all in one site, and is being extensively utilized for retrieval of information by various researchers across the world.

5.5 Different Molecular Markers in the Evaluation of Potato Genetic Diversity

Utilization of molecular markers in the potato breeding programs offer new opportunities for the selection of genotypes. To date, many markers linked to useful traits have been mapped to the potato genome, which are now available for rapid and efficient molecular-marker-assisted selection of potato genotypes (Barone 2004). Various non-commercial varieties of potatoes harbor genetic potential for the improvement of disease resistance and abiotic stress tolerance in commercial potato cultivars, however, introducing traits from non-commercial varieties to breeding stock can be extremely labor intensive (Ramakrishnan et al. 2015). Molecular genetic markers closely associated with a trait can be used to decrease the time spent for phenotyping varieties. In the section below, we summarize various studies on the utilization of RAPD, RFLP, AFLP, SSR, ISSR, EST-SSR and SNPs for the evaluation of genetic diversity in potato.

(A) Randomly Amplified Polymorphic DNA (RAPD) markers in potato genetic diversity evaluation

RAPD is a PCR-based technique, which involves the use of single arbitrary short primers (8–12 nucleotides), resulting in the amplification of many discrete DNA. The segments of DNA are randomly amplified, and prior knowledge of the genome sequence is not required in the RAPD method (Welsh and McClelland 1990; Williams et al. 1990; Kumari and Thakur 2014). The RAPD marker system is characterized by speed, simplicity and independence from prior sequence knowledge and has been extensively used for the identification of the genetic variations or polymorphisms in a population of species (Demeke et al. 1996).

The genetic diversity of 16 potato cultivars was assessed using RAPD, culminating into the establishment of a molecular profile of these potato cultivars (Rocha et al. 2010). The cultivars were classified according to their geographical origin and morphological traits of their parent variety using a dendrogram based on RAPD markers. Hoque et al. (2013) used eight RAPD primers to explore the genetic diversity and polymorphism of 12 potato varieties in Bangladesh. The use of RAPD markers for potato cultivar identification has been well documented in molecular characterization studies, as indicated by Bianchi et al. (2004). Sosinski and Donches (1996) pioneered a similar strategy in North America in 1996, employing RAPD markers for fingerprinting, identification and characterization of potato cultivars. Shamsuzzaman et al. (2021) deployed six RAPD markers to analyze genetic diversity in a set of nine potato somaclonal variants and three check varieties. A total of 54 distinct amplicons were obtained with a size range of 88–3265 bp, out of which 47 amplicons (87%) were observed to be polymorphic in nature. Nei's genetic distance among all the 12 potato genotypes ranged from 0.0972 to 0.6217, while genetic identity was in the range of 0.5370 to 0.9074. The UPGMA dendrogram divided potato genotypes into two clusters, each comprising eight and four genotypes, respectively. The highest genetic distance was observed between SUP-6 and SUP- 68 genotype pairs. Cluster A comprised five somaclonal variants and three source materials, while cluster B comprised the remaining four somaclonal variants. It had been inferred that the somaclonal variants of cluster B might have possessed comparatively larger genetic variations than that of cluster A (Shamsuzzaman et al. 2021). Islam et al. (2022) carried out molecular characterization and genetic diversity evaluation among ten local potato genotypes of Bangladesh using five RAPD primers. A total of 33 PCR amplicons were obtained with the size range of 131–1188 bp, out of which 28 amplicons were found to be polymorphic. Gene frequency ranged between 0.3 to 1.0 and gene diversity was in the range of 0 to 0.7. UPGMA dendrogram based on Nei's genetic distance grouped all the potato genotypes into two main groups.

(B) RFLP markers in potato genetic diversity evaluation

RFLP method relies on variations in the lengths of DNA fragments generated by specific restriction enzymes at recognition sites. The majority of RFLP markers are characterized by co-dominance inheritance, which allows them to identify both alleles in a heterozygous sample, and they possess remarkable locus-specificity (Beckmann and Soller 1983). Early research studies demonstrated that RFLP patterns are effective in distinguishing diploid and tetraploid potato genotypes (Gebhardt et al. 1989). In total, 38 diploid and 20 tetraploid genotypes were differentiated by utilizing a combination of two highly variable RFLP markers (Gorg et al. 1992). Later, Gorg et al. (1992) classified a significant majority of 136 tetraploid varieties using a single RFLP marker and a single restriction enzyme. Phenetic investigations using RFLP molecular markers have also been carried out on both wild and cultivated relatives of *S. tuberosum* ssp. *tuberosum*, as indicated by Debener et al. (1990). These studies involved the construction of phenetic trees for accessions from 14 wild and three cultivated *Solanum species*. The data utilized in these studies encompassed information from as many as 777 RFLP fragments.

(C) AFLP markers in potato genetic diversity evaluation

AFLP is a PCR-based molecular technique that uses selective amplification of a subset of digested DNA fragments from any source to generate and compare unique fingerprints of genomes (Sheeja et al. 2021). In potato, the AFLP marker system had been utilized for analysis of genetic diversity, taxonomy studies and phylogenetic analysis. Kim et al. (1998) used AFLP markers to characterize 12 potato cultivars and obtained 84 polymorphic bands. Bamberg and Del Rio (2016) used AFLP markers to analyze the genetic diversity among three potato species cultivars in the US potato GenBank. They concluded that there had been redundant deposits of potato cultivars in the GenBank, as a result of which the genetic diversity had not been enriched with the increasing number of cultivar deposits over the time (Bamberg and Del Rio 2016). In the following year, both SSR and AFLP markers were used to evaluate genetic diversity and population structure in a collection of 288 potato germplasm accessions procured from eight countries and International Potato Centre (CIP) (Jian et al. 2017). All the 20 SSR markers resulted in polymorphic amplicons with a total of 190 alleles, ranging from 2 to 23, with 9.5 alleles as the average number of alleles per locus, and 0.7312 as the mean PIC value. Ten AFLP marker combinations amplified a total of 988 alleles, of which 99.49% alleles were polymorphic and the allele number ranged from 91–116, with 98.3 average number of alleles per marker, along with a mean PIC value of 0.987. All the 288 potato germplasm accessions were divided into seven sub groups and an admix group based on Bayesian analysis. It was inferred from the results obtained from Bayesian analysis, cluster and principal co-ordinate analysis that there was a lack of geographical differentiation among country-wise collections, indicating towards the possibility of common introduction of potato germplasm to these countries.

(D) ISSR markers in potato genetic diversity evaluation

Inter-simple sequence repeat (ISSR) markers have evolved as a valuable tool for investigating natural populations of various organisms, including plants, fungi, insects and vertebrates. ISSRs are PCR amplified, employing microsatellite core sequences as primers, supplemented with a small number of selective nucleotides that anchor them to adjacent non-repeating regions (16–18 bp). This method enables the simultaneous generation of approximately 10–60 fragments from multiple genetic loci, which are subsequently separated through gel electrophoresis and scored based on the presence or absence of fragments of specific sizes (Kumar et al. 2009). During early studies, Allainguillaume et al. (1997) employed ISSR-PCR to detect gene flow from *Solanum phureja* into potato dihaploids. Thereafter, the effectiveness of ISSR-PCR was assessed for creating genetic fingerprints in potato by Prevost and Wilkinson (1999). They employed four primers on a set of 34 cultivars for the development of these genetic fingerprints in potato (Prevost and Wilkinson 1999). Bornet et al. (2002) utilized three primers to analyze the genetic diversity in 28 potato accessions. It has been concluded that potato grown in Argentina and Europe exhibited distinct genetic profiles, with a greater degree of genetic diversity among the Argentina potato (Bornet et al. 2002). In another study, five ISSR primers were employed to characterize 35 potato accessions originating from Mexico, Europe and the United States. The primary goal was to gather information on genetic diversity and relationships among these genotypes to enhance their utilization in breeding programs. The ISSR markers effectively differentiated between various potato accessions and grouped the genotypes based on their geographic origin. This information regarding genetic diversity and relationships holds significant value in selecting parental lines and conducting mapping studies (Onamu et al. 2016). In summary, ISSR markers serve as a valuable tool in assessing the genetic diversity within potato populations, and aiding further understanding of potato genetics, breeding and conservation. They provide insights into the genetic relationships among different potato varieties and populations, which can be crucial for sustainable potato cultivation and crop improvement (Onamu et al. 2016).

(E) SSR and EST-SSR markers in potato genetic diversity evaluation

SSRs are repetitive, short, tandem repeat motifs ranging from one to six base pairs, which are found throughout coding as well as non-coding regions of genomes of eukaryotes with high repeatability (Tiwari et al. 2019). Among the various classes of molecular markers, there is a growing significance attributed to SSRs or microsatellites in the realms of plant breeding and genetics. These repetitive DNA motifs, ranging from one to six nucleotides in length, have garnered attention due to their co-dominant inheritance, multi-allelic characteristics, relatively abundant presence, remarkable variability, reproducibility, capacity of thorough genome coverage, including organellar genomes, specific chromosomal locations, adaptability to automation, and suitability for high-throughput genotyping (Walunjkar et al. 2013). SSRs could be efficiently utilized for analyzing the phylogenetic relationships and carrying out comparative genome mapping in many crops, because they are derived from conserved sequences of the genome (Ghislain et al. 2004).

Milbourne et al. (1997) used three types of PCR-based markers, viz. RAPD, SSR, AFLP and compared their efficacy in the variability estimation among 16 potato cultivars. Among the three marker types, SSRs were found to be the best in establishing genetic relationships and analyzing the genetic distances among potato cultivars. In the following year, Milbourne et al. (1998) identified a set of 112 SSR primer sequences located on all the 12 chromosomes of potato genome. This information equipped potato molecular biologists with a potential set of SSR markers, which could be utilized in genetic mapping and genetic diversity evaluation programs. In another study, SSR markers proved helpful in establishing phylogenetic relationships among wild and domesticated potato species (Ashkenazi et al. 2001). Subsequently, Raker and Spooner (2002) used SSR markers to distinguish *S. tuberosum* subsp. *tuberosum* from *S. tuberosum* subsp. *andigena*. Later, Ghislain et al. (2004) used 22 SSR markers to analyze 900 native potato accessions from the Andes, which belonged to eight different *Solanum* species. Barandalla et al. (2006) characterized 41 potato cultivars from ten locations of Tenerife Island of Spain using 19 SSR markers. They reported a total of seven alleles, out of which 12 alleles were present in all the cultivars. However, they observed some discrepancies while deciphering the genetic relationships among various potato cultivars on the basis of SSR allelic data and proposed for the need of a more comprehensive study. Spooner et al. (2007) used SSR markers for genetic diversity studies in potato accessions and deciphered their genetic structure classification. Ghislain et al. (2004, 2009) developed a potato genetic identification (PGI) kit consisting of the best 24 SSR markers, after evaluating 148 SSRs among a panel of 742 potato landraces. Galarreta et al. (2011) deciphered relationships among 105 local potato cultivars from Spain comprising 41 cultivars from Tenerife Island, 19 from La Palma Island and 45 local cultivars from peninsular Spain using 19 SSR markers. A total of 76 alleles were identified with seven alleles common in all the varieties under study. Another study carried out for evaluation of genetic diversity among a panel of 47 potato genotypes of Europe and America origin, by Salimi et al. (2016) deployed 25 genomic-SSR and six EST-SSR primer pairs. They reported two–nine alleles with an average of 6.22 alleles/locus. UPGMA dendrogram based on Jaccard's similarity coefficient grouped the American and European potato genotypes according to their geographical distribution. It was further observed that the expected heterozygosity values were higher for the American genotypes in comparison to the European genotypes. This trend might be due to the fact that European potato genotypes have originated from a limited number of introductions and this might have contributed to the narrow genetic base, depicting relatively less genetic diversity (Salimi et al. 2016).

Anoumaa et al. (2017) evaluated genetic variability among 138 potato cultivars from the western Highlands region of Cameroon using 12 SSR primers pairs. They reported 0.74 average PIC value and 7.08 average number of alleles per SSR marker, which inferred a high extent of genetic diversity amongst potato cultivars. Further, they reported that genetic diversity estimates were significantly lower for cultivars from low elevation in comparison to the potato cultivars from medium and higher elevations. Cluster analysis grouped all the 138 cultivars into three distinct groups, whereas two gene pools were observed using structure analysis. A higher level of admixture had been found between gene pools within locations and elevations. On the basis of SSR-allelic data, a core collection of potato cultivars was also developed, which consisted of 27 highly diverse potato genotypes, which captured 99.15% of the total alleles and represented 19.57% of the whole collection. Tiwari et al. (2018) analyzed allelic variations in 48 Indian potato varieties using 12 SSR markers with an aim to develop a dataset of SSR markers for varietal identification. Allele number ranged from four to 35 and PIC values were in the range of 0.53 to 0.92. They could identify a set of a minimum of two SSR markers viz. STIKA and STU65SNRN for potato varietal identification. It has been felt over the years that there is a constant need for development of new potato varieties with better attributes like resistance to biotic and abiotic stresses and improved nutritional quality traits to feed the ever-increasing population. Also, the newly developed cultivars should be well-registered for protection of breeders' rights using morphological and physiological traits. However, the advent of molecular markers in general and SSR markers in particular has really expedited the varietal registration process by developing unique DNA fingerprints of these cultivars. In an attempt to develop a DNA-fingerprinting profile, Tillault and Yevtushenko (2019) characterized 20 potato varieties developed in Alberta, Canada using ten SSR markers. They obtained 4.4 average number of alleles per SSR marker and the allele number ranged from 2–6. The PIC values ranged from 0.477 to 0.802 with 0.675 as the mean PIC value per SSR locus. All the ten SSR markers developed 77 unique DNA-fingerprinting patterns and three SSRs viz. STM0037, STM1016 and STM1104 were reported as the best marker systems for distinguishing all the varieties. It was also inferred that the potato varieties used in this study were quite different from their respective maternal parents at genetic level. Tiwari et al. (2019) studied allelic variations in 82 accessions belonging to 22 wild potato species using 14 SSR markers. They obtained four to 13 alleles with allele size ranging from 104 to 304 bp. The PIC value was in the range of 0.66 to 0.91. The NJ-clustering method based on the Dice similarity coefficient divided all the wild potato genotypes into five major clusters. A range of distinct allelic variation had been observed among the genotypes irrespective of their origin, series and species. In an attempt to evaluate genetic diversity and population structure of potato landrace and varieties in China, Wang et al. (2019) used 30 SSR markers covering the entire genome of potato in a diverse population of potato comprising 292 genotypes including foreign elite lines, local landraces and cultivars. A total of 174 alleles were reported with 5.5 average number of alleles per SSR marker. Structure analysis divided all the potato genotypes into two main groups P1 and P2. Group P1 consisted of a higher number of commercial varieties and revealed less genetic diversity as compared to group P2, which consisted of wild relatives, thus depicting a higher amount of genetic diversity due to accumulation of new and unique alleles from wild species. It was further inferred that the potato landraces in group P1 might have been derived from ancient European and International Potato Centre genotypes, while group P2 potato landraces might have been derived from the modern population from these regions. Patil et al. (2020) evaluated genetic diversity among 47 different late blight responsive Indian potato cultivars using 24 SSR markers. A total of 193 alleles were obtained with an average of 7.04 alleles per locus and 4.1 per genotype. Moderately resistant potato cultivars depicted higher genetic diversity than susceptible and resistant cultivars. Cluster analysis grouped all the genotypes into two main and four sub-clusters; however, the late blight resistant cultivars were found distributed across all the clusters.

In an attempt to understand the genetic characteristics and distinguish closely related potato varieties, recently, Muhammad et al. (2022) used 35 SSR markers to characterize 25 potato varieties cultivated in Pakistan. A total of 214 alleles were obtained, of which 89.2% alleles exhibited polymorphism. PIC values were in the range of 0.0–0.87 and the amplicon/allele size ranged from 30–1000 bp. Three groups were obtained on the basis of cluster analysis. Group 1 revealed the maximum of genetic diversity, followed by group 2 and group 3 varieties. A total of 29 polymorphic SSRs demonstrated unique DNA fingerprints for 16 potato varieties. DNA fingerprinting of potato cultivars using molecular markers offers an easy and efficient way for protecting plant breeders' rights. Rahman et al. (2022) developed a DNA-fingerprinting profile of 12 potato cultivars grown in Punjab province of Pakistan using 214 SSR markers. A total of 1720 alleles were obtained and the number of alleles ranged from 1 to 31 with 8.04 average number of alleles per SSR locus. The PIC value was in the range of 0 to 0.96, with an average PIC value of 0.73. A total of 72 SSRs exhibited unique allelic patterns for DNA fingerprinting. Fifteen SSR markers clearly identified potato varieties – Rubby and Sadaf, while 12 SSRs could identify Faisalabad Red and SH-5 cultivars. The SSR-DNA fingerprint developed in this study would be very useful for registration of these varieties for protection of plant breeders' rights. Tiwari et al. (2022) carried out molecular profiling of 84 multi-colored flesh potato hybrids and 81 inter-specific somatic hybrids using two SSR markers. A higher degree of polymorphism was obtained with the STIKA SSR marker than STU6SNRN. Jaccard's dissimilarity coefficient based on the NJ tree method using Darwin software clearly distinguished the two groups of potato hybrids. Behn et al. (2023) evaluated genetic diversity in a panel of 290 germplasm accessions of *S. tuberosum* subsp. *tuberosum* belonging to the *Chilotanum* group using a set of 22 microsatellite markers and correlated them with flesh and skin color, total phenolic content and anthocyanin content. Structure analysis divided all the potato accessions into two groups, whereas seven clusters were obtained using UPGMA. UPGMA-based clusters exhibited common phenotypic traits as well as anthocyanin content, thus supporting a strong correlation between phenotypic traits and DNA fingerprint. These SSR markers will assist plant breeders to breed for potatoes with high polyphenol and anthocyanin content using molecular marker- assisted breeding.

Scattered efforts had been made for genetic diversity evaluation in potato. All of these studies utilized a very limited number of SSR markers in a limited panel of different potato genotypes. SSR markers have been used for characterization of potato gene pools and identification of potato cultivars in several countries and SSR-based DNA fingerprinting has been suggested and recommended as the most reliable procedure for cultivars' certification process. In India too, ICAR-New Delhi has made it a mandatory process to have the SSR-based DNA-fingerprinting profile of any elite material for its subsequent registration and commercial release in the form of a variety. So far, no comprehensive study has been conducted using a large and a common set of SSR markers to characterize potato accessions and varieties. So, the results of those labs always remained incomparable. In the way to identify cultivars and assessment of genetic variability or complete genome genetic-mapping studies, a complete spectrum of representative SSRs is needed, because it is not always sensible and financially suitable to test multiple numbers of SSRs. The development of a portable set of molecular markers is urgently needed for molecular characterization of potato germplasm accessions and varieties, and in potato improvement programs to lend precision and speed. A universal and reliable set of SSR markers is required for fingerprinting and genetic diversity profiling among different collections of potato by different laboratories and research groups.

(F) SNP markers in potato genetic diversity evaluation

Single nucleotide polymorphisms (SNPs), represent small variations in nucleotides found throughout genomes. These SNPs serve various purposes, including (1) assessing genetic diversity, (2) phylogenetic analyses, (3) identifying genes of agricultural significance using methods like QTLs, and genome-wide association studies (GWAS) mapping, (4) facilitating marker-assisted selection (MAS), (5) enabling varietal identification (Felcher et al. 2012). Efficient high-throughput SNP genotyping platforms have been developed to identify genetic diversity in potato. Specifically, 8K and 20K SNP arrays have been developed for potato (Felcher et al. 2012; Vos et al. 2015). The 8K array comprises 8303 SNPs chosen from a pool of over 69,000 markers identified through transcriptomics and expressed sequence tag (EST) data from six North American potato cultivars: Premier Russet, Bintje, Kennebec, Shepody, Snowden and Atlantic (Hamilton et al. 2011). This 8K SNP matrix has been instrumental in investigating the genetic diversity within potato germplasm collections of European origin, as well as those from North and South America (Hirsch et al. 2013). The genetic assessment of the University of Nariño collection involved studying the 8303 SNPs contained in the SolCAP version 1 SNParray (Hamilton et al. 2011). This comprehensive analysis aided in genotype selection, resulting in a polymorphism rate of 57.2%. This specific set of SNPs has been utilized across diverse potato populations with varying origins for evaluation. Several research endeavors have employed this SNP panel to examine different potato populations.

SNP studies identified the 77% polymorphism rate among 250 monoploid, diploid, and tetraploid genotypes in the United States (Hirsch et al. 2013). The phenomenon of high heterozygosity in potatoes primarily correlated with their allogamous nature, diverse ploidy levels and inherent genetic variability (Spooner et al. 2010). Additionally, Colombian diploid potato collections have been assessed using SSR and SNP markers, revealing elevated heterozygosity levels (Juyó et al. 2015). Similarly, Stich et al. (2013) observed a 74% polymorphism rate among 44 diploid and tetraploid genotypes of European-grown varieties. Differences in polymorphism percentages across these studies can be attributed to varying sample sizes, ranging from 44 (Stich et al. 2013) to 809 (Berdugo-Cely et al. 2017) samples. Hardigan et al. (2015) identified 61% polymorphism among 287 diploid, tetraploid, and hexaploid genotypes from various *Solanum* sect. Petota species and elite genotypes in the United States. Furthermore, these studies encompassed different ploidy levels, including monoploids (Hirsch et al. 2013) to hexaploids (Hardigan et al. 2015). Kolech et al. (2016) observed a 44.5% polymorphism rate among 109 tetraploid genotypes spanning the United States, Europe, Peru and Ethiopia. The substantial count of polymorphic markers identified in this investigation signifies the effectiveness of the SolCAP 8K matrix for genetic analysis within the potato breeding collection at the Universidad de Nariño in Colombia. Within this collection, both diploid and tetraploid populations exhibited considerable genetic diversity and linkage disequilibrium (LD) among markers. Notably, genetic diversity levels were lower in diploid genotypes than in tetraploids. Berdugo-Cely et al. (2017) carried out genetic diversity analysis among 809 *Andigenum* group accessions using 5968 SNPs and predicted the genetic structure of *Andigenum* germplasm and conferred that all the accessions can be divided into two major groups showing strong association with their ploidy level viz. *Phureja* (diploid) and *Andigena* (tetraploid). Esnault et al. (2016) found 61% polymorphism in 48 tetraploid genotypes from the National Institute for Agronomic Research in France. Similarly, Endelman et al. (2017) reported a 61% polymorphism rate among 719 tetraploid genotypes in the United States. Previous research using SNP markers in potato germplasm has reported heightened heterozygosity (Esnault et al. 2016; Kolech et al. 2016) and LD (Vos et al. 2017), with diploid genotypes displaying reduced genetic diversity and greater LD compared to tetraploid genotypes (Igarashi et al. 2019). In summary, SNP markers play a critical role in identifying and characterizing the genetic diversity of potato populations. This knowledge is valuable for the breeding programs aiming to develop improved potato varieties that are resilient, productive and adapted to changing environmental conditions (Berdugo-Cely et al. 2021).

5.6 Conclusions and Future Implications

To address the concerns of global food security, there is an urgent need to develop high yielding and stress resistance potato varieties. This book chapter discusses how the developed molecular markers have been efficiently utilized for studies involving potato diversity analysis and studying the phylogenetic relationships. Various abiotic and biotic stresses are known to hamper potato yield potentials. Molecular markers have been employed for tagging of various genes contributing to the resistance to different abiotic and biotic stress tolerance and selection for these traits by marker-assisted selection. Molecular marker-based genetic studies have significantly contributed to the development of improved stress-tolerant potato varieties. Additionally, DNA fingerprinting of potato cultivars using molecular markers offers an easy and efficient way for protecting plant breeders' rights. In future, the advanced markers' techniques may be utilized on larger genome coverage of potato by utilizing a newer class of DNA elements like retrotransposons, mitochondrial and chloroplast-based microsatellites for deciphering the genetic variations. Also, to reveal the genetic basis of biological responses, markers like RAPD and AFLP could be applied to cDNA-based templates in potato. Application of newer methods with modifications to basic techniques of molecular markers will help in achieving improved sensitivity and resolution in deciphering the genetic distinctiveness among various potato cultivars.

Acknowledgements

The authors wish to thank the Director, ICAR-CPRI, Shimla and acknowledge the CABIN Project for necessary support.

References

Allainguillaume, J., Wilkinson, M.J., Clulow, S.A. and Barr, S.N.R., 1997. Evidence that genes from the male parent may influence the morphology of potato dihaploids. *Theoretical and Applied Genetics*, *94*, pp.241–248.

Anoumaa, M., Yao, N.K., Kouam, E.B., Kanmegne, G., Machuka, E., Osama, S.K., Nzuki, I., Kamga, Y.B., Fonkou, T. and Omokolo, D.N., 2017. Genetic diversity and core collection for potato (*Solanum tuberosum* L.) cultivars from Cameroon as revealed by SSR markers. *American Journal of Potato Research*, *94*, pp.449–463.

Ashkenazi, V., Chani, E., Lavi, U., Levy, D., Hillel, J. and Veilleux, R.E., 2001. Development of microsatellite markers in potato and their use in phylogenetic and fingerprinting analyses. *Genome*, *44*(1), pp.50–62.

Bamberg, J. and Del Rio, A., 2016. Accumulation of genetic diversity in the US Potato Genebank. *American Journal of Potato Research*, *93*, pp.430–435.

Barandalla, L., De Galarreta, J.R., Rios, D. and Ritter, E., 2006. Molecular analysis of local potato cultivars from Tenerife Island using microsatellite markers. *Euphytica*, *152*(2), pp.283–291.

Barone, A., 2004. Molecular marker-assisted selection for potato breeding. *American Journal of Potato Research*, *81*, pp.111–117.

Beckmann, J.S. and Soller, M., 1983. Restriction fragment length polymorphisms in genetic improvement: methodologies, mapping and costs. *Theoretical and Applied Genetics, 67*, pp.35–43.

Behn, A., Lizana, C., Zapata, F., Gonzalez, A., Reyes-Díaz, M. and Fuentes, D., 2023. Phenolic and anthocyanin content characterization related to genetic diversity analysis of Solanum tuberosum subsp. tuberosum Chilotanum Group in southern Chile. *Frontiers in Plant Science, 13*, p.5469.

Berdugo-Cely, J., Valbuena, R.I., Sánchez-Betancourt, E., Barrero, L.S. and Yockteng, R., 2017. Genetic diversity and association mapping in the Colombian Central Collection of *Solanum tuberosum* L. Andigenum group using SNPs markers. *PloS One, 12*(3), p.e0173039.

Berdugo-Cely, J.A., Martínez-Moncayo, C. and Lagos-Burbano, T.C., 2021. Genetic analysis of a potato (*Solanum tuberosum* L.) breeding collection for southern Colombia using Single Nucleotide Polymorphism (SNP) markers. *PLoS One, 16*(3), p.e0248787.

Bianchi, D.W., 2004. Circulating fetal DNA: its origin and diagnostic potential-a review. *Placenta, 25*, pp.S93–S101.

Birch, P.R., Bryan, G., Fenton, B., Gilroy, E.M., Hein, I., Jones, J.T., Prashar, A., Taylor, M.A., Torrance, L. and Toth, I.K., 2012. Crops that feed the world 8: potato: are the trends of increased global production sustainable?. *Food Security, 4*, pp.477–508.

Bonierbale, M.W., Plaisted, R.L. and Tanksley, S., 1988. RFLP maps based on a common set of clones reveal modes of chromosomal evolution in potato and tomato. *Genetics, 120*(4), pp.1095–1103.

Bornet, B., Goraguer, F., Joly, G. and Branchard, M., 2002. Genetic diversity in European and Argentinian cultivated potatoes (*Solanum tuberosum* subsp. tuberosum) detected by inter-simple sequence repeats (ISSRs). *Genome, 45*(3), pp.481–484.

Chand, R., Kumar, P., Kumar, S., 2011. Total Factor Productivity and Contribution of research investment to Agricultural Growth in India. National Centre for Agricultural Economics and Policy Research, New Delhi.

Debener, T., Salamini, F. and Gebhardt, C., 1990. Phylogeny of wild and cultivated Solanum species based on nuclear restriction fragment length polymorphisms (RFLPs). *Theoretical and Applied Genetics, 79*, pp.360–368.

de Galarreta, J.I.R., Barandalla, L., Rios, D.J., Lopez, R. and Ritter, E., 2011. Genetic relationships among local potato cultivars from Spain using SSR markers. *Genetic Resources and Crop Evolution, 58*, pp.383–395.

Demeke, T., Laroche, A. and Gaudet, D.A., 1996. A DNA marker for the Bt-10 common bunt resistance gene in wheat. *Genome, 39*(1), pp.51–55.

de Santayana, M.P., Pieroni, A. and Puri, R.K. eds., 2010. *Ethnobotany in the new Europe: people, health, and wild plant resources* (Vol. 14). Berghahn Books.

Diwan, N., Bouton, J.H., Kochert, G.A.R.Y. and Cregan, P.B., 2000. Mapping of simple sequence repeat (SSR) DNA markers in diploid and tetraploid alfalfa. *Theoretical and Applied Genetics, 101*, pp.165–172.

Earle, R., 2017. Promoting potatoes in eighteenth-century Europe. *Eighteenth-Century Studies, 51*(2), pp.147–162.

Endelman, J.B., Schmitz Carley, C.A., Douches, D.S., Coombs, J.J., Bizimungu, B., De Jong, W.S., Haynes, K.G., Holm, D.G., Miller, J.C., Novy, R.G. and Palta, J.P., 2017. Pedigree reconstruction with genome-wide markers in potato. *American Journal of Potato Research, 94*, pp.184–190.

Esnault, F., Pellé, R., Dantec, J.P., Bérard, A., Le Paslier, M.C. and Chauvin, J.E., 2016. Development of a potato cultivar (*Solanum tuberosum* L.) core collection, a valuable tool to prospect genetic variation for novel traits. *Potato Research, 59*(4), pp.329–343.

Felcher, K.J., Coombs, J.J., Massa, A.N., Hansey, C.N., Hamilton, J.P., Veilleux, R.E., Buell, C.R. and Douches, D.S., 2012. Integration of two diploid potato linkage maps with the potato genome sequence. *PloS One, 7*(4), p.e36347.

Gebhardt, C., Blomendahl, C., Schachtschabel, U., Debener, T., Salamini, F. and Ritter, E., 1989. Identification of 2n breeding lines and 4n varieties of potato (*Solanum tuberosum*, ssp. tuberosum) with RFLP-fingerprints. *Theoretical and Applied Genetics, 78*, pp.16–22.

Ghislain, M., Núnez, J., del Rosario Herrera, M., Pignataro, J., Guzman, F., Bonierbale, M. and Spooner, D.M., 2009. Robust and highly informative microsatellite-based genetic identity kit for potato. *Molecular Breeding, 23*, pp.377–388.

Ghislain, M., Spooner, D.M., Rodríguez, F., Villamón, F., Nunez, J., Vásquez, C., Waugh, R. and Bonierbale, M., 2004. Selection of highly informative and user-friendly microsatellites (SSRs) for genotyping of cultivated potato. *Theoretical and Applied Genetics, 108*, pp.881–890.

Gorg, R., Schachtschabel, U., Ritter, E., Salamini, F. and Gebhardt, C., 1992. Discrimination among 136 tetraploid potato varieties by fingerprints using highly polymorphic DNA markers. *Crop Science, 32*(3), pp.815–819.

Govindaraj, M., Vetriventhan, M. and Srinivasan, M., 2015. Importance of genetic diversity assessment in crop plants and its recent advances: an overview of its analytical perspectives. Genetics research international, 2015(1), p.431487.

Hamilton, J.P., Hansey, C.N., Whitty, B.R., Stoffel, K., Massa, A.N., Van Deynze, A., De Jong, W.S., Douches, D.S. and Buell, C.R., 2011. Single nucleotide polymorphism discovery in elite North American potato germplasm. *BMC Genomics, 12*, pp.1–12.

Hardigan, M.A., Bamberg, J., Buell, C.R. and Douches, D.S., 2015. Taxonomy and genetic differentiation among wild and cultivated germplasm of Solanum sect. Petota. *The Plant Genome, 8*(1), pp.1–16.

Hirsch, C.N., Hirsch, C.D., Felcher, K., Coombs, J., Zarka, D., Van Deynze, A., De Jong, W., Veilleux, R.E., Jansky, S., Bethke, P. and Douches, D.S., 2013. Retrospective view of North American potato (*Solanum tuberosum* L.) breeding in the 20th and 21st centuries. *G3: Genes, Genomes, Genetics* , *3*(6), pp.1003–1013.

Hoque, M.E., Huq, H. and Moon, N.J., 2013. Molecular diversity analysis in potato (*Solanum tuberosum* L.) through RAPD markers. *SAARC Journal of Agriculture, 11*(2), pp.95–102.

Huang, C.H., Zhang, C., Liu, M., Hu, Y., Gao, T., Qi, J. and Ma, H., 2016. Multiple polyploidization events across Asteraceae with two nested events in the early history revealed by nuclear phylogenomics. *Molecular Biology and Evolution, 33*(11), pp.2820–2835.

Igarashi, T., Tsuyama, M., Ogawa, K., Koizumi, E., Sanetomo, R. and Hosaka, K., 2019. Evaluation of Japanese potatoes using single nucleotide polymorphisms (SNPs). *Molecular Breeding, 39*, pp.1–19.

Islam, M.S., Hoque, M.E., Alauddin, M. and Shamsuzzaman, M., 2022. Molecular characterization and diversity analysis of some local potato (*Solanum tuberosum*

L.) genotypes of Bangladesh by RAPD markers: molecular characterization and diversity analysis of local potato. *SAARC Journal of Agriculture, 20*(2), pp.43–55.

Jian, W.A.N.G., Lu, H.O.U., Wang, R.Y., He, M.M. and Liu, Q.C., 2017. Genetic diversity and population structure of 288 potato (*Solanum tuberosum* L.) germplasms revealed by SSR and AFLP markers. *Journal of Integrative Agriculture, 16*(11), pp.2434–2443.

Jacobs, J.M.E., Van Eck, H.J., Arens, P., Verkerk-Bakker, B., te Lintel Hekkert, B., Bastiaanssen, H.J.M., El-Kharbotly, A., Pereira, A., Jacobsen, E. and Stiekema, W.J., 1995. A genetic map of potato (*Solanum tuberosum*) integrating molecular markers, including transposons, and classical markers. *Theoretical and Applied Genetics, 91*, pp.289–300.

Juyó, D., Sarmiento, F., Álvarez, M., Brochero, H., Gebhardt, C. and Mosquera, T., 2015. Genetic diversity and population structure in diploid potatoes of *Solanum tuberosum* group Phureja. *Crop Science, 55*(2), pp.760–769.

Kim, J.H., Joung, H., Kim, H.Y. and Lim, Y.P., 1998. Estimation of genetic variation and relationship in potato (*Solanum tuberosum* L.) cultivars using AFLP markers. *American Journal of Potato Research, 75*, pp.107–112.

Koch, M., Busse, M., Naumann, M., Jákli, B., Smit, I., Cakmak, I., Hermans, C. and Pawelzik, E., 2019. Differential effects of varied potassium and magnesium nutrition on production and partitioning of photoassimilates in potato plants. *Physiologia Plantarum, 166*(4), pp.921–935.

Kolech, S.A., Halseth, D., Perry, K., Wolfe, D., Douches, D.S., Coombs, J. and De Jong, W., 2016. Genetic diversity and relationship of Ethiopian potato varieties to germplasm from North America, Europe and the International Potato Center. *American Journal of Potato Research, 93*, pp.609–619.

Kumari, N. and Thakur, S.K., 2014. Randomly amplified polymorphic DNA-a brief review. *American Journal of Animal and Veterinary Sciences, 9*(1), pp.6–13.

Kumar, P., Gupta, V.K., Misra, A.K., Modi, D.R. and Pandey, B.K., 2009. Potential of molecular markers in plant biotechnology. *Plant Omics, 2*(4), pp.141–162.

Liu, J., Sun, Z., Zou, Y., Li, W., He, F., Huang, X., Lin, C., Cai, Q., Wisniewski, M. and Wu, X., 2021. Pre-and postharvest measures used to control decay and mycotoxigenic fungi in potato (*Solanum tuberosum* L.) during storage. *Critical Reviews in Food Science and Nutrition, 62*(2), pp.415–428.

Majeed, A., Muhammad, Z., Ullah, Z., Ullah, R. and Ahmad, H., 2017. Late blight of potato (*Phytophthora infestans*) I: fungicides application and associated challenges. *Turkish Journal of Agriculture-Food Science and Technology, 5*(3), pp.261–266.

McKenzie, F.C. and Williams, J., 2015. Sustainable food production: constraints, challenges and choices by 2050. *Food Security, 7*, pp.221–233.

McNeill, W.H., 1999. How the potato changed the world's history. Social Research, 66, pp.67–83.

Meksem, K., Leister, D., Peleman, J., Zabeau, M., Salamini, F. and Gebhardt, C., 1995. A high-resolution map of the vicinity of the R1 locus on chromosome V of potato based on RFLP and AFLP markers. *Molecular and General Genetics MGG, 249*, pp.74–81.

Milbourne, D., Meyer, R., Bradshaw, J.E., Baird, E., Bonar, N., Provan, J., Powell, W. and Waugh, R., 1997. Comparison of PCR-based marker systems for the analysis of genetic relationships in cultivated potato. *Molecular Breeding, 3*, pp.127–136.

Milbourne, D., Meyer, R.C., Collins, A.J., Ramsay, L.D., Gebhardt, C. and Waugh, R., 1998. Isolation, characterisation and mapping of simple sequence repeat loci in potato. *Molecular and General Genetics MGG, 259*, pp.233–245.

Mohammadi, S.A. and Prasanna, B.M., 2003. Analysis of genetic diversity in crop plants—salient statistical tools and considerations. *Crop Science, 43*(4), pp.1235–1248.

Marhadour, S. and Prodhomme, C., 2023. Recent trends in genetics studies and molecular breeding of potato. In *Potato production worldwide* (pp. 273–301). Academic Press.

Muhammad, A., Noor, S., Hussain, I., Ali, K., Shahzad, A., Numan, M., Zeshan, M., Ijaz ul Hassan, S. and Ali, G.M., 2022. Simple sequence repeats-based genetic characterization and varietal identification of potato varieties frown in Pakistan. *Sustainability*, 14(18), p.11561.

Onamu, R., Legaria, J., Rodríguez, J.L., Sahagùn, J. and Pèrez, J., 2016. Molecular characterization of potato (*Solanum tuberosum* L.) genotypes using random amplified polymorphic DNA (RAPD) and inter simple sequence repeat (ISSR) markers. *African Journal of Biotechnology, 15*(22), pp.1015–1025.

Parita, B., Kumar, S.N., Darshan, D. and Karen, P., 2018. Elucidation of genetic diversity among ashwagandha [*Withania somnifera* (L.) Dunal] genotypes using EST-SSR markers. *Research Journal of Biotechnology, 13*(10), pp.52–59.

Patil, P.G., Singh, N.V., Parashuram, S., Bohra, A., Mundewadikar, D.M., Sangnure, V.R., Babu, K.D. and Sharma, J., 2020. Genome wide identification, characterization and validation of novel miRNA-based SSR markers in pomegranate (*Punica granatum* L.). *Physiology and Molecular Biology of Plants, 26*(4), pp.683–696.

Patil, V.U., Sharma, N.N. and Chakrabarti, S.K., 2017. High-Throughput Sequencing of the Potato Genome. In: Kumar Chakrabarti, S., Xie, C., Kumar Tiwari, J. (eds) The Potato Genome. Compendium of Plant Genomes. Springer, Cham, pp.95-107.

Pham, G.M., Hamilton, J.P., Wood, J.C., Burke, J.T., Zhao, H., Vaillancourt, B., Ou, S., Jiang, J. and Buell, C.R., 2020. Construction of a chromosome-scale long-read reference genome assembly for potato. *Gigascience, 9*(9), p.giaa100.

Prevost, A. and Wilkinson, M.J., 1999. A new system of comparing PCR primers applied to ISSR fingerprinting of potato cultivars. *Theoretical and Applied Genetics, 98*, pp.107–112.

Provan, J., Powell, W. and Waugh, R., 1996. Microsatellite analysis of relationships within cultivated potato (*Solanum tuberosum*). *Theoretical and Applied Genetics, 92*, pp.1078–1084.

Rahman, S.U., Jamil, S., Shahzad, R., Yasmeen, E., Sattar, S. and Iqbal, M.Z., 2022. Genetic diversity and DNA fingerprinting of potato varieties using simple sequence repeat (SSR) markers. *Journal of Animal & Plant Sciences, 32*(3), pp.775–783.

Raker, C.M. and Spooner, D.M., 2002. Chilean tetraploid cultivated potato, *Solanum tuberosum*, is distinct from the Andean populations: microsatellite data. *Crop Science, 42*(5), pp.1451–1458.

Ramakrishnan, A.P., Ritland, C.E., Blas Sevillano, R.H. and Riseman, A., 2015. Review of potato molecular markers to enhance trait selection. *American Journal of Potato Research, 92*, pp.455–472.

Rocha, E.A., Paiva, L.V., Carvalho, H.H.D. and Guimarães, C.T., 2010. Molecular characterization and genetic diversity of potato cultivars using SSR and RAPD markers. *Crop Breeding and Applied Biotechnology, 10*, pp.204–210.

Salimi, H., Bahar, M., Mirlohi, A. and Talebi, M., 2016. Assessment of the genetic diversity among potato cultivars from different geographical areas using the genomic and EST microsatellites. *Iranian Journal of Biotechnology, 14*(4), p.270.

Sandweiss, D.H. and Richardson, J.B. 2008. Central Andean Environments. In: Silverman, H., Isbell, W.H., (eds) The Handbook of South American Archaeology. Springer, New York, NY, pp.93–104

Schneider, K. and Douches, D.S., 1997. Assessment of PCR-based simple sequence repeats to fingerprint North American potato cultivars. *American Potato Journal, 74*, pp.149–160.

Scott, G.J. and Suarez, V., 2012. The rise of Asia as the centre of global potato production and some implications for industry. *Potato Journal*, 39(1), pp.1-22.

Shamsuzzaman, M., Bhattacharjya, D.K., Islam, M.S. and Hoque, M.E., 2021. Molecular diversity analysis of somaclonal variants of potato (*Solanum tuberosum* L.) by random amplified polymorphic DNA markers. *Annual Research & Review in Biology*, pp.63–76.

Sheeja, T.E., Kumar, I.P.V., Giridhari, A., Minoo, D., Rajesh, M.K. and Babu, K.N., 2021. Amplified Fragment Length Polymorphism: Applications and Recent Developments. In: Besse, P. (eds) Molecular Plant Taxonomy. Methods in Molecular Biology, vol 2222. Humana, New York, NY, pp.187–218.

Sosinski, B. and Douches, D.S., 1996. Using polymerase chain reaction-based DNA amplification to fingerprint North American potato cultivars. *HortScience, 31*(1), pp.130–133.

Spooner, D.M., 2005. *Molecular markers for genebank management* (No. 10). Bioversity International.

Spooner, D.M., Núñez, J., Trujillo, G., del Rosario Herrera, M., Guzmán, F. and Ghislain, M., 2007. Extensive simple sequence repeat genotyping of potato landraces supports a major reevaluation of their gene pool structure and classification. *Proceedings of the National Academy of Sciences, 104*(49), pp.19398–19403.

Spooner, D.M., T. Gavrilenko, S.H. Jansky, A. Ovchinnikova, E. Krylova, S. Knapp, and R. Simon (2010). Ecogeography of ploidy variation in cultivated potato (Solanum sect. Petota). *American Journal of Botany, 97*, pp. 2049–2060

Stich, B., Urbany, C., Hoffmann, P. and Gebhardt, C., 2013. Population structure and linkage disequilibrium in diploid and tetraploid potato revealed by genome-wide high-density genotyping using the SolCAP SNP array. *Plant Breeding, 132*(6), pp.718–724.

Swarup, S., Cargill, E.J., Crosby, K., Flagel, L., Kniskern, J. and Glenn, K.C., 2021. Genetic diversity is indispensable for plant breeding to improve crops. *Crop Science*, 61(2), pp.839–852.

Tanksley, S.D., Ganal, M.W., Prince, J.P., De Vicente, M.C., Bonierbale, M.W., Broun, P., Fulton, T.M., Giovannoni, J.J., Grandillo, S. and Martin, G., 1992. High density molecular linkage maps of the tomato and potato genomes. *Genetics, 132*(4), pp.1141–1160.

Tillault, A.S. and Yevtushenko, D.P., 2019. Simple sequence repeats analysis of new potato varieties developed in Alberta, Canada. *Plant Direct, 3*(6), p.e00140.

Tiwari, J.K., Ali, N., Devi, S., Kumar, V., Zinta, R. and Chakrabarti, S.K., 2018. Development of microsatellite markers set for identification of Indian potato varieties. *Scientia Horticulturae, 231,* pp.22–30.

Tiwari, J.K., Ali, N., Devi, S., Zinta, R., Kumar, V. and Chakrabarti, S.K., 2019. Analysis of allelic variation in wild potato (Solanum) species by simple sequence repeat (SSR) markers. *3 Biotech, 9,* pp.1–9.

Tiwari, J.K., Buckseth, T., Challam, C., Natarajan, N., Singh, R.K. and Kumar, M., 2022. Omics approaches for biotic, abiotic, and quality traits improvement in potato (Solanum tuberosum L.). In: Sharma, P., Yadav, D. and Gaur, R.K., (eds), Bioinformatics in Agriculture. Academic Press, pp. 341–359.

Torres-Rendon, A., Stewart, R., Craig, G.T., Wells, M. and Speight, P.M., 2009. DNA ploidy analysis by image cytometry helps to identify oral epithelial dysplasias with a high risk of malignant progression. *Oral Oncology, 45*(6), pp.468–473.

Van Eck, J.O.Y.C.E., Conlin, B.R.I.A.N., Garvin, D.F., Mason, H., Navarre, D.A. and Brown, C.R., 2007. Enhancing beta-carotene content in potato by RNAi-mediated silencing of the beta-carotene hydroxylase gene. *American Journal of Potato Research, 84,* pp.331–342.

Visser, R.G., Bachem, C.W., de Boer, J.M., Bryan, G.J., Chakrabati, S.K., Feingold, S., Gromadka, R., van Ham, R.C., Huang, S., Jacobs, J.M. and Kuznetsov, B., 2009. Sequencing the potato genome: outline and first results to come from the elucidation of the sequence of the world's third most important food crop. *American Journal of Potato Research, 86,* pp.417–429.

Vos, P.G., Paulo, M.J., Voorrips, R.E., Visser, R.G., van Eck, H.J. and van Eeuwijk, F.A., 2017. Evaluation of LD decay and various LD-decay estimators in simulated and SNP-array data of tetraploid potato. *Theoretical and Applied Genetics, 130,* pp.123–135.

Vos, P.G., Uitdewilligen, J.G., Voorrips, R.E., Visser, R.G. and van Eck, H.J., 2015. Development and analysis of a 20K SNP array for potato (*Solanum tuberosum*): an insight into the breeding history. *Theoretical and Applied Genetics, 128,* pp.2387–2401.

Walunjkar, B., Parihar, A., Chaurasia, P., Pachchigar, K. and Chauhan, R.M., 2013. Genetic analysis of wild and cultivated germplasm of pigeonpea using random amplified polymorphic DNA (RAPD) and simple sequence repeats (SSR) markers. African Journal of Biotechnology, 12(40), pp. 5823-5832.

Wang, X., Chen, W., Luo, J., Yao, Z., Yu, Q., Wang, Y., Zhang, S., Liu, Z., Zhang, M. and Shen, Y., 2019. Development of EST-SSR markers and their application in an analysis of the genetic diversity of the endangered species *Magnolia sinostellata. Molecular Genetics and Genomics, 294,* pp.135–147.

Watanabe, K., 2015. Potato genetics, genomics, and applications. *Breeding Science, 65*(1), pp.53–68.

Welsh, J. and McClelland, M., 1990. Fingerprinting genomes using PCR with arbitrary primers. *Nucleic Acids Research, 18*(24), pp.7213–7218.

Williams, J.G., Kubelik, A.R., Livak, K.J., Rafalski, J.A. and Tingey, S.V., 1990. DNA polymorphisms amplified by arbitrary primers are useful as genetic markers. *Nucleic Acids Research, 18*(22), pp.6531–6535.

6

Soilless Cultivation for the Production of Spice and Essential Oil-Bearing Plants

Amit Kumar, Mahinder Partap, Shiv Rattan, Pankaj Kumar and Ashish R. Warghat

6.1 Introduction

In the 21st century world, the increasing demand for spice and essential oil-bearing plants is a major challenge faced by countries, with limited water and land resources (Savvas and Gruda 2018). Spices are available in the form of seeds, fruits, roots, leaves or other plant material, mainly used during the cooking process. Spices provide color, flavor, taste, aroma and garnishing to the food (Partap et al. 2021). Additionally, spices are also used in the drugs, cuisine and beverages industries. In the market, the spices are available in two forms; fresh, and full dried or pre-ground dried (Vimolmangkang et al. 2010). The full dried spices have a long shelf-life, so they can be bought and stored in more quantities, making them less-priced per serving. Fresh spices often have a stronger flavor than their dried counterparts; they are also costlier and have a considerably shorter shelf-life. However, the seeds are often used in whole or in powder. The aroma and flavor properties of spices are majorly due to the presence of essential oil. Basically, essential oil (EO) is a concentrated hydrophobic mixture of aromatic terpenes in the secretory tissues of plants (Sharifi-Rad et al. 2017). The EO is "essential" in the sense that it carries the aroma of the plant from which it is produced. Generally, the EOs are often extracted by hydro-distillation with the use of steam from the plants and their material. However, several other processes are also available for the extraction of EOs, including ram press, solvent extraction, slow-folding process, absolute-oil extraction, resin-tapping, wax embedding and cold-pressing (Aziz et al. 2018). Essential oils are extensively utilized for their aromatic and flavoring properties in foods and cosmetics. However, EOs also possess various pharma-logical effects against various diseases (Bakkali et al. 2008). However, most of the cultivation of spice and essential oil-bearing plants relies on conventional agriculture systems. Due to the increase in

DOI: 10.1201/9781003625674-6

urbanization, a more effective approach or cultivation strategy is required to fulfil the unmet demand and supply gap of economically important herbs.

In the current scenario, producers and consumers from across the world are very interested in more environmentally friendly and sustainable plant cultivation methods. Spice and essential oil-bearing plants are abundant sources of bioactive substances that promote human health. These substances are regulated by a number of variables, including light, temperature, humidity, CO_2, genotype, environmental factors and farming techniques (Aires 2018). In the last few years, a number of soilless growing techniques has been devised for the large-scale, environmentally controlled production of economically significant herbs (Partap et al. 2021). In the agricultural and horticultural industries, soilless systems are significantly advertised for quality biomass production in significantly less time duration. These farming techniques also provide a stable and sustainable platform for the development of a wide range of commercially significant crops, including medicinal, blooming and vegetable crops, as well as spices, and aromatics crops (Thakur et al. 2019; Partap et al. 2020). Compared to soil-cultivated plants, these cultivation methods produce more metabolite yield, and biomass in a regulated, and nourished environment (Thakur et al. 2019; Partap et al. 2020).

The chapter discusses up-to-date information on the soilless cultivation of spice and essential oil-bearing plants. The chapter also discusses the influences of soilless techniques on growth and metabolite accumulation in spice and essential oil-bearing plants.

6.2 Literature Review

For the book chapter, the literature has been searched through PubMed, Google Scholar, SciFinder, Web of Science, ScienceDirect, Education Resources Information Center, National Medicinal Plant Board (NMPB, India) and Scopus databases. The search keywords in the databases were soilless cultivation of spice crops, essential oil-bearing plants, hydroponic, aeroponic, aquaponics, vertical farming, secondary metabolites, soilless cultivation market, etc. The data presented in this book chapter were obtained from published articles, review articles, book chapters and reports in the English language only.

6.3 Soilless Cultivation

Instead of soil, soilless cultivation is an approach to cultivating plants in a growth box containing nutrient-rich solution or mist (Partap et al. 2022). In this cultivation system, plants and their roots are physically anchored with neutral materials like pebbles, gravels, rock-wools, peat-moss, sawdust, fibers, vermiculites and perlite. However, based on the cultivation system, soilless agriculture is divided into two types: (1) open and (2) closed (Fussy et al. 2022).

6.3.1 Open-Soilless Cultivation

A dripping framework is generally utilized in an open soilless system to provide dissolved nutrients to plants for their growth and development. Generally, these practices must include a water/nutrient solution run-off system at one end. Additionally, on the other side, the solution is filled at regular intervals to ensure the optimum level to irrigate the root zone of plants. The nutrient solution is not re-circulated or recycled in this type of cultivation. Due to the frequent changes in the nutritional solution, the major benefit of this method is that there will not be any chance of contamination in the plants. This practice is classified into the following techniques: (1) root-dipping, (2) floating, and (3) capillary action (Gul et al. 1997).

6.3.2 Closed-Soilless Cultivation

In this cultivation, the nutrient solution is re-circulated, managed, and regulated periodically in accordance with plants. The system is regulated by an automated or manual control system to maintain environmental conditions. The closed-soilless system is further categorized in the following subheadings.

6.3.2.1 Hydroponic System

Hydroponic is a type of soilless cultivation technique in which plants are growing in the growth chambers containing mineral nutrient solutions without using the soil in an environmental control system. The growing solutions contain precise amounts of nutrients, which are required for the growth of the plant (Mok et al. 2014). Generally, the hydroponic system is used to grow leafy herbs, vegetables, spices, essential oil-bearing plants, aromatic and medicinal herbs, flowers and fruit crops (Asao 2012). In addition, the spacing between the cultivated plants is considerably less in this system, thus increasing the overall yield from the lesser space (Thakur et al. 2019). Therefore, the agro-industries adopt these cultivation practices more frequently to increase crop productivity. Also, the concentration of nutrients can be monitored easily to identify nutritional deficits or toxicity in the growth solutions (Partap et al. 2020). Additionally, this system can also couple with a UV sterilization tube to eradicate the microorganism contaminations in the solutions (Partap et al. 2020). As a result, a hydroponic system is more effective compared to soil-based cultivation systems, when all the advantages are considered. The following categories further categorize this method (Sharma et al. 2018): (1) nutrient film technique (NFT), (2) wick system, (3) water culture or deep-water culture (DWC), (4) ebb and flow systems and (5) drip hydroponic system. These types of techniques are mostly utilized for better growth of above-ground or shoot parts of the plants.

6.3.2.2 Aeroponic System

In an aeroponic system, plant roots are grown in a closed chamber containing nutrients' solution spray or mist to supply extra aeration. However, the plant roots are anchored with neutral supporting material to sustain the growth (Partap et al. 2020). Therefore, it differs from both conventional and hydroponic cultivation systems. The nutrient solution is sprayed on plants in an atomized form using a nozzle and fogger system (Mbiyu et al. 2012). The solution spray intermittently sprinkles plant roots with nutrients for a set amount of time rather than misting continuously (Lakhair et al. 2018). In general, aeroponics technology can be a good approach for growing underground food storage crops, such as mini tubers or potatoes (Christie et al. 2003). The system comprises three bases; the first does not typically use a water pump to spray the solution mist. In a low-pressure framework aeroponic, water pumps are usually required to spray the nutrient mist from the sprinklers. The third structure consists of an ultrasonic fogger that releases a high-pressure mist (Domingues et al. 2012). In an aeroponic system, the following functional nozzles/foggers are used to spray the mist: (1) ultrasonic-atomization, (2) high-pressure atomization and (3) pressurized airless.

6.3.2.3 Aquaponics System

In the aquaponics system, the aqua-culture is merged with hydroponic. In this approach, both fish and plants are cultivated in a mutualistic way (Savidov et al. 2005). In this, the hydroponic system receives the processed, nutrient-rich fish water. Before that, nitrifying bacteria such as *Nitrosomonas* and *Nitrobacter* species transform ammonia released by fish into nitrites and nitrates through metabolic activity, respectively (Azad et al. 2016). Further, the processed water is supplied to plant growth chambers and after being used by plants. Moreover, it can be cleaned and enriched with oxygen before being returned to the aquaculture system, and re-cycles. Hydroponics such as DWC, and NFT are commonly utilized in aquaponics cultivation (Maharana and Koul 2011).

6.3.2.4 Vertical Farming

Vertical farming can produce a variety of crops in a climate-resilient manner, by using the hydro-aeroponic and aquaponics concept with less space and water consumption than conventional farming in multi-storey racks or buildings. These buildings are frequently tall with multi-levels of growth beds that are often equipped with a controlled light system. Different across the world, urban vertical farms of all forms and sizes are emerging. On the rooftop of both old and modern buildings, including commercial and private ones, as well as cafes and shopping centres, this kind of vertical farming is practiced (Touliatos et al. 2016).

6.4 National and International Status

The Netherlands, France, England, Canada, Australia, Israel and the United States are the leading global players in soilless farming. Fifty percent of the edible crops in the Netherlands are produced through commercial soilless farming. The hydroponic sector in Australia is estimated to be worth between USD 300 and USD 400 million dollars, or up to 20% of the value of all the vegetable and cut flowers produced in Australia (RIRDC). Australia currently produces more hydroponically grown lettuce than any other country in the world, as well as more strawberries and cut flowers than the US. Spain and Canada are also emerging countries in the production of soilless crops and vegetables. Israel uses a soilless technique to grow vast amounts of berries, citrus fruits and bananas due to the country's dry and arid environment. Major cities like Delhi, Bangalore, Mumbai and Chandigarh are already employing soilless farming to produce a number of leafy green, herbs and spices for fresh utilization on their terraces and balconies. The global market of soilless agriculture was 9.5 billion USD in 2020. It is predicted to have a value of 17.9 billion USD by 2026, increasing at a compound annual growth rate (CAGR) of 11.3% (Global Forecast, 2026). The hydro-aeroponic market in India is anticipated to expand at a CAGR of 13.53% from 2022 to 2027 (www.datami ntelligence.com/). India, Israel, China and the Netherlands are developing soilless cultivation leading countries. This technology's worldwide market value was estimated at USD 662.49 million in 2019 and is anticipated to expand at a CAGR of 12.5% to reach USD 1.29 billion in 2026 (www.industry arc.com/). According to projections, the worldwide market for herbal plants would be worth around 72 billion US dollars in 2020 and 7 trillion by 2050, growing at a CAGR of 14.88% (Thakur et al. 2019). In 2021, the Indian spice markets had a value of INR 142,569.3 crores. Looking forward, the IMARC Group anticipates the market to develop at a compound annual growth rate (CAGR) of 11.15% from 2022 to 2027, reaching INR 270,928.4 crores (Partap et al. 2020). India's essential oil market was estimated to be worth USD 149.63 million in 2020 and is anticipated to expand to USD 239.85 million by 2028, with a CAGR of 6.97% from 2021 to 2028. The output of essential oils was predicted to have tripled since 1990 (45,000 tonnes) and to have reached over 150,000 tons in 2017 valued at around 6 billion USD. Half of this increase has taken place since 2007 (Thakur et al. 2019). The Indian essential oils market is predicted to reach an output of ~50,400 tons by the end of 2030.

6.5 Plant Cultivation in the Soilless System

The details about various commercially important spices and essential oil-bearing plants cultivated using the soilless system are represented in Table 6.1. The common protocol used for the cultivation of plants in the soilless system is highlighted in Figure 6.1.

6.5.1 *Valeriana jatamansi* Jones. (Tagar)

Valeriana jatamansi Jones. is commonly known as Indian valerian, Muskbala, Sugandhbala (Hindi) and Tagar (Sanskrit). It is an essential oil-bearing medicinal herb belonging to the family *Valerianaceae*. It is used to treat a number of illnesses, including leprosy, hysteria, asthma and sleeplessness (Joseph et al. 2016; Jugran et al. 2019). In the roots, rhizomes and subterranean portion, 0.4–0.5% aromatic essential oil and secondary metabolites are accumulated (Navarrete et al. 2006; Singh et al. 2010; Jugran et al. 2019; Partap et al. 2020).

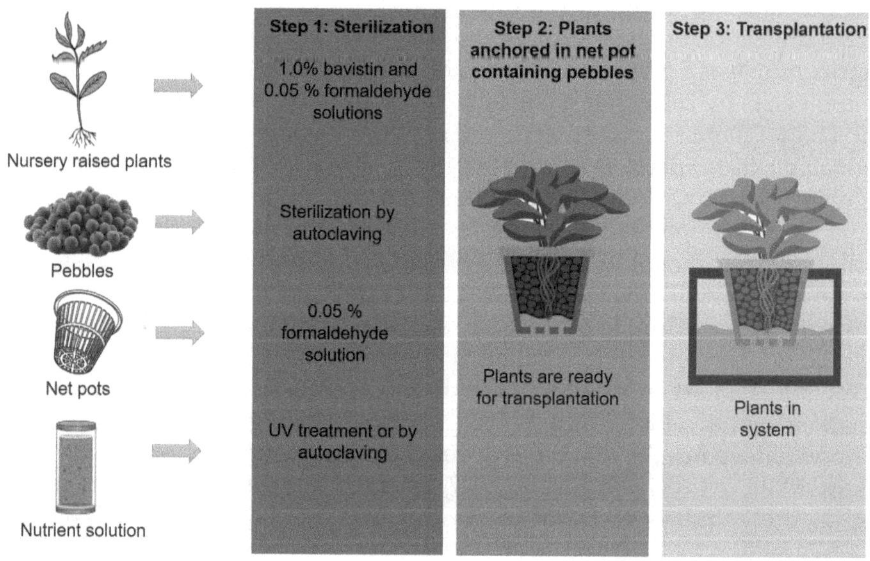

FIGURE 6.1
Plant cultivation process in a hydroponic system.

TABLE 6.1

Soilless Cultivation of Spice and Essential Oil-Bearing Plants

Spice and Essential Oil-Bearing Crops	Family	Trade Name*	Useful Part*	Cultivation Technique	Annual Trade (MT)*	Specialized metabolites	References
Anethum graveolens L.	Apiaceae	Sowa	Seed, and fruit	Hydroponics	200–500	Carvone	Currey et al. (2019)
Artemisia annua L.	Asteraceae	Seeme davana	Whole plant	Hydroponics	1000–2000	Artemisinin	Jehnsen et al. (2010)
Brassica juncea L.	Brassicaceae	Kaduku	Seed	Hydroponics	100–200	Glucosinolates	Dou et al. (2020)
Chrysanthemum balsamita L.	Asteraceae	Costmary	Shoot	Hydroponics	<10	Carvone	Hassanpouraghdam et al. (2007)
Conobea scoparioides (Cham. & Schltdl.) Bentham	Scrophulariaceae	Pataqueira	Aerial part	Hydroponics	1000–2000	Methylthymol	Costa et al. (2014)
Coriandrum sativum L.	Apiaceae	Dhaniya	Leaves and seeds	Hydroponics	500–1000	Linalool	Hazeri et al. (2012)
Crocus sativus L.	Iridaceae	Kesar	Flower	Hydroponics, Aeroponics	50–100	Picrocrocin, Crocin	Souret and Weathers et al. (2000)
Echinacea purpurea L.	Asteraceae	Eastern purple coneflower	Root	Hydroponics	<10	Germacrene D	Ahmadi et al. (2021)
Mentha piperita L.	Lamiaceae	Peppermint	Aerial Part	Hydroponics	2000–5000	Menthol	Surendran et al. (2017)
Ocimum basilicum L.	Lamiaceae	Sweet Basil	Whole plant	Hydroponics	200–500	Methyl chavicol and linalool	Kiferle et al. (2011)
Origanum vulgare L.	Lamiaceae	Sattar pati	Aerial part	Hydroponics and aeroponics	<10	thymol and carvacrol	Yeritsyan et al. (2001)
Petroselinum crispum (Mill.) Fuss	Apiaceae	Parsley	Leaf, and seed	Hydroponics	350–600	Myristicin	Rattan et al. (2021)
Rosmarinus officinalis L.	Lamiaceae	Rosemary	Flower	Aeroponics	50–100	Carnosol and carnosic acid	Loera-Muro et al. (2021)
Valeriana jatamansi Jones ex Roxb.	Valerianaceae	Tagar	Rhizome and Root	Hydroponics and aeroponics	1000–2000	Valerenic acid and derivatives	Partap et al. (2020)
Valeriana officinalis L.	Valerianaceae	Valerian	Root	Hydroponics	<10	Valepotriate and valerenic acid	Mousavi et al. (2021)
Zingiber officinale Roscoe	Zingiberaceae	Soonth	Root	Aeroponics	2000–5000	Gingerols	Hayden et al. (2004)

*The National Medicinal Plants Board (NMPB), Ministry of AYUSH, Government of India

Valerenic acid (VA), acetoxy valerenic acid (AVA) and hydroxy valerenic acid (HVA) are the main constituent metabolites in *Valeriana* species and these metabolites are mainly present in the underground part of *Valeriana* (Navarrete et al. 2006). Many pharmaceutical companies advertised their therapeutic products comprising valerenic acid derivatives (Partap et al. 2020). Valerenic acid 3.0 mg/g in the *Valeriana* sp. sample has been suggested as a top-notch ingredient for making medicines (Bos et al. 1998). The traditional agricultural methods and the plant's natural environment are the only sources of supply for the large-scale market demand for *Valeriana* phytochemicals. Due to the enormous demand-supply imbalance, using high-tech/modern agricultural techniques, including hydro-aeroponics, provides an alternative method for producing high-quality biomass in accordance with industry requirements (Dorais et al. 2001; Tabatabaei 2008; Thakur et al. 2019). The aeroponic system provides an alternative, commercially feasible, agro-technique for increased phytochemical production in *V. jatamansi*, according to the latest report of Partap et al. (2020). The study demonstrated the impact of soil culture, aeroponics farming and different elicitor treatments, on plant development, yield, morphology and secondary metabolite accumulation in *Valeriana* species. In this experiment, the plant was grown for 12–14 weeks in controlled parameters in an aeroponic system. The parameters include temperature (25°C), photoperiod (16 hrs light/8 hrs dark), relative humidity (65%), and electrical conductivity (0.5–1.5 mS/cm) with photosynthetic photon flux density (150 μmol m^{-2} s^{-1}) were set. The pH (6.8–7.0), and chiller temperature (10°C) was set for Hoagland nutrient solution. Further, the treatment of yeast extract (0.5 mg/L) resulted in the maximum leaves, plant height and rootlet length. Whereas, yeast extract (1.5 mg/L) and methyl jasmonate (150 μM) levels showed the highest leaf width and leaf length, respectively. Photosynthesis has been reported maximum at 0.5 mg/L and 1.5 mg/L yeast extract. However, 150 μM methyl jasmonate (MJ) treatment showed the highest transpiration rate. Additionally, 150 μM and 100 μM of MJ levels accumulated the highest content of valerenic acid and hydroxy valerenic acid. However, acetoxy valerenic acid has been reported to be highest in root at 150 μM MJ. Whereas the leaves treated with 1.5 mg/L yeast extract accumulated maximum acetoxy valerenic acid. In GC-MS analysis, isovaleric acid, patchouli alcohol and baldrinal have been reported as major compounds in roots (Partap et al. 2020). A total of 28 volatile compounds has been identified in elicitor-treated and control plants. The study also highlighted that besides the cultivation system, elicitors performed an important role in the enhancement of volatile compounds in *V. jatamansi*. The study also showed that the aerial part of *V. jatamansi* would be employed as a substitute source for metabolites and essential oil extraction. Hence, aeroponic cultivation offers an easy-to-access successive biomass production without damaging the endangered plants compared to soil-based cultivation.

6.5.2 *Zingiber officinale* Roscoe (Ginger)

Ginger (*Zingiber officinale* Roscoe), is a bulbous, herbaceous, perennial crop, belonging to the family *Zingiberaceae* (Surh et al. 1999). Ginger is one of the most widely used spices and is obtained from the ginger plant's underground stem. The spicy aroma of this plant is mainly due to the presence of ketones, especially gingerols (Jiang et al. 2005). Ginger is economically proliferated from "seed pieces" of the rhizome. In field cultivation, ginger is prone to many pathogens and soil-borne diseases, like Cucumber Mosaic Viruses, bacterial wilt, Erwinia soft rot, and root-knot nematode (Inden et al. 1988; Stirling 2002; Vilsoni et al. 1976). Soilless cultivation can be an alternative agriculture technology for crops that are more prone to soil-borne pathogens. The unchanging and controlled cultivation environment in a greenhouse may produce crops with a higher level of secondary metabolites. Unfortunately, there are limited hydroponics or aeroponics cultivation systems appropriate for rhizomatous crops. In aeroponic cultivation roots of the plant are suspended in a closed compartment and sprayed intermittently with a nutritional solution by using a timer and pumps. In the aeroponic system, roots are simply manageable for checking, collecting and harvesting (Hayden et al. 2004). In this experiment, the plant was cultivated under controlled atmosphere conditions like a temperature of 25°C, an electrical conductivity (EC) of 2.0 mS/cm, and a pH (5.7). An external timer-controlled pump circulated the solution, spraying the roots for one minute and then stopping for three minutes at a rate of roughly 0.03 L s^{-1}. The nutrition solution was changed once a week, and no spraying was applied at night. Two treatments were used in the experiments: (1) the growing medium used in the rhizome compartment (RC) and (2) the heat vs. no heat in the reservoir below the spray chamber. The growing medium is made up of either (a) perlite, (b) peat moss or (c) no aggregate medium (NAM). The study highlighted that perlite provides a higher growth condition over the NAM medium. Those plants growing in a heated nutrient medium root grow quicker and produce more rhizomes compared to the same medium without a heated nutrient medium. To conclude, ginger rhizomes were effectively produced in aeroponic cultivation.

6.5.3 *Petroselinum crispum* (Mill.) Nyman (Parsley)

Parsley, *Petroselinum crispum* (Mill.) Nyman is a culinary plant belonging to the (*Umbelliferae*) *Apiaceae* family. It is also known as baweri in Hindi and garden parsley in English. Parsley is an upright, much-branched plant, which can reach up to 30–100 cm long, with green leaves and yellow-greenish flowers. The leaves of parsley are an abundant source of vitamins, flavonoids, carotenoids and ascorbic acid and are consumed as a good source of food (Maodaa et al. 2016). Nowadays, parsley is used commercially in both dried and fresh forms in a wide range of cuisine preparations. The essential oils found in parsley's roots and leaves, are responsible for their anti-inflammatory and therapeutic properties (Farzaei et al. 2013). Due to its therapeutic, cosmetics, condiments and preservation qualities, parsley essential oil is widely used (Dong et al. 2017). Myristicin has been detected as one of the main volatile constituents in the essential oil of parsley, which has a number of medicinal uses and strong antioxidant characteristics (Dong et al. 2017). Martins et al. (2019) investigated the impact of brackish water and electrical conductivity (EC) on the development and biomass of parsley in a hydroponics growing system. In this study, it was shown that an increase in electrical conductivity (EC) in a hydroponic system led to an increase in the shoot and root dry biomass of parsley. The level of EC was adjusted and resulted in six isosmotic levels (ECns =1.7 to 6.7 dS m^{-1}) with the four types of salt: $CaCl_2$, NaCl, KCl and $MgCl_2$. In this study, two strategies were used to restore the amount of consumption of parsley plants. The urban water source was made a substitute under the first strategy and the respective brackish water under the second. The study showed that hydroponics is an efficient and effective technology for the production of parsley by using wastewater for large-scale biomass production and growth. In another study, a deep flow technique (DFT)-based hydroponic system was used for the cultivation of parsley (Rattan et al. 2022). In this experiment, the plants were grown in two different nutrient media, as described by Hoagland and Arnon (1950) (code; T1) and Thakur et al. (2019) (code; T2). Moreover, *P. crispum* plants showed an improved morpho-physiological response, growth, antioxidant property and volatile oil quantity in nutrient recipe T1 treated hydroponic compared to pot cultivation. In the present research, the study also showed the EC level of nutrient mixture needed for vegetative (0.5–1.5 mS cm^{-1}) and production stage (2.0–2.5 mS cm^{-1}) of parsley. Hoagland solution treatment on parsley plants during hydroponic cultivation enhanced morpho-physiological growth, myrısticın content and growth characteristics. Myristicin, apiol, β-phellandrene and elemicin were identified as the main constituents of the volatile compound in GC-MS analysis of volatile oil. Any disease symptoms, nutrient deficiencies, and morphological abnormalities were not detected in plants grown hydroponically. In conclusion, hydroponics could be a good alternative system for efficient biomass and crop production in a variety of related species.

6.5.4 *Mentha piperita* L. (Peppermint)

Peppermint, botanically known as *Mentha piperita* L. is an aromatic perennial herb belonging to the family *Lamiaceae* (Iscan et al. 2002). Leaves of peppermint are used globally, in the form of an infusion, tincture, tea, extract and essential oil. It has pharmacological activities like anesthetic, antihelmintic, antimicrobial, antifungal and antioxidant activities (de Oliveira Hashimoto et al. 2016). The distinctive flavors of peppermint leaves include a warm, pungent aroma, a sweetish flavor, and a cooling aftertaste. Industrial-scale cultivation for essential oil extraction must be carried out as the market demand for peppermint essential oils is continuously increasing (Vimolmangkang et al. 2010). The research was conducted to examine the viability of growing peppermint hydroponically and to compare its biochemical characteristics with those of plants cultivated in the soil. The results showed that the peppermint was more productive in terms of morphology, physiology, yield, and its impact on biochemical characteristics. In comparison to conventional soil cultivation, peppermint had a 61% higher yield percentage. Therefore, it is proposed that growing these plants hydroponically is a feasible option in urban regions and eliminates the issues that are inherent in growing these plants conventionally. Surendran et al. (2017) conducted a study to compare hydroponically produced plants with those grown in the soil. In this study, the hydroponic set-up is made up of PVC pipe, and Hoagland nutrient solution is used as a growth medium. A drip irrigation technology was used to apply the nutritional solution. The supporting media, which included cocopeat and vermiculite, is used, and pH, TDS, EC and salt concentration were measured regularly. The nutrient solution was changed once a week. The enhanced aeration and nutrient retention capacity of the medium may have a positive impact on shoot growth and results in the maximum plant height recorded in cocopeat in this study (Putra and Henry 2015). The outcome also demonstrates that the total chlorophyll content was greater in the hydroponically grown plant. When compared to alternative media and typical soil-grown plants, cocopeat media gained the highest yield. The maximum value of proline (0.022 mg/100 g) and phenol (0.51 mg/100 g) was observed in soil-grown plants, which is an indication that the plants are under some type of biotic and abiotic stress. In conclusion, this investigation of the hydroponic cultivation of peppermint and comparison of its morphological and physiological characteristics with those of soil-grown plants showed that it might be successfully and profitably produced under hydroponic systems. Based on these results, it is suggested that hydroponics is a feasible and effective substitute production system for plant growth and development.

6.5.5 *Ocimum basilicum* L. (Sweet Basil)

Ocimum basilicum L. (sweet basil) is one of the most important culinary herbs belonging to the family *Lamiaceae*. It is also known as basilica and alfavaca. A variety of bioactive metabolites in basil has been shown to have positive effects on human health (Ciriello et al. 2022). Due to its high quantity of essential oil and rosmarinic acid, sweet basil is utilized in pharmacological and aesthetic preparations (Javanmardi 2002; Makri and Kintzios 2007; Juliani et al. 2008; Zheljazkov et al. 2008). Due to large market demand, basil is grown all over the world for the production of essential oils (Favorito et al. 2011). Additionally, growing basil takes up a lot of space, especially close to populated areas. The efficient option for meeting the enormous demand for this crop is hydroponic cultivation. The hydroponic method of plant production is an efficient alternative for countries short on water supply and limited in agricultural soil. Stress and salinity, hinder basil growth and cause morpho-physiological abnormalities that affect nutritional upregulation, and decrease stomatal activity, photosynthesis and transpiration rate (Maia et al. 2017). In this study, the electrical conductivity (EC) of the medium was adjusted to 1.55–1.80 dS m^{-1}, and pH 5.5–7.0 (Kiferle et al. 2011). In the hydroponic system, plants grew vigorously and abundantly within one month after planting. The accumulation of fresh and dry biomass in the leaves, shoots and roots of hydroponically grown plants was much more during the blooming stage than during the vegetative stage. In comparison to the vegetative stage, basil plant leaves and roots had more rosmarinic acid (RA) during the blooming stage. The concentration of RA was maximum in the roots (on average, 17.64 mg g^{-1} DW) as compared to leaves (10.69 mg g^{-1} DW). According to earlier studies, the hydroponic farming technique was shown to be an effective and viable method for the large-scale production of plant material (Pardossi et al. 2005). The major advantage of the hydroponics farming approach is the ability to harvest root tissues, which comprised about 10% of the total dry biomass and had a much greater RA concentration than shoot tissue. The maximum amount of RA was spotted at full flowering in both leaves and roots (Juliani et al. 2008). In conclusion, hydroponics is an advantageous and effective cultivation system for growth and biomass production in significantly less time.

6.5.6 *Coriandrum sativum* L. (Dhania)

Coriander (also called cilantro) is native to the Mediterranean, but cultivated worldwide as a culinary annual herb belonging to the family *Apiaceae* (Laribi et al. 2015). It is also known as cilantro, dhania or Chinese parsley. It is widely used to enhance the flavor and aroma of many cuisines (Rashed and Darwesh 2015). Coriander plants under the natural environment experience different abiotic stress, such as salinity, water deficit, temperature and other environmental stress. Additionally, the lower availability of groundwater in many countries' region has led to limiting the availability of crops. To minimize such problems, studies have proposed the hydroponic technique (soilless cultivation). This research compared the biochemical compositions of the essential or volatile oil of coriander grown in hydroponics with those growing in the field and under a greenhouse environment (Hazeri et al. 2012). These experiments were marked as follows: S1; coriander cultivated in hydroponic cultivation, S2; plant cultivated in the field under natural habitat, and S3; plants cultivated in a greenhouse. In this experiment, Hoagland nutrient solution is used as a growing medium, and the nutrient medium consists of distilled water. The air-dried aerial parts of coriander were used for oil extraction. Essential oil is extracted by hydro-distillation using a clevenger-type instrument. The oil extracted by hydro-distillation from samples S1, S2 and S3 is a colorless or pale-yellow liquid. These oils are analyzed by the GC-MS technique. The major constituents, such as aliphatic aldehyde and decanal (S1; 25.8%, S2; 31.0%, and

S3; 34.7%) have been detected in all samples. Twenty-four compounds were identified in the oil of hydroponically grown *C. sativum* L (S1). Twenty-four compounds were characterized in the oil isolated from the field cultivated (S2). Twenty-three compounds were characterized in the oil isolated from the greenhouse cultivated (S3). These chemical variations indicated that environmental conditions, different cultivation methods and other parameters affect the essential oil composition of coriander.

6.5.7 *Conobea scoparioides* Benth. (Pataqueira)

Conobea scoparioides Benth. (*Scrophulariaceae*), popularly known as "pataqueira", is a popular herbaceous and aromatic medicinal plant found in rainy zones of the Brazilian Amazon (Maia et al. 2000). It is used to cure leishmaniasis, pain and beriberi (de Lima et al. 2020). Leaves of pataqueira contains an essential oil that is used in various industries, pharmaceutical companies and cures to treat diseases. Additionally, the oil has strong antioxidant properties, and many industries of aesthetics and perfume have shown huge interest in the demand for the "pataqueira" oil. But its commercial utilization is obstructed by the incapability to sustain a cultivation technology in rainy areas of the region. So, it is important to maintain a yearly and systematic cultivation system without changing the qualitative and medicinal characteristics of pataqueira essential oils. Numerous studies have demonstrated that aromatic plants grown hydroponically have increased production and yield. Additionally, when compared to field-grown plants, acquires more essential oil per unit. In this study, the NFT hydroponics system was used for the cultivation of pataqueira (Costa et al. 2014). Following experimental conditions were maintained as the pH of 6.0±1.0, the EC 1.6±0.4 mS/cm, the temperature of 27°C, and reduced light intensity by 30%. The oils were analyzed by GC and GC-MS techniques. The major components of the essential oil were methylthymol, thymol, and α-phellandrene. The NFT hydroponic farming technology was very well adapted by pataqueira. After the first two months of cultivation, the species more than doubled in size, reaching a height of 30 cm; after 4 months, it reached an 85 cm height. The growth of the root biomass of pataqueira in the NFT system has been found to be very good after 4 months of cultivation. Seedlings of pataqueira showed excellent growth and biomass production in the NFT hydroponics system.

6.6 Conclusion

Global recognition of soilless farming is providing farmers and consumers opportunities in the field of contemporary agriculture. The book chapter delivered a comprehensive discussion on soilless cultivation technology for growth and secondary metabolites production in spice and essential oil-bearing plants. Based on the literature review, it is suggested that the soilless cultivation system is well-appropriate for the cultivation of upright parts of essential oil-bearing plants like a leaves, shoots and seeds. Whereas aeroponic is most preferable for the production of tuberous, fibrous roots and rhizomatous plant production. Modern soil-less agro-practices have the potential to produce large amounts of plant biomass and its essential oil production to satisfy producers' and consumers' demands. Contrary to soil-based cultivation, soilless cultivation is heavily utilized in safe agriculture to support eco-friendly and boost the agro-economy. Further soilless farming can be scaled-up by utilizing the vertical farming system to fulfill the demand of the urban population with respect to spices and essential oils.

Acknowledgments

The authors acknowledge the Council of Scientific and Industrial Research (CSIR), Government of India, under the project "MLP-0201 and MLP-0151" for providing funding. AK, MP and SR, acknowledge the Academy of Scientific and Innovative Research (AcSIR), Ghaziabad, India, for PhD enrolment. AK and MP acknowledge the CSIR, New Delhi, for providing the Junior/Senior Research Fellowship. The authors are grateful to the Director, CSIR-IHBT, Palampur, for providing the necessary facilities.

Author Contributions

ARW conceived the concept and framed the layout of book chapter. AK and MP: data collection, literature collection and interpretation, original figures and table preparation and original draft writing SR: data collection and original draft writing. PK and ARW edited the book chapter and approved the final version of the draft.

References

Ahmadi, F., Samadi, A., Sepehr, E., Rahimi, A. and Shabala, S., 2021. Optimizing hydroponic culture media and NO_3−/NH^{4+} ratio for improving essential oil compositions of purple coneflower (*Echinacea purpurea* L.). *Scientific Reports*, 11(1): 1–10.

Aires, A., 2018. Hydroponic production systems: impact on nutritional status and bioactive compounds of fresh vegetables. In: Asaduzzaman, M.D. and Asao, T. (eds.) *Vegetables-importance of quality vegetables to human health*. IntechOpen. https://doi.org/10.5772/intechopen.73011

Asao, T. ed., 2012. *Hydroponics: A standard methodology for plant biological researches*. BoD–Books on Demand.

Azad, K.N., Salam, M.A. and Azad, K.N., 2016. Aquaponics in Bangladesh: current status and future prospects. *Journal of Bioscience and Agriculture Research*, 7 (2): 669–677.

Aziz, Z.A., Ahmad, A., Setapar, S.H.M., Karakucuk, A., Azim, M.M., Lokhat, D., Rafatullah, M., Ganash, M., Kamal, M.A. and Ashraf, G.M., 2018. Essential oils: extraction techniques, pharmaceutical, and therapeutic potential-a review. *Current Drug Metabolism*, 19(13): 1100–1110.

Bakkali, F., Averbeck, S., Averbeck, D. and Idaomar, M., 2008. Biological effects of essential oils–a review. *Food and Chemical Toxicology*, 46(2): 446–475.

Bos, R., Hendriks, H., Scheffer, J. and Woerdenbag, H., 1998. Cytotoxic potential of valerian constituents and valerian tinctures. *Phytomedicine*, 5: 219–225.

Christie, C.B. and Nichols, M.A., 2003, February. Aeroponics-a production system and research tool. *Acta Horticulturae*, 648: 185–190.

Ciriello, M., Formisano, L., Soteriou, G. A., Kyratzis, A., De Pascale, S., Kyriacou, M. C. and Rouphael, Y., 2022. Differential response to NaCl osmotic stress in sequentially harvested hydroponic red and green basil and the role of calcium. *Frontiers in Plant Science*, 13, 799213. https://doi.org/10.3389/fpls.2022.799213

Costa, R.G., de Faria, L.J.G., de Gusmao, S.A.L., et al., 2014. Essential oil of pataqueira (*conobea scoparioides* benth.): from natural occurrence and produced by hydroponics. *Advances Plants Agricultural Research*, 1: 390–394. https://doi.org/10.15406/apar.2014.01.00015

Currey, C.J., Walters, K.J. and Flax, N.J., 2019. Nutrient solution strength does not interact with the daily light integral to affect hydroponic cilantro, dill, and parsley growth and tissue mineral nutrient concentrations. *Agronomy*, 9: 7–389.

de Lima, E.J., Fontes, S.S., Nogueira, M.L., ahmsure, V.R., Santos, L.D.S., D'Elia, G.M., Dias, R.B., Sales, C.B., Rocha, C.A.G., Vannier-Santos, M.A. and Soares, M.B., 2020. Essential oil from leaves of *Conobea scoparioides* (Cham. & Schltdl.) Benth. (*Plantaginaceae*) causes cell death in HepG2 cells and inhibits tumor development in a xenograft model. *Biomedicine & Pharmacotherapy*, 129: 110402.

de Oliveira Hashimoto, G.S., Neto, F.M., Ruiz, M.L., Acchile, M., Chagas, E.C., Chaves, F.C.M. and Martins, M.L., 2016. Essential oils of Lippia sidoides and Mentha piperita against monogenean parasites and their influence on the hematology of Nile tilapia. *Aquaculture*, 450 : 182–186.

Domingues, D.S., Takahashi, H.W., Camara, C.A. and Nixdorf, S.L., 2012. Automated system developed to control pH and concentration of nutrient solution evaluated in hydroponic lettuce production. *Computers and Electronics in Agriculture*, 84: 53–61.

Dong, X., Jiang, Z.T., Jiang, S. and Li, R., 2017. Composition comparison of essential oils extracted by hydrodistillation and microwave-assisted hydrodistillation from *Petroselinum crispum* grown in China. *Journal Essential Oil Bearing Plants*, 20(2): 368–374.

Dorais, M., Papadopoulos, A. P., Luo, X., Leonhart, S., Gosselin, A., Pedneault, K., et al., 2001. Soilless greenhouse production of medicinal plants in North Eastern Canada. *Acta Horticultrae,* 554: 297–303. https://doi.org/10.17660/ActaHortic.554.32

Dou, H., Niu, G., Gu, M. and Masabni, J., 2020. Morphological and physiological responses in basil and brassica species to different proportions of red, blue, and green wavelengths in indoor vertical farming. *Journal of the American Society for Horticultural Science,* 145(4): 267–278.

GÃl, A., TÃzel, I.H., Tuncay, Ã., Eltez, R. Z. and Zencirkiran, E., 1997. Soilless culture of cucumber in glasshouses: I. A comparison of open and closed systems on growth, yield and quality. *International Symposium Greenhouse Management for Better Yield & Quality in Mild Winter Climates,* 491: 389–394.

Farzaei, M.H., Abbasabadi, Z., Ardekani, M.R.S., Rahimi, R. and Farzaei, F.V., 2013. Parsley: a review of ethnopharmacology, phytochemistry and biological activities. *Journal of Traditional Chinese Medicine,* 33(6): 815–826.

Favorito, P.A., Echer, M.M., Offemann, L.C., Schlindwein, M.D., Colombare, L.F., Schineider, R.P. and Hachmann, T.L., 2011. Productive characteristics of basil (*Ocimum basilicum* L.) as a function of spacing between plants and between rows. *Revista Brasileira de Plantas Medicinais,* 13: 582–586.

Fussy, A. and Papenbrock, J., 2022. An overview of soil and soilless cultivation techniques—chances, challenges and the neglected question of sustainability. *Plants,* 11(9): 1153.

Hassanpouraghdam, M.B., Tabatabaie, S.J., Nazemiyeh, H., Vojodi, L. and Aazami, M.A., 2008. Essential oil composition of hydroponically grown *Chrysanthemum balsamita* L. *Journal of Essential Oil Bearing Plants,* 11(6): 649–654.

Hayden, A.L., Brigham, L.A. and Giacomelli, G.A., 2004, March. Aeroponic cultivation of ginger (*Zingiber officinale*) rhizomes. *VII International Symposium on Protected Cultivation in Mild Winter Climates: Production, Pest Management and Global Competition,* 659: 397–402.

Hazeri, N., Valizadeh, J., Shakeri, A. and Rajabpour, M., 2012. Evaluation of essential oil and mineral composition of coriander (*Coriandrum sativum* L.) among growth conditions of hydroponic, field and greenhouse. *Journal of Essential Oil Bearing Plants,* 15(6): 949–954.

Hoagland, D.R. and Arnon, D.I., 1950. The water-culture method for growing plants without soil. Circular, California Agricultural Experiment Station, 347, 1–32.

Inden, H., Asahira, T. and Hirano, A., 1988, May. Micropropagation of *ginger. Symposium on High Technology in Protected Cultivation,* 230: 177–184.

İşcan, G., Kirimer, N., Kürkcüoğlu, M., Başer, H.C. and Demirci, F., 2002. Antimicrobial screening of Mentha piperita essential oils. *Journal of Agricultural and Food Chemistry,* 50(14): 3943–3946.

Javanmardi, J., Khalighi, A., Kashi, A., Bais, H.P. and Vivanco, J.M., 2002. Chemical characterization of basil (*Ocimum basilicum* L.) found in local accessions and used in traditional medicines in Iran. *Journal of Agricultural and Food Chemistry,* 50(21): 5878–5883.

Jehnsen, J., 2010. *Effects on Artemisia annua growth with the addition of gibberellic acid, salicylic acid, and methyl jasmonate in hydroponic systems.* UC. Berkeley Spring, 1–23.

Jiang, H., Sólyom, A.M., Timmermann, B.N. and Gang, D.R., 2005. Characterization of gingerol-related compounds in ginger rhizome (*Zingiber officinale* Rosc.)

by high-performance liquid chromatography/electrospray ionization mass spectrometry. *Rapid Communications in Mass Spectrometry: An International Journal Devoted to the Rapid Dissemination of Up-to-the-Minute Research in Mass Spectrometry*, 19(20): 2957–2964.

Joseph, L., Puthallath, R.E. and Rao, S.N., 2016. Acute and chronic toxicity study of *Valeriana wallichii* rhizome hydro-ethanolic extract in Swiss albino mice. *Asian Journal of Medicine Science*, 72: 49–54.

Jugran, A.K., Rawat, S., Bhatt, I.D. and Rawal, R.S., 2019. Valeriana jatamansi: An herbaceous plant with multiple medicinal uses. *Phytotherapy Research*, *33*(3), 482-503.

Juliani H.R., Koroch A.R. and Simon J.E., 2008. Basil: A new source of rosmarinic acid. In: Ho C.T., Simon J.E., Shahidi F. and Shao Y. (eds.) *Dietary Supplements, American Chemical Society Symposium Series 987*. American Chemical Society, 129–143.

Kiferle, C., Lucchesini, M., Mensuali-Sodi, A., Maggini, R., Raffaelli, A. and Pardossi, A., 2011. Rosmarinic acid content in basil plants grown in vitro and in hydroponics. *Central European Journal of Biology*, 6(6): 946–957.

Lakhiar, I.A., Jianmin, G., Syed, T.N., Chandio, F.A., Buttar, N.A. and Qureshi, W.A., 2018. Monitoring and control systems in agriculture using intelligent sensor techniques: a review of the aeroponic system. Journal of Sensors. 2018(1), 8672769.

Laribi, B., Kouki, K., M'Hamdi, M. and Bettaieb, T., 2015. Coriander (*Coriandrum sativum* L.) and its bioactive constituents. *Fitoterapia*, 103: 9–26.

Loera-Muro, A., Troyo-Diéguez, E., Murillo-Amador, B., Barraza, A., Caamal-Chan, G., Lucero-Vega, G. and Nieto-Garibay, A., 2021. Effects of vermicompost leachate versus inorganic fertilizer on morphology and microbial traits in the early development growth stage in mint (*Mentha spicata* L.) and rosemary (*Rosmarinus officinalis* L.) plants under closed hydroponic system. *Horticulturae*, 7: 5–100.

Maharana, L. and Koul, D.N., 2011. The emergence of Hydroponics. *Yojana (June)*, 55: 39–40.

Maia, J.G.S., Zoghbi, M.D.G.B., Andrade, E. and da Silva, M.H.L., 2000. Essential oils from *Conobea scoparioides* (Cham. & Schltdl.) Benth. *Flavour and Fragrance Journal*, 15(64): 13–414.

Maia, S.S., da Silva, R.C., Oliveira, F.D.A.D., Silva, O.M.D.P., Silva, A.C.D. and Candido, W.D.S., 2017. Responses of basil cultivars to irrigation water salinity. *Revista Brasileira de Engenharia Agrícola e Ambiental*, 21: 44–49.

Makri, O. and Kintzios, S., 2007. *Ocimum* sp. (Basil): botany, cultivation, pharmaceutical properties, and biotechnology. *Journal of Herbs, Spices & Medicine Plants*, 13: 123–150.

Maodaa, S.N., Allam, A.A., Ajarem, J., Abdel-Maksoud, M.A. and Al-Basher, G.I., 2016. Effect of parsley (*Petroselinum crispum, Apiaceae*) juice against cadmium neurotoxicity in albino mice (*Mus Musculus*). *Behaviour Brain Function*, 12: 1–6.

Martins, J.B., Júnior, J.A.S., Júnior, F.J.S., Silva, G.F. and Medeiros, S.S., 2019. Production of parsley in hydroponic conditions under isosmotic brackish nutrient solutions. *Cienc e Agrotecnologia*, 43: e023418.

Mbiyu, M.W., Muthoni, J., Kabira, J., Elmar, G., Muchira, C., Pwaipwai, P., Ngaruiya, J., Otieno, S. and Onditi, J., 2012. Use of aeroponics technique for potato (Solanum tuberosum) minitubers production in Kenya. Journal of Horticulture and Forestry. 4(11), 172-177.

Mok, H.F., Williamson, V.G., Grove, J.R., Burry, K., Barker, S.F. and Hamilton, A.J., 2014. Strawberry fields forever Urban agriculture in developed countries: a review. *Agronomy for Sustainable Development*, 34(1): 21–43.

Mousavi, S.A., Dalir, N., Rahnemaie, R. and Ebadi, M.T., 2021. Phosphate concentrations and methionine application affect quantitative and qualitative traits of valerian (*Valeriana officinalis* L.) under hydroponic conditions. *Industrial Crops and Products*, 171: 113821.

Navarrete, A., Avula, B. and Choi, Y.W., 2006. Chemical fingerprinting of Valeriana species: Simultaneous determination of valerenic acids, flavonoids, and phenylpropanoids using liquid chromatography with ultraviolet detection. *Journal of AOAC International*, 89: 8–15.

Papadopoulos, A.P., Luo, X., Leonhart, S., Gosselin, A., Pedneault, K., Angers, P.,

Pardossi, A., Malorgio, F., Incrocci, L. and Tognoni, F., 2005. Hydroponic technologies for greenhouse crops. In: *Crops: Quality, Growth and Biotechnology*. WFL Publisher Science & Technology, pp. 360-378

Partap, M., Kumar, P., Kumar, A., Joshi, R., Kumar, D. and Warghat, A.R., 2020. Effect of elicitors on morpho-physiological performance and metabolites enrichment in *Valeriana jatamansi* cultivated under aeroponic conditions. *Frontiers in Plant Science*, 11:01263.

Partap, M., Rattan, S., Sood, A., Kumar, P. and Warghat, A.R., 2021. Hydroponic and aroponic cultivation of economically important crops for production of quality biomass. In: Kumar Srivastava, D., Kumar Thakur, A., Kumar, P. (eds), Agricultural Biotechnology: Latest Research and Trends. Springer, 573–585.

Putra, P.A. and Yuliando, H., 2015. Soilless culture system to support water use efficiency and product quality: a review. *Agriculture and Agricultural Science Procedia*, 3: 283–288.

Rashed, N.M. and Darwesh, R.K., 2015. A comparative study on the effect of microclimate on planting date and water requirements under different nitrogen sources on coriander (*Coriandrum sativum*, L.). *Annals of Agricultural Sciences*, 60(2): 227–243.

Rattan, S., Partap, M., Kumar, S. and Warghat, A.R., 2022. Nutrient feeding approach enhances the vegetative growth biomass, volatile oil composition, and myristicin content in hydroponically cultivated *Petroselinum crispum* (Mill.) Nyman. *Journal of Applied Research on Medicinal and Aromatic Plants*, 26: 100–359.

Savidov, N.A., Hutchings, E. and Rakocy, J.E., 2005, September. Fish and plant production in a recirculating aquaponic system: a new approach to sustainable agriculture in Canada. *International Conference and Exhibition on Soilless Culture: ICESC 2005*, 742: 209–221.

Savvas, D. and Gruda, N., 2018. Application of soilless culture technologies in the modern greenhouse industry—a review. *European Journal of Horticultural Sciences*, 83(5): 280–293.

Sharifi-Rad, J., Sureda, A., Tenore, G.C., Daglia, M., Sharifi-Rad, M., Valussi, M., Tundis, R., Sharifi-Rad, M., Loizzo, M.R., Ademiluyi, A.O. and Sharifi-Rad, R., 2017. Biological activities of essential oils: from plant chemoecology to traditional healing systems. *Molecules*, 22: 70.

Sharma, N., Acharya, S., Kumar, K., Singh, N. and Chaurasia, O.P., 2018. Hydroponics as an advanced technique for vegetable production: An overview. *Journal of Soil and Water Conservation*, 17(4): 364–371.

Singh, R.D., Meena, R.L., Sharma, B., Singh, B., Kaul, V.K. and Ahuja, P.S., 2010. Seasonal variation of bioactive components in *Valeriana jatamansi* from Himachal Pradesh, India. *Industrial Crops and Products*, 32: 292–296.

Souret, F.F. and Weathers, P.J., 2000. The growth of saffron (*Crocus sativus* L.) in aeroponics and hydroponics. *Journal of Herbs, Spices & Medicinal Plants*, 7(3): 25–35.

Stirling, A.M., 2002. *Erwinia chrysanthemi*, the cause of soft rot in ginger (*Zingiber officinale*) in Australia. *Australasian Plant Pathology*, 31(4): 419–420.

Surendran, U., Chandran, C. and Joseph, E.J., 2017. Hydroponic cultivation of *Mentha spicata* and comparison of biochemical and antioxidant activities with soil-grown plants. *Acta physiologiae plantarum*, 39: 11–14.

Surh, Y.J., 1999. Molecular mechanisms of chemopreventive effects of selected dietary and medicinal phenolic substances. *Mutation Research/Fundamental and Molecular Mechanisms of Mutagenesis*, 428(1–2): 305–327.

Tabatabaei, S. J., 2008. Effects of cultivation systems on the growth, and essential oil content and composition of Valerian. *Journal of Herbs, Spices & Medicinal Plants*, 14(1–2): 54–67. https://doi.org/10.1080/10496470802341219

Thakur, K., Partap, M., Kumar, D. and Warghat, A.R., 2019. Enhancement of picrosides content in Picrorhiza kurroa Royle ex Benth. mediated through nutrient feeding approach under aeroponic and hydroponic system. *Industrial Crops and Products*, 133: 160–167.

Touliatos, D., Dodd, I.C. and McAinsh, M., 2016. Vertical farming increases lettuce yield per unit area compared to conventional horizontal hydroponics. *Food and Energy Security*, 5(3): 184–191.

Vilsoni, F., McClure, M.A. and Butler, L.D., 1976. Occurrence, host range, and histopathology of *Radopholus similis* in ginger (*Zingiber officinale*). *Plant Disease Reporter*, 60(5): 417–420.

Vimolmangkang, S., Sitthithaworn, W., Vannavanich, D., Keattikunpairoj, S. and Chittasupho, C., 2010. Productivity and quality of volatile oil extracted from *Mentha spicata* and *M. arvensis* var. piperascens grown by a hydroponic system using the deep flow technique. *Journal of Natural Medicines*, 64: 131–135.

Yeritsyan, N. and Economakis, C., 2001, July. Effect of nutrient solution's iron concentration on growth and essential oil content of oregano plants grown in solution culture. *International Conference on Medicinal and Aromatic Plants. Possibilities and Limitations of Medicinal and Aromatic Plant*, 576: 277–283.

Zheljazkov, V.D., Cantrell, C.L., Tekwani, B. and Khan, S.I., 2008. Content, composition, and bioactivity of the essential oils of three basil genotypes as a function of harvesting. *Journal of Agricultural and Food Chemistry*, 62(5): 380–385.

7

PGPR-Biotechnological Interventions for Improvement of Crop Plants

Neerja Rana, Arti Ghabru, Shivani Chauhan and Himani Sharma

7.1 Introduction

The Green Revolution, which began in the latter part of the 20th century, led to a global boom in the agricultural industry. The Green Revolution greatly increased plant productivity and crop yields by introducing new high-yielding seed varieties and increasing the use of synthetic fertilizers, insecticides and other agrochemicals. The biological and physicochemical health of the arable soil has deteriorated due to the widespread overuse of synthetic agrochemicals to increase crop yield, which has resulted in a recent decline in agricultural productivity around the world (Kumar et al., 2019). In the current situation, land resources are becoming scarcer, and biological diversity is diminishing. The yield and productivity of agricultural crops must be raised at the same time as the production of goods related to agriculture in order to meet the growing need for sustainable agriculture. An effective strategy to stop the rapid environmental deterioration while ensuring high agricultural productivity and better soil health is to promote sustainable agriculture with a gradual decrease in the use of synthetic agrochemicals and a more prominent utilization of biowaste-derived substances as well as the biological and genetic potential of crop plants and microorganisms (Majeed et al., 2017). Certain members of the soil microbial community, especially those found in the plant rhizosphere, may help plants in minimizing or partially overcoming environmental challenges in addition to genetic manipulation of crop physiology and metabolism for yield enhancement. Plants have been subjected to various biotic and abiotic stressors ever since the terrestrial environment was colonized. Stress is typically understood to be an external factor that influences a plant's growth. These circumstances limit the plants' ability to grow and develop as well as their capacity to reproduce and pass on their DNA to the following generation. In the worst-case circumstances, stress is quantified by crop production, plant survival, biomass buildup or CO_2 and mineral uptake. Stress conditions in nature can be brought on by

DOI: 10.1201/9781003625674-7

both biotic and abiotic sources (Shameer & Prasad, 2018). Plant survival may be threatened by biotic stress brought on by nematodes, viruses, bacteria, fungus, insects like white flies, and several other types of living things.

The development of biofertilizers and other microbial-based products, such as organic extracts and vermicompost teas, as well as their subsequent use, were sparked by the search for environmentally friendly alternatives to reduce the negative effects of toxic agrochemicals and cope with biotic and abiotic stresses. These microbial compounds can be used to promote plant development and manage diseases because they are non-toxic and safe for the environment (Kumar et al., 2021). As a viable alternative for reducing the use of synthetic agrochemicals in crop production, the use of efficient plant growth-promoting rhizobacteria (PGPR) as biofertilizers and biological control agents is being considered. Additional characteristics that may help promote plant growth include those that cause systemic resistance or enzymes that break down ethylene, hydrogen cyanide or reactive oxygen species, which are created by stressed plants (Kumar et al., 2019; Bhatt et al., 2020). The expression of each of these features can also be influenced by the phenomena of quorum regulation as PGPR interacts with the local microbial community. This chapter offers a clearer explanation of the various PGPR-based biofertilizers' features, their possibilities and limitations, and finally the road map for their commercialization.

7.2 Plant Growth-Promoting Rhizobacteria (PGPR)/ Biofertilizer

Due to the remarkable advancements made in the research of the relationship between microorganisms and plants over the past 20 years, the terms "biofertilizer" and "bioinoculant" have taken on a variety of meanings. According to the most popular definition, a biofertilizer is "a substance that contains living microorganisms that, when applied to seed, plant surfaces or soil, colonizes the rhizosphere or the interior of the plant and promotes growth by increasing the supply or availability of primary nutrients to the host plant" (Glick, 2012). The microorganisms found in biofertilizers use a variety of techniques to benefit agricultural plants. They may be proficient at all three processes—nitrogen fixation, phosphate solubilization and plant growth promotion—or they may combine all three. Via the biological nitrogen fixation (BNF) process, biofertilizers can fix atmospheric N_2, as well as solubilize plant-necessary minerals like phosphate, zinc and potassium, and secrete compounds that encourage plant growth, such as different hormones. Biofertilizers can also multiply, take part in nutrient cycling and aid in crop development for sustainable farming when used as seed or soil inoculants (de Bruijin, 2015). They fight against abiotic stressors and display antagonistic action against some agricultural diseases. Based on their capacity to draw nutrients from the soil, fix atmospheric N_2, promote the solubilization of nutrients and serve as biocontrol agents, a variety of microbial taxa have been utilized commercially as effective biofertilizers (Thakur et al., 2023). The confined area of soil that surrounds the root systems of developing plants, known as the plant rhizosphere, serves as a hub for microbial activity in the soil. Prokaryotes (archaea, bacteria and viruses) and eukaryotes (fungi, oomycetes, nematodes, protozoa, algae and arthropods), of which bacteria and fungi constitute the most prevalent groups showing fundamental ecological activities, inhabit the rhizosphere. Plant growth-promoting rhizobacteria (PGPR), a name coined by Kloepper et al. (1980), are free-living soil bacteria that flourish in the rhizosphere, aggressively colonize plant roots, and promote plant development. There have been reported cases of PGPRs from the genera *Acinetobacter*, *Aeromonas*, *Agrobacterium*, *Allorhizobium*, *Arthrobacter*, *Azoarcus*, *Azorhizobium*, *Azospirillum*, *Azotobacter*, *Bacillus*, *Bradyrhizobium*, *Burkholderia*, *Caulobacter*, *Chromobacterium*, *Delftia*, *Enterobacter*, *Flavobacterium*, *Frankia* (Mercado-Blanco, 2015). If a rhizobacterial strain exhibits particular characteristics that promote plant growth and can boost plant growth after inoculation, it is regarded as a suspected PGPR. The following qualities should be present in a good PGPR strain: (1) it must be extremely eco-friendly and rhizosphere-competent, (2) upon inoculation, it ought to colonize the plant roots in appreciable quantities, (3) it ought to be capable of encouraging plant development, (4) a wide range of actions should be visible, (5) it ought to coexist peacefully with other rhizosphere bacteria, (6) it should be capable of withstanding physical and chemical conditions like heat, desiccation, radiation and oxidants and (7) it ought to outperform the current rhizobacterial communities in terms of competitive abilities.

7.2.1 Mechanisms of PGPR Action

PGPRs are either actively or inactively involved in promoting plant growth since they are the predominant rhizosphere microbial community. By facilitating the tolerance to biotic and abiotic stress and promoting the nutrition of the host plants, they can serve as biofertilizers that encourage plant growth and development. Through their diverse modes of action, which include root colonization, favorable effects on plant physiology and growth, biofertilization, induced systemic resistance, biocontrol of phytopathogens, etc., these advantageous bacterial groups protect plants and facilitate the promotion of plant growth (Spence & Bais, 2013). Direct and indirect mechanisms occurring inside and outside of the plant, respectively, have historically been used to categorize the ways by which PGPRs promote plant development (Figure 7.1).

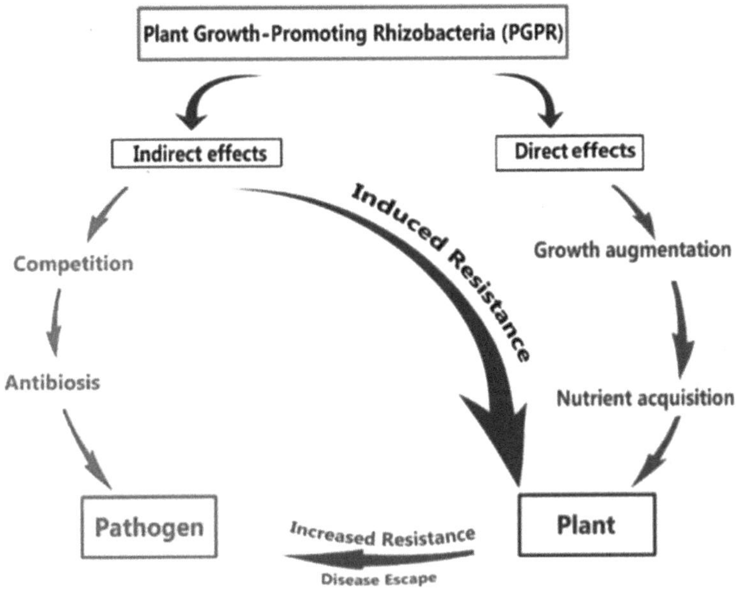

FIGURE 7.1
Overview of interactions between plant growth-promoting rhizobacteria (PGPR), plants and pathogens.

Improved plant nutrition can be achieved by supplying phytonutrients like fixed nitrogen or solubilized minerals from the soil (P, K, Zn, Fe and other crucial mineral nutrients), or plant growth and development can be accelerated by controlling phytohormone levels (like auxins, cytokinins, gibberellins, abscisic acid and ethylene). The indirect effects of PGPRs include affecting plant health by inhibiting phytopathogens and other harmful microorganisms through parasitism, competing for nutrients and niches in the rhizosphere, producing antagonistic substances (like hydrogen cyanide, siderophores, antibiotics and antimicrobial metabolites) and lytic enzymes (like chitinases, glucanases and proteases), and inducing systemic resistance in plants against a wide range of pathogens (Bhasan et al., 2014; Chouhan et al., 2021). PGPRs are good candidates to be developed and marketed as bioinoculants and phytoprotective microbial products because of the direct and indirect impacts they have on host plants. The host plant type affects the PGPR's mode and method of activity. Abiotic factors including soil type, composition, past soil management practices and existing environmental conditions as well as biotic factors like plant genotype, developmental phases, plant defense systems and the presence of other members of the microbial community can influence PGPR action (Olivares et al., 2013).

7.3 Contribution of Rhizobacteria to Plant Nutrition

Microorganisms with processes that improve nutrient availability, promote nutrient uptake or promote plant development. Certain PGPR can mineralize organic molecules, fix nitrogen and solubilize mineral nutrients (Table 7.1). The most well-researched PGPRs that are regarded as biofertilizers involve the fixing of nitrogen and the use of phosphorus in insoluble forms (Caryalhais et al., 2013; Rana et al., 2020).

7.3.1 Nitrogen Fixation

Nitrogen fixation is constrained by the poor availability of N due to large losses through emission or leaching. Bacteria that can make atmospheric N available to plants play a crucial role. Symbiotic and non-symbiotic biological fixation are the two forms. The first is the most significant process by which the majority of atmospheric nitrogen is fixed, however, it is only used by diverse tree and shrub species that produce actinorhizal roots with Frankia and legume plant species (Hassan & Mathesius, 2012). This action takes place in clearly defined nodular formations (Figure 7.2). The symbiotic bacteria *Rhizobium*, *Bradyrhizobium*, *Sinorhizobium* and *Mesorhizobium* have received

TABLE 7.1

Terms Adopted for Classified Mechanisms by Which Plant Growth-Promoting Bacteria Stimulate Plant Growth

Term	Definition	Mechanisms
Biofertilizer	A substance that contains live microorganisms which when applied on the seed, plant surface or the soil, colonizes the rhizosphere or the interior of the plant and promotes growth through increased supply or availability of primary nutrients for the host plant	Biological nitrogen fixation Utilization of insoluble forms of phosphorous
Phytostimulants	Microorganisms with the ability to produce or change the concentration of growth regulators, such as indole acetic acid, gibberellic acid, cytokinin and ethylene	Production of phytohormones (auxins, cytokinins and gibberellins) Decreased ethylene concentration (in the interior of the plant)
Biopesticides or biocontrol agents	Microorganisms that promotes plant growth through the control of phytopathogenic agents, mainly for the production of antibiotics and antifungal metabolites	Production of antibiotics (siderophores, HCN, antifungal metabolites) Competitive exclusion Acquired and induced systemic resistance

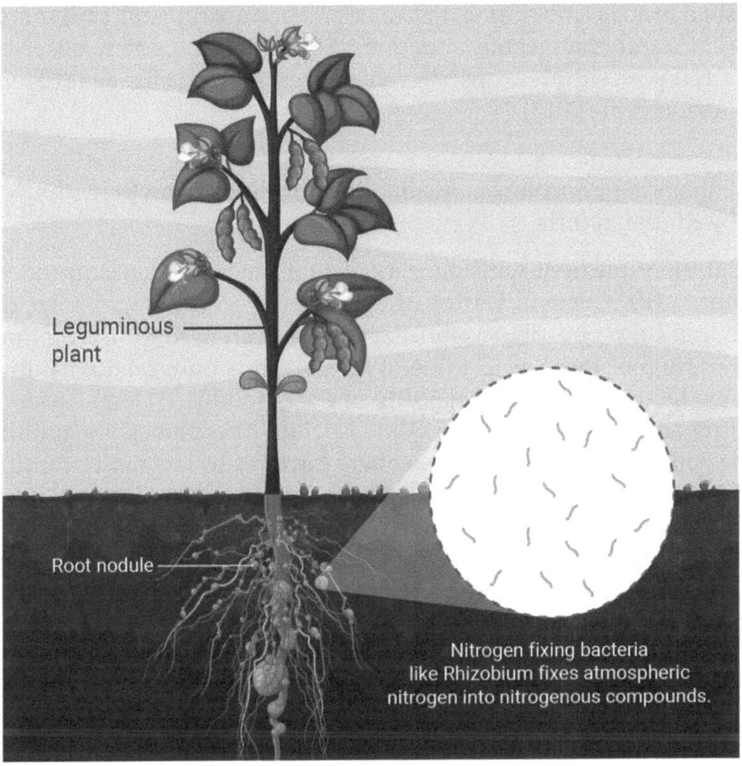

FIGURE 7.2
Leguminous plant with roots showing the formation of root nodules by rhizobium.

the most research attention. Rhizobia are not classified PGPR unless they are symbiotically connected with non-legume plants, despite the fact that the benefits of this relationship with legume plants are well documented. On the other hand, free-living diazotrophic organisms carry out non-symbiotic biological N fixation, which can promote the growth of non-legume plants. Studies have shown that N-fixing bacteria, both free-living and Rhizobium strains, can promote the development of non-legumes like radish and rice, reducing the need for N-based fertilizers (Levy et al., 2018). Many soils have yielded non-symbiotic N-fixing rhizospheric bacteria from the genera *Azoarcus*, *Azospirillum*, *Burkholderia*, *Gluconacetobacter* and *Pseudomonas*.

The capacity of non-symbiotic bacteria to fix considerable amounts of N is constrained due to the high energy demand for N fixation and the comparatively low metabolic activity of free-living organisms that must compete for root exudates outside of a nodule environment. It is no longer believed that a diazotrophic bacterium's presence in the rhizosphere of a particular plant indicates that it significantly contributes to N fixation and N supply for plant growth. Even while the N-fixing ability of some bacteria can be easily established in an *in vitro* setting, field and greenhouse research presents a more challenging and changeable situation. According to certain observations, rhizobacteria can supply crops with sizable amounts of nitrogen (Barea et al., 2013; Carrion et al., 2019).

7.3.2 Enhancing Phosphorus Availability for Plant Growth by Rhizobacteria

A crucial plant nutrient with poor availability in many agricultural soils is phosphorus (P). Currently, a lot of agricultural soils contain a lot of total phosphorus since phosphorus fertilizers have been used for a long time. Yet, a large portion of this P is in mineral forms and is only gradually absorbed by plants. In acidic soils, the majority of the insoluble P forms are found as aluminum and iron phosphates, while calcium phosphates are abundant in alkaline soils. The power of rhizosphere bacteria to lower pH through the excretion of organic acids (such as gluconate, citrate, lactate and succinate) and protons (during the assimilation of NH_4^+) has been related to their ability to solubilize insoluble P minerals (Schreiter et al., 2015). In numerous studies of P solubilizing bacteria from various rhizospheric soils, these bacteria have been identified as representatives of the genera *Bacillus, Burkholderia, Enterobacter, Klebsiella, Streptomyces* and *Pseudomonas*. These microorganisms thrive in media that solely contain tricalcium phosphate or comparable insoluble materials as a source of phosphate, and they not only ingest the element but also solubilize amounts that exceed their dietary needs to make it available to plants (Wu et al., 2013). On the other hand, organic P can make up between 30 and 50% of the soil's total P, with a significant amount of it being phytate. In this situation, some microorganisms can produce the phytase enzymes needed to mineralize phytates. There are currently very few studies describing rhizobacteria that can mineralize the phytate. The most prevalent culturable rhizobacteria among those that produce phytase are those from the genera *Bacillus, Burkholderia, Enterobacter, Pseudomonas, Serratia* and *Staphylococcus*. Numerous of these microorganisms are incredibly effective. Similar to bacteria that fix nitrogen, the development of plant growth hormones that increase root surface area can indirectly affect a plant's capacity to effectively remove phosphorus from soil (Xiang et al., 2019). It is therefore likely that many so-called biofertilizers have multiple action effects that are mediated by direct inorganic phosphorus solubilization, organic phosphorus mineralization and stimulatory effects on plant root growth or mycorrhizal fungi production.

7.3.3 Production of Phytohormones

It is now believed that one of the most significant processes by which many rhizobacteria enhance plant growth is through the generation of phytohormones by PGPR. Phytohormones are signal molecules that function as chemical messengers and are essential for controlling plant growth and development. Organic substances called phytohormones influence biochemical, physiological and morphological processes in plants at very low quantities, and their synthesis is tightly controlled. Many bacterial and fungal species have the ability to create phytohormones (Barea, 2015; Yang et al., 2019). The ability to produce phytohormones is broadly spread among soil and plant-associated microorganisms. According to studies, the PGPR can either control the high amounts of endogenous ethylene in the plant or induce the formation of auxins (indole acetic acid), gibberellines and cytokinins (Li et al., 2018).

7.3.3.1 Indole Acetic Acid (IAA) Producing Rhizobacteria

The synthesis of auxins, with IAA being the primary plant auxin, is a key component of many significant plant–microbial interactions. The IAA promotes root extension and is responsible for the division, growth and differentiation of plant cells and tissues. Many rhizobacteria, as well as pathogenic, symbiotic and free-living bacterial species, have been found to be able to synthesis IAA (Malusa et al., 2016). The manufacturers of phytohormones that have received the most research to date are rhizobacteria that synthesize auxin. Several rhizobacteria use various routes to convert tryptophan into IAA, while it can also be produced in smaller amounts through tryptophan-independent processes. The primary method used by phytopathogenic bacteria to produce IAA, which has been linked to plant tumor induction, is the indole acetamide pathway. It is unclear if it is utilized by good bacteria (Zhang et al., 2017). In contrast, the acid indole pyruvic route seems to be the primary pathway present in helpful bacteria that promote plant growth. *Azospirillum* is one of the most extensively researched IAA producers among PGPR species. Many rhizosphere soils have yielded IAA-producing bacteria from the genera *Aeromonas, Azotobacter, Bacillus, Burkholderia, Enterobacter, Pseudomonas* and *Rhizobium*. PGPR that produces IAA has been injected into plants to promote seed germination, speed up root development, alter the architecture of the root system, and boost root biomass (Du Jardin et al., 2015; Wang et al., 2018).

7.3.3.2 Regulation of Plant Ethylene by Rhizobacteria

Ethylene is necessary for the growth and development of plants, however depending on the concentration in the root tissues, it has various impacts on plant growth. It can be dangerous at high doses because it causes defoliation, cellular processes that limit stem and root growth and premature senescence, all of which diminish crop performance. Plants respond to several environmental stresses by producing 1-aminocyclopropane-1-carboxylate (ACC), which is the precursor for ethylene, including cold, drought, flooding, pathogen infections and the presence of heavy metals (Kumar et al., 2019).

A portion of the ACC is released into the rhizosphere, where it is readsorbed by the roots and transformed into ethylene. This buildup of ethylene causes a negative cycle effect because it makes it harder for plants to get the water and nutrients they need, which in turn makes them more stressed. As a result, PGPR, which has the potential to degrade ACC in the rhizosphere, can aid in breaking this downward spiral and re-establishing a strong root system, which is necessary to handle environmental stress (Malusa et al., 2016). The principal method by which rhizobacteria break down ethylene is the ACC deaminase enzyme's degradation of ethylene (EC 4.1.99.4). Some of the negative consequences of the high quantities of ethylene can be mitigated or avoided by this enzyme. The immediate ethylene precursor ACC, which is present in higher plants, is broken down by the enzyme ACC deaminase into alpha-ketobutyrate and ammonium (Singhal et al., 2016; Bona et al., 2019). Several soils have yielded isolates of rhizosphere bacteria with ACC deaminase activity from the genera *Achromobacter, Azospirillum, Bacillus, Enterobacter, Pseudomonas* and *Rhizobium*. Studies have shown that plants treated with PGPR bacteria that produce ACC deaminase are more resilient to environmental stress than untreated plants (Khan et al., 2020).

7.3.4 Rhizobacteria Act as Biocontrol Agents

Phytopathogenic microorganisms significantly reduce agricultural productivity as well as plant health and crop quality (Liu & Murray, 2016). The conventional approach to controlling phytopathogens involves using chemical pesticides, but this approach has sparked worries about environmental contamination and given rise to the pesticide treadmill, in which pathogens gradually become more resistant to particular chemical controls over time, necessitating the ongoing creation of new pesticides (Berendsen et al., 2018). Since they are capable of providing biocontrol of disease or insect pests, rhizobacteria are seen as an alternative to chemical pesticides in this context (biopesticides). Biocontrol involves a wide range of mechanisms, including direct antagonism through the production of antibiotics, siderophores, HCN and hydrolytic enzymes (chitinases, proteases, lipases, etc.), as well as indirect mechanisms in which the biocontrol organisms act as probiotics by competing with the pathogen for a niche (infection and nutrient sites). Induced systemic resistance (ISR) responses in plants, activation of acquired systemic resistance (ASR) and alterations to the hormone levels in plant tissues can all function as mediating mechanisms for biocontrol (Lyu et al., 2020; Basu et al., 2022).

7.3.5 Antibiotic-Producing Rhizobacteria

One of the most effective and extensively researched biocontrol techniques for battling phytopathogens is the generation of antibiotics. A diverse range of microbes create antibiotics, which are a large and heterogeneous category of low molecular weight chemical organic chemicals (Meena et al., 2017; Ismail et al., 2021). Many varieties of antibiotics made by PGPR have demonstrated efficacy against phytopathogenic agents in laboratory settings. Butyrolactones, zwittermycin A, kanosamine, oligomycin A, oomycin A, phenazine-1-carboxylic acid, pyoluteorin, pyrrolnitrin, viscosinamide, xanthobaccin and 2,4-diacetyl phloroglucinol are among the antibiotics made by PGPR (2,4-DAPG). The last one, which can be produced by different strains of *Pseudomonas*, one of the most prevalent bacterial species in the rhizosphere, is one of the most effective antibiotics for controlling plant infections (Barnawal et al., 2017). The 2,4-DAPG is antifungal, antibacterial and antihelmintic, giving it a broad range of characteristics. It inhibits the growth of the pathogenic fungus *Gaeumannomyces graminis* var. *tritici*, which infects wheat (Bhat et al., 2020).

7.3.6 Hydrogen Cyanide (HCN)-Producing Rhizobacteria

In addition to 2,4-DAPG synthesis, some rhizobacteria can also produce HCN (hydrogen cyanide, also known as cyanide). A volatile secondary metabolite called HCN inhibits the growth of microorganisms and adversely affects plant growth and development. Many metal enzymes are effectively inhibited by HCN, particularly cytochrome C oxidases (Smith et al., 2015). The HCN synthetase enzyme, which is connected to the plasma membrane of certain rhizobacteria, converts glycine into HCN. To date, a wide range of bacterial genera, including species of *Alcaligenes*, *Aeromonas*, *Bacillus*, *Pseudomonas* and *Rhizobium*, have demonstrated the ability to produce HCN (Behm et al., 2014). According to numerous research, HCN has a disease-protective function, preventing nematodes *Meloidogyne javanica* and *Thielaviopsis basicota* from causing "root-knot" and black rot in tobacco and tomato roots, respectively. HCN also manages the subterranean termite *Odontotermes obesus*, a significant pest in Indian agricultural and forestry crops (Malusa et al., 2016). Investigations have revealed negative impacts on plants, a reduction in lettuce's root growth, and an obstruction of the energy metabolism of potato root cells. The main growth of roots in Arabidopsis is also inhibited by HCN generated by *Pseudomonas* in the rhizosphere due to the inhibition of an auxin-sensitive gene (Du Jardin, 2015).

7.3.7 Siderophore-Producing Rhizobacteria

Low molecular weight substances known as siderophores are synthesized and used as iron (Fe) chelating agents by bacteria and fungi. Due to low iron solubility at high pH, different types of bacteria manufacture these chemicals in response to iron deficiency, which often occurs in neutral to alkaline pH soils (Kloepper et al., 1980). Since iron is necessary for cellular growth and metabolism, bacteria's ability to colonize plant roots and compete for iron with other microbes in the rhizosphere depends critically on their ability to acquire Fe through the creation of siderophores. By securing Fe^{3+} in the vicinity of the root, siderophore-producing PGPR can stop the spread of harmful microbes. While low Fe concentrations occur at microsites with strong microbial activity during the development of the pathogen, Fe depletion in the rhizosphere has no effect on the plant. Many bacterial siderophores can be used as iron sources by many plants, although the overall concentrations are probably too low to significantly affect plant iron intake (Malusa et al., 2016). Additionally, plants use their own processes to obtain iron, either by the synthesis of phytosiderophores in the case of monocots or the action of a protein called root membrane reductase, which changes the insoluble Fe^{3+} ion into the more soluble Fe^{2+} ion. From the rhizosphere, siderophore-producing bacteria from the *Bradyrhizobium, Pseudomonas, Rhizobium, Serratia* and *Streptomyces* genera have been isolated in a number of studies (Barea, 2015). Yet, numerous studies demonstrate that the PGPR contributes significantly to iron sequestration through the generation of chelating agents, increasing the iron availability for both plants and the organism itself. Most of the iron in soil is in the inaccessible state Fe^{3+}. In the presence of oxygen and a neutral pH, the Fe^{2+} undergoes rapid oxidation, converting to Fe^{3+} and decreasing its solubility in soil, rendering it biologically inaccessible to plants and rhizobacteria (Khan et al., 2020). Rhizobacteria, on the other hand, create siderophores, which are low-molecular-weight peptides (400–1000 Da) that function as chelating agents and bind to Fe^{3+} ions with a high affinity.

Siderophores, which can be classified as extracellular and intracellular, are water-soluble substances. Siderophores form a compound with Fe^{3+} ions that are then transferred to the cytoplasm. Nevertheless, this complex is transported differently by Gram-positive and Gram-negative bacteria. The siderophore-Fe^{3+} combination is delivered to the periplasm by Gram-negative bacteria after binding to OMTs (outer membrane transporters). However, in order to transport the complex into the periplasm, OMTs must be activated, which is accomplished by TonB machinery. TonB-ExbB-ExbD complexes, which are part of the TonB machinery, are tethered to the cytoplasmic membrane (CM). Siderophore-Fe^{3+} complex is trapped in its binding site by OMT when the siderophore-Fe^{3+} complex binds to it and causes a conformational shift. Transporters are activated by Ton to move the complex to the periplasm in the following phase (Barret et al., 2013; Thomas et al., 2019). A crucial component of the Fe transport mechanism in bacteria is the siderophore-Fe^{3+} complex, which is bound in the periplasm by periplasmic SBPs (siderophore binding proteins). ABC transporters are responsible for the siderophore-Fe^{3+} complex's ultimate translocation in the cytoplasm (Carvalhais et al., 2013; Schreiter et al., 2015). The siderophore-Fe^{3+} complex can be transported into the cytoplasm by the siderophore permeases-ATPase system, which is an ATP-dependent mechanism. Ferric iron reductase releases the Fe^{2+} inside the cell after reducing the Fe^{3+} inside the cytoplasm (Figure 7.3A). As Gram-positive

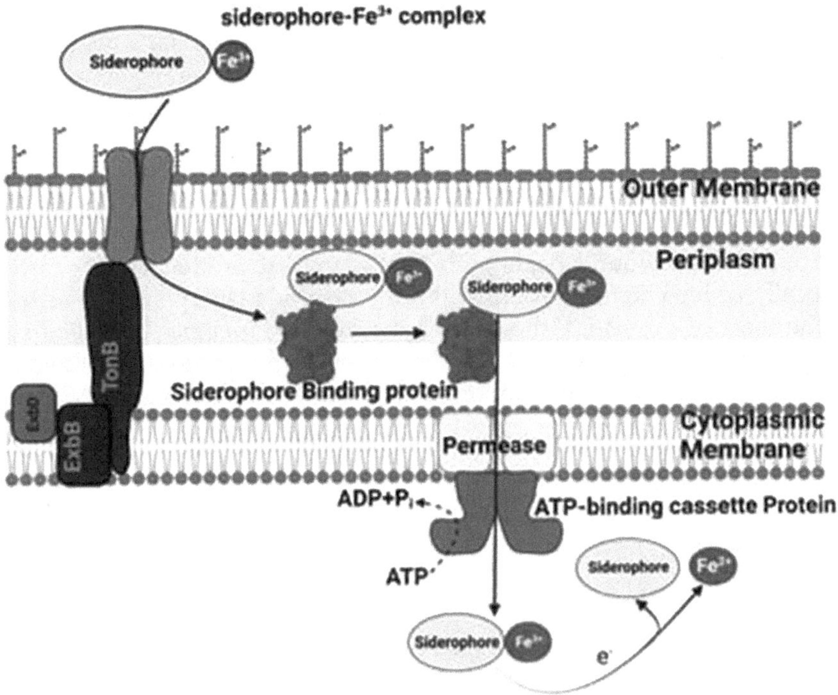

(A)

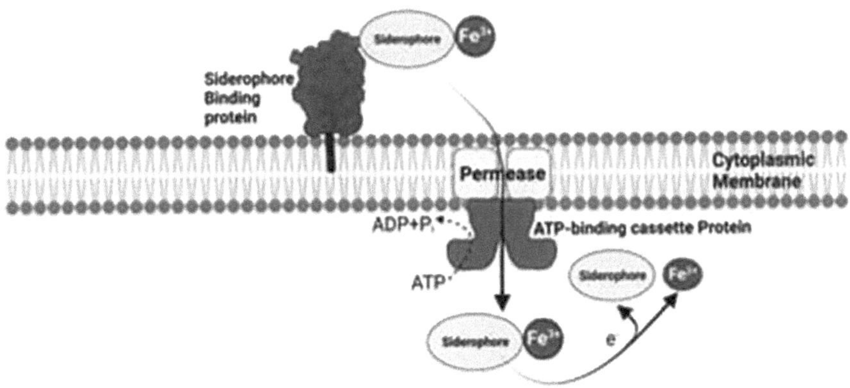

(B)

FIGURE 7.3
(A) Iron uptake mechanism by Gram-negative bacteria and (B) iron uptake mechanism by Gram-positive bacteria.

bacteria do not have an outer membrane, OMTs or TonB machinery, their process for absorbing iron is different. The SBP attached to fatty acid groups on the cell membrane is directly bound by the siderophore-Fe^{3+} complex (Jat et al., 2021). ABC transporters deliver the iron complex straight into the cytoplasm (Figure 7.3B).

Several investigations have revealed that PGPRs make iron accessible to both themselves and plants. *Centella asiatica*, an Indian pennywort, is used in traditional Chinese and Indian remedies due to its potential for neuroprotection, wound healing and anti-aging. It produces the pharmaceutically active chemical asiaticoside (trisaccharide triterpene). By settling on the roots of *C. asiatica*, the endophytic fungus *Piriformospora indica* successfully increases the production of asiaticoside. When different strains of *Bacillus* are used to cure siderophore under salinity stress, *Arabidopsis thaliana* produces more biomass.

7.4 Rhizobial Biosphere Engineering

Global demand for increased food output is anticipated given the effects of climate change and population growth. Plant-microbial mutualism will aid in sustainable agriculture by providing resilience to biotic and abiotic pressures by improving plant health and productivity. New research opportunities in the search for superior rhizobial inoculants are made possible by recent technological advancements and the burgeoning interest in the microbiome approach (de Bruijin, 2015). Particularly, a set of techniques called microbiome engineering has recently been put out as a way to comprehend and put together microbial communities. Yet, this methodology was also suggested as a tool to discover changes to the abundance of microbial species and changes to allele frequencies in microbial populations of microbiomes. Microbiome engineering has primarily focused on the assembly of beneficial microbiomes for plant hosts (Olivares et al., 2013). Initially, the microbial community that the host interacts with is shaped by plant host processes. Second, the host microbiome associated with an interesting attribute in the host (like fruit production) is found and applied to fresh, microbe-free hosts. The final constructed microbiome has been proven to affect intentionally selected plant features like biomass and blooming by regularly transferring microbial communities from hosts with an interesting trait to microbe-free hosts (i.e., passaging). Strong host regulatory systems that mould microbial populations are essential for microbiome engineering (Singhal et al., 2016). Legumes in particular have been shown to significantly modify microbial populations both in planta and ex-planta, making them particularly receptive to this process (Figure 7.4). The four stages of rhizobial transmission—root

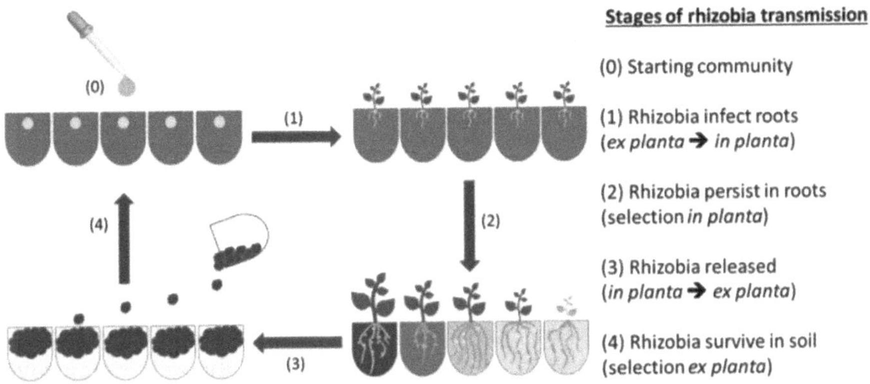

FIGURE 7.4
The four stages of rhizobia transmission.

infection, persistence in host tissue, release from host roots and survival in the soil—reflect the obstacles that result from rhizobial engineering. We can create methods that can quickly spot alterations at any of these four stages of transmission; thanks to microbiome engineering and the accessibility of next-generation sequencing. To comprehend the rhizobial competition issue, we first give an overview of designing legume microbiomes. Then, to further our comprehension of the rhizobial competitive population, we investigate three distinct approaches: (1) a commercially accessible inoculum approach, (2) a hyper-competitive inoculum strategy and (3) a synthetic community approach (Spence & Bais, 2013). Adding 1-aminocyclopropane-1-carboxylate (ACC) deaminase to the banana endospheric bacterial cell walls (*Enterobacteriaceae*) is a promising way to give bananas resistance against Fusarium wilt progression. To explore the priority effects and keystone species in the phyllosphere of the gnotobiotic Arabidopsis model system, a synthetic microbial community has been used. Such a useful method aids in putting to the test the underlying ideas that influence community structure in the phyllosphere (Mercado-Blanco, 2015). To better understand the structure and gene activities of the microbiota, it is therefore helpful to manipulate the enormous diversity of bacteria that are connected to the host plant. Here, the approaches indicated below have been used to create and enhance beneficial microbiome functions that guarantee plant productivity for several generations. Rhizobia infection and persistence should be controlled by legume mechanisms, according to microbiome engineering, yet selection forces associated to released rhizobia and their survival in the soil are vulnerable to passaging techniques used during microbiome engineering (Glick, 2012). By altering rhizobial species abundance and rhizobia allele frequencies, the microbiome engineering methods we describe aim to examine the rhizobial competition dilemma.

7.5 Soil Amendments

The physicochemical characteristics of the soil are collectively controlled by the interplay of climatic, edaphic and biotic circumstances. The distribution and functional activities of microbial species are supported by factors such as soil structure, organic matter, pH, texture, temperature, quality and quantity of carbon inputs (root exudates). Plant metabolites in the root region have a positive feedback effect on which bacteria are chosen to colonize (Schreiter et al., 2015). This causes several reactions that improve the functionality of the ecosystem. Such traits benefit both plant health and microbial assembly that is specific to a certain soil type. Organic additions to the soil can improve productivity and promote stable agroecosystem operation. By modifying the heterogeneous distribution of microbial species, plant characteristics and soil biophysical properties, organic amendments can trigger a chain of events that affects various trophic levels in an ecosystem (Thomas et al., 2019). The distribution, composition and organization of the microbial population are affected by the application of organic amendments. With notable changes in soil fertility, plant and microbial biomass, root characteristics, tissue elemental composition, substrate consumption and decomposability, they provide a variety of advantages (Barea et al., 2013; Meena et al., 2017). Some scientists have possibly looked into the crucial role that soil systems handled organically play in comparison to conventional systems. Using 16S rRNA gene-based next-generation sequence approaches, they evaluated the diversity and composition of microbial species. Their research offers a thorough understanding of how different soil health treatments alter the phylogenetic richness, diversity and heterogeneity of soil microbiota (Malusa et al., 2016). This idea is that organic additions have a positive impact on crop production, soil chemistry and microbial composition. For the sustainable management of soil fertility, frequent addition of organic substrates such as charcoal, alfalfa hay and glucose offers a proven and efficient method (Yang et al., 2019). Application of organic amendments generally controls the soil nitrogen cycle, enhances soil water retention, enhances carbon sequestration and reduces the effects of climate change, which in turn affects plant health and productivity.

7.6 Microbe-Based Uses

The synergistic interactions and population-based behavior that the host-associated microbial community exhibits have significant effects on human food security, biodiversity and agricultural output. By resolving pressing environmental problems, these interactions affect the host's physiology. As a result, it makes for a desirable target for *in situ* microbial engineering. For

TABLE 7.2

Beneficial Microbes Helping in Plant Growth-Promoting Traits (Barea, 2015; Liu et al., 2018; Kumawat et al., 2022)

Plant	Beneficial Microbe	Plant Growth-Promoting Trait
Bioinoculants/biofertilizer		
Lens culinaris var. *PL-*	Chryseobacterium sp. PSR10	Phosphate solubilization
Sarraceria species	Endophytic diazotrophic bacteria	Nitrogen fixation
Zea mays	Bacterial endophytes (specifically Bacillus and Brevbacillus)	Enhanced plant growth promotion abilities
Stress resilience		
Black pepper	*Bacillus subtilis, Bacillus siamensis, Brevibacillus gelatini, Pseudomonas geniculate, Pseudomonas beteli, Burkholderia ubonensis* and *Burkholderia territorii*	Antagonistic to soil-borne *Fusarium solani*
Cajanus cajan and *Festuca*	*Azotobacter chroococcum, Bacillus megaterium and Pseudomonas fluorescens*	Impaired growth of Listeria monocytogenes
Phaseolus vulgaris	*Aneurinibacillus aneurinilyticus and Paenibacillus sp.*	1-Aminocyclopropane-1-carboxylic acid (ACC) deaminase activity Alleviating salinity stress
Oryza Sativa	*Bacillus pumilus strain JPVS11*	Stress tolerance
Avicennia marina	*Halomonas sp.*	Heavy metal stress

agricultural settings to promote plant growth and development, the integration of reductionist methods by vertical transmission of microbial species is crucial. Several examples of this have been enumerated (Table 7.2). This is frequently related to modifying the genetic, biochemical, physical and metabolic aspects of the plant through navigating the plant system (Wang et al., 2016). There has been a recent increase in interest in expediting the development of microbiome research, which has significant impacts on environmental changes. Synthetic biology has become a useful tool for filling in knowledge gaps regarding the dynamics of microbe–microbe interaction. With an emphasis on particular applications, it is a field for planning, projecting and controlling the behavior of natural microbial populations (Xiang et al., 2019). A top-down strategy for imaginatively generating synthetic microbial consortia has been made possible by recent developments in the next-generation sequencing (NGS) platform, gene-editing technologies, culture-independent analyses, meta-genomics and bioinformatics tools. By adding genes from different photosynthetic bacteria or by building synthetic metabolic pathways *in vivo*, synthetic biologists are honing their skills to transform agricultural production (Liu et al., 2018). Synthetic microbial communities (SynComs), a recent innovation, are based on tailoring plant–microbe interactions by enlisting and modifying microbial consortia with improved colonization, prevalence and stress-resilient features. Such artificially created microbial inoculants combine the genetics and ecology of microbial communities to produce sustainable agricultural results. These microbial-based interventions have a tremendous deal of potential for addressing the global standstill in agricultural output, biomass demand and climate uncertainty (Xu et al., 2019).

7.7 Plant-Based Manipulations

There is a need to explore holistic and sustainable solutions that assure the worldwide conservation of food supplies to handle issues like climatic fluctuations, desertification, salinization, the introduction of new illnesses, and resistance to biotic and abiotic pressures. In this respect, advanced molecular biology techniques, biochemical methods, mutagenesis, selection and breeding of crop types have all contributed to the major dismantling of the plant system against diverse environmental changes (Table 7.3). High-throughput transcription profiling, large-scale identification of a particular protein network, molecular modeling and their correlation with environmental changes are further strategies that can be used to maintain the famed Green Revolution's success and improve agricultural traits and yields (Qiu et al., 2019).

TABLE 7.3

Unraveled the Plant System Against Various Environmental Changes

S. No	Methods	Strategy
1	Synthetic selection	Designing a microbiome comprising of a few culturable microbial strains serving as a streamlined model for unraveling the fundamental principles of plant–microbe interaction
2.	Artificial microbiome selection	Synthetically selecting multispecies assemblages (culturable and nonculturable) with desirable traits for optimizing the functional dynamics of microbial communities
3.	Single-cell genomics, Whole-community metagenomics and metaproteomics	Orchestrating genomic data with transcriptomic and metabolomic data for improving individual microbiome function and network interactions
4.	Metabolic network monitoring	Designing models at the species level for predicting the dynamic shifts in microbe–microbe interaction in response to substrate gradients, metabolic dependence, competition and spatial heterogeneity

The discovery of novel features, trait development, and site-specific genome changes have been made possible across the plant kingdom; thanks to the advent of site-specific nucleases (TALENs), targeted mutagenesis, and genome-editing tools like CRISPR/Cas systems. It has been extensively researched how CRISPR/Cas systems can be used to specifically create crop features (Foo et al., 2017). Using molecular scissors with the ability to introduce site-specific double-stranded breaks is a requirement for focused enhancement of characteristics (DSBs). The DNA repair machinery in the cell is stimulated by these nuclease-induced breaks to repair these double-strand breaks (DSBs) either through nonhomologous end-joining (NHEJ) or homology-directed repair (HDR). Frameshift mutations caused by NHEJ lead to functional knockouts of a gene (Backer et al., 2018). In contrast to NHEJ, HDR accurately modifies/repairs the genome using an oligonucleotide template with homologous ends or arms to the DSB flanking region, introducing particular point mutations or a new gene sequence. Such site-specific programmable strategies are applied to crop enhancement. As a result, using eukaryotic gene variations that have been created will lead to a large increase in plant yield and food security. Also, significant developments in functional genomics will foresee future synergies between genomics and plant breeding methods for enhanced crop genetic resources, the production of new cultivars, and targeted crop improvement (Kumawat et al., 2022). From an agronomic perspective, key gene players of biosynthetic and metabolic pathways can be targeted at the transcriptional, and post-transcriptional levels for the manipulation and better understanding of pathways. Gene clusters like Phenylalanine ammonia-lyase (PALs), Flowering Locus T-Like (FTL), Calcium-dependent protein kinases (CDKs), and other regulatory networks have undergone substantial editing with better performance-tailored plants (Foo et al., 2017). Hence, a new avenue for balancing the intricate arrangements of phytomicrobiome will undoubtedly open up a profound grasp of multiple strategies at the soil, microbe and plant levels. Also, the integrated study will serve as a deciding factor in shaping the inherently complex, heterogeneous and dynamic microbiome assembly. Nevertheless, the technical limitations that exist today prevent these manipulation tactics from being widely adopted and commercialized (Thomas et al., 2019). These include host genotype, lack of *in situ* manipulation tools, standardization protocols, competition by native soil microflora at the root–soil interface, poor microbial colonization, environmental stress conditions, limited persistence, inability of mutual recognition and root exudate composition and architecture (Backer et al., 2018; Lyu et al., 2021). QTL mapping and the incorporation of other meta-omics methods can be used to fill in these information gaps (metagenomics, meta-transcriptomics and metabolomics). The formation and success of healthy microbiomes may be aided by the application of synthetic biology techniques for the identification of novel phages, plasmids and transposable elements for *in situ* microbiome engineering (Lee et al., 2012).

7.8 Challenges and Limitations with PGPR-Based Biofertilizers

Presently, there is an escalating interest in the use of microbial-based products as bioinoculants. Still, their use is associated with several challenges moving from the lab to the field. The preliminary use of these bioinoculants has been made on crop plants such as legumes and cereals. For developing a new PGPR strain as an effective bioinoculant, an initial laboratory screening is required, which depends on specific direct and indirect mechanisms of plant growth promotion by PGPRs (Saad et al., 2020). Mere primary screening of axenic culture isolates for PGPR traits does not guarantee efficacious plant growth promotion under field conditions. Parallelly, those pure culture isolates that exhibit less *in vitro* growth-promoting activities might possess different plant growth promotion strategies (Thomas et al., 2019). These mechanisms are not thoroughly known, making it challenging to screen for such isolates under normal circumstances. As a result, because of their subpar *in vitro* performance, such beneficial strains exhibiting these pathways may get thrown out (Lyu et al., 2021). The widespread use and implementation of PGPRs calls for many crucial concerns to be addressed and some difficulties and limitations to be overcome (Figure 7.5).

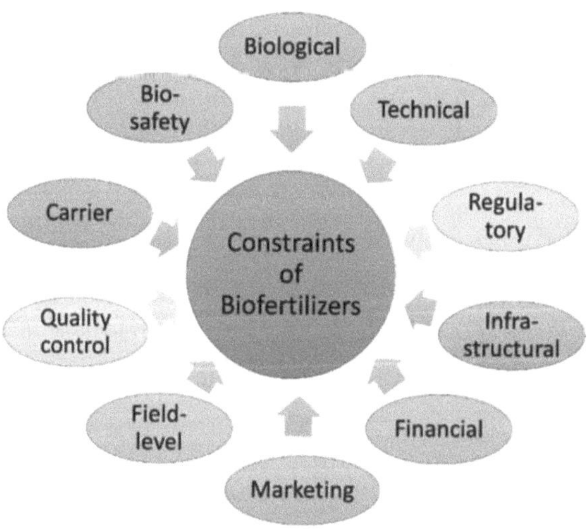

FIGURE 7.5
Limitations in the utilization, production and commercialization of PGPR-based biofertilizers.

7.8.1 Biological Limitations

It can be difficult to choose the right PGPR strain(s) for biofertilizer development. The strain(s) should have a wide range of hosts and not be particularly selective or targeted (to certain crops). Their selectivity is one of the key challenges that limits them. In contrast to PGPRs, conventional agrochemicals have a tendency to affect the entire resident microbiota. Yet, because different microbes are present in the field, the quality and effectiveness of these PGPRs inevitably fluctuates (Barnawal et al., 2017). Possible isolates should be chosen based on how well they function in the field with a variety of crops in various soil types and environmental settings. The strains must successfully replace the natural, ineffective strains and must not compete with other helpful microbial species (Xiang et al., 2019). In order to be effective as biofertilizers, PGPRs need to be able to sufficiently colonize the roots of the host plant, develop a healthy rhizosphere and boost the bioavailability of N, P, K and antagonistic characteristics. PGPR dynamics, which primarily alters with the host crop, the mid- and long-term effects, the impact of crop rotation, and site variation, is a significant element in PGPR colonization (Smith et al., 2015). While not all rhizobacteria have the same methods of action for promoting plant growth, employing PGPRs presents a further barrier due to their various modes of action. It is recognized that a number of Gram-negative rhizobacteria have biocontrol capability. Due to their incapacity to create spores, they are challenging to formulate, which poses a constraint. Moreover, their formulations have a short shelf-life, and desiccation is likely to destroy the bacteria (Malusa et al., 2016).

7.8.2 Technical Constraints

One of the considerable challenges encountered through the improvement of a biofertilizer and the commercialization of an effective PGPR strain is its shelf existence. Biofertilizers with a short shelf existence deliver the danger of recycling if they are not used or offered before expiry resulting in an internet monetary loss to the advertising corporation. Because biofertilizers comprise live microbial cells, their storage and transportation require more care and precaution (Meena et al., 2017). The technical constraints involve the risk of degradation of the product due to shorter shelf lifestyles or spontaneous mutations bobbing up for the duration of fermentation or garage. The mutations result in bioinoculant effectiveness and in extreme trouble that raises the cost of production and quality of the bioinoculant. Inadequate availability of soil-specific strains location-smart extensively limits the vast use of bioinoculants (Liu & Murray, 2016).

7.8.3 Regulatory Limitations

Regulatory limitations encompass the demanding situations in product registration and patent filing. The regulations frequently vary among distinctive areas and nations and are not regular. Further, the regulatory tactics are quite complex, and the fees, although variable, are totally on the higher side. The documentation approaches for product registration are complicated. The absence of a standardized felony and regulatory definition for "plant biostimulants" is the number one motive at the back of the dearth of a globally coordinated uniform regulatory policy (Bona et al., 2019). The system of registering the biocontrol agent within a country is commonly in phases and is quite prolonged and complex. Generally, the active ingredient inside a biofertilizer has to get an authorization certificate from the Directorate of Health and Purchaser Affairs, and after that, the formulated product has to be nationally approved. For that reason, the entire method starting from registration to commercializing a capacity biofertilizer is prolonged and can stretch to several years (Singhal et al., 2016). The international locations have their own norms to reply of their unique language, and the registering employer also can require even extra data.

7.8.4 Infrastructural Limitations

Biofertilizer production and quality assurance require sophisticated technology as well as qualified and skilled labor. The primary infrastructure constraints are a lack of advanced technology, necessary technical support and equipment, trained labor, and skilled technical personnel (Bhat et al., 2020).

7.8.5 Marketing Limitations

One of the major limitations for developing the product in the market is the unavailability of proper transportation services along with storage facilities. Farmers possess little or inadequate knowledge regarding the advantages of biofertilizers over hazardous agrochemicals for sustainable agriculture. Thus, the demand for such eco-friendly products is reduced. The establishment of extension centers does not help in creating awareness among farmers due to the lack of well-qualified technical staff. The biofertilizer developers face a significant problem because the agricultural crops are grown under various physicochemical and environmental conditions, including diverse ranges of temperature, rainfall, soil type and crop variety. These conditions tend to change from farm to farm or even within a single field (Liu & Murray 2016; Basu et al., 2022). The lack of appropriate transportation and storage services is one of the main obstacles for developing a product for the market. Farmers are not well-informed about the benefits of using biofertilizers for sustainable agriculture as opposed to dangerous agrochemicals. As a result, there is less of a demand for such eco-friendly goods. Due to a lack of technically skilled staff, the establishment of extension centers does not aid in raising awareness among farmers. The fact that agricultural crops are grown in a variety of physicochemical and environmental conditions, including different ranges of temperature, rainfall, soil type and crop variety, presents a significant challenge for the developers of biofertilizers. These circumstances frequently vary from farm to farm, sometimes even within a single field. The effectiveness of PGPR-based biofertilizers is therefore inconsistent as a result of these variances (Li et al., 2018; Rana et al., 2020). Before any microbial goods get to the point of commercialization, a broad plan is followed in every state within a nation. The agriculture ministry or department gives the all-clear to place orders mostly from their own production facilities. Biofertilizer packets are shipped from here to a number of districts. Before these packets are delivered to the field, a chain of extension workers is engaged. The microorganisms used as bioinoculants are exposed to high temperatures (over 40°C) during this course, which could cause them to become inactive or die, making them low- or poor-quality biofertilizers. These packets will henceforth be detrimental to the farmers as well as the crop production (Xiang et al., 2019).

7.8.6 Field-Level Limitations

Since it takes time for the inoculum to build up its concentration and begin root colonization, the response of crops to applied biofertilizers is very gradual and occasionally ineffective. As a result, farmers only accept biofertilizers to a limited extent. In field application, the purity of the inoculants and the inoculation methods are extremely important. Because of the negative aftereffects of synthetic chemicals and the current unfavorable abiotic circumstances, biofertilizers lose some of their efficiency (Foo et al., 2017). Another significant factor in the decline of biological activity is environmental stress, such as salt and drought in some regions. There are biotic and abiotic stressors on the inoculants. The soil's acidity and alkalinity, the use of pesticides, and excessive concentrations of nitrate in the soil, which restrict the ability of the bioinoculants to fix nitrogen, are other factors that collectively lead to poor performance of the bioinoculants. Many soils have dangerous levels of heavy metals like Cd, Hg and Cr as well as a lack of other crucial minerals like P, Cu, Mo and Co, which lower the biological potential of fertilizers based on PGPR (Arif et al., 2020).

7.8.7 Quality-Control Limitations

Quality control is the most crucial feature that farmers look for in a biofertilizer. Living microorganisms have an extremely limited shelf-life because they are natural products. Any microbial-based product's failure in the field could be attributed to the availability of low-quality goods. There are no quality checks for biofertilizers currently available (Chen et al., 2018; Rana et al., 2020). Setting up quality-control criteria for biofertilizers is therefore very important to demonstrate the effectiveness of these fertilizers in boosting plant development in the fields.

7.9 Biofertilizer Carrier

The limited shelf-life of the bioinoculant agent necessitates the use of a suitable carrier for the biofertilizer when applied in the field. Thus, one of the main obstacles to its widespread usage in fields is the lack of a suitable carrier. Peat, charcoal, lignite and other suitable carriers are utilized in the manufacturing of biofertilizers. Due to the fact that the majority of these carriers are not available in developing nations like India, they once again present technical challenges. These carriers are insufficiently plentiful and have a desirable property. Only charcoal can be utilized as a forming agent because it is easily accessible on the Indian market (Thomas et al., 2019). Although it is acknowledged that among the possible carriers, peat is the most suitable, the problem is that it has a shelf-life of less than 6 months. Biochar can be employed as a good carrier for biofertilizers due to its capacity to enhance soil and plant health. The bioinoculant should have a number of other qualities to demonstrate its effectiveness as a carrier. It should be inexpensive, have a high concentration of organic matter and water-holding ability, and have a longer period of organism retention (Lyu et al., 2021). For a biofertilizer to be of high quality, it should be nearly sterile, have no moisture, be toxicn-free, non-polluting and have a pH that is nearly neutral.

7.10 PGPR's Role Against Abiotic Stresses

PGPRs work by means of a number of mechanisms. The first step in promoting plant growth is the microbe's colonization of the plant roots. This complex process necessitates the capacity of bacteria to compete in the rhizosphere soil for a suitable niche in order to promote a beneficial plant–microbe interaction. Abiotic factors, such as soil type, temperature, pH, radiation, oxygen concentration, availability of nutrients and the degree of interaction with the native soil microbiota, also have a significant impact on the plant–microbe interaction, which in turn affects the existence and survivability of the microbes within the host plant (Hart et al., 2018; Arif et al., 2020). Consequently, the climatic conditions necessary for a specific type of cultivated crop determine whether the field application of PGPRs will be

TABLE 7.4
Different Mechanisms Employed by PGPRs to Overcome Abiotic Stress and Enhance the Growth of Plants

Abiotic Stress	
Salinity, Drought, Nutrition, Heavy Metal Toxicity	
PGPR	ACC deaminase VOC synthesis Phytohormones Osmoprotectectants Priming Siderophores Exopolysacchrides Biostimulants

successful. PGPRs are frequently applied directly to host plants as an inoculum without first being mixed with a suitable carrier. However, due to competition with the pre-existing soil micro- and macro-biota, their amounts are insufficient to support effective rhizosphere colonization in a field. Typically, soils linked to high-value crops are fumigated using broad-spectrum biocidal fumigants. The microbial community of these soils is altered as a result of these fumigants. Long-term fumigation has a significant negative impact on the soil microbial community and the helpful interactions that assist host plants mobilize and acquire nutrients (Woo et al., 2014; Lyu et al., 2021). As a result, the PGPR inoculant's ability to colonize the rhizosphere is reduced. In order to combat stressful situations and encourage plant growth and development, PGPR uses a variety of techniques (Table 7.4).

In response to abiotic stresses, PGPR uses a variety of mechanisms to assist plants in surviving environmental changes. These mechanisms include (i) the production of ACC deaminase, which lowers ET levels in plants; (ii) the secretion of osmolytes (proline, choline and trehalose), which serve as osmoprotectants; (iii) the secretion of bacterial volatiles to induce stress tolerance. In addition to these well-known pathways, PGPR can also cause physiological occurrences or processes that are controlled by a complicated web of signaling activities, leading to increased stress tolerance and priming (Qiu et al., 2019; Saad et al., 2020). As an appealing technique to control plant stress, the use of PGPR to produce abiotic stress tolerance in plants is being investigated more and more. Many processes by which these organisms induce stress tolerance have been elucidated (Figure 7.6). In addition to the previously described plant growth methods, PGPR can also help plants tolerate abiotic stress by altering phytohormonal activity, preserving iron homeostasis and restoring osmotic equilibrium (Yang et al., 2019).

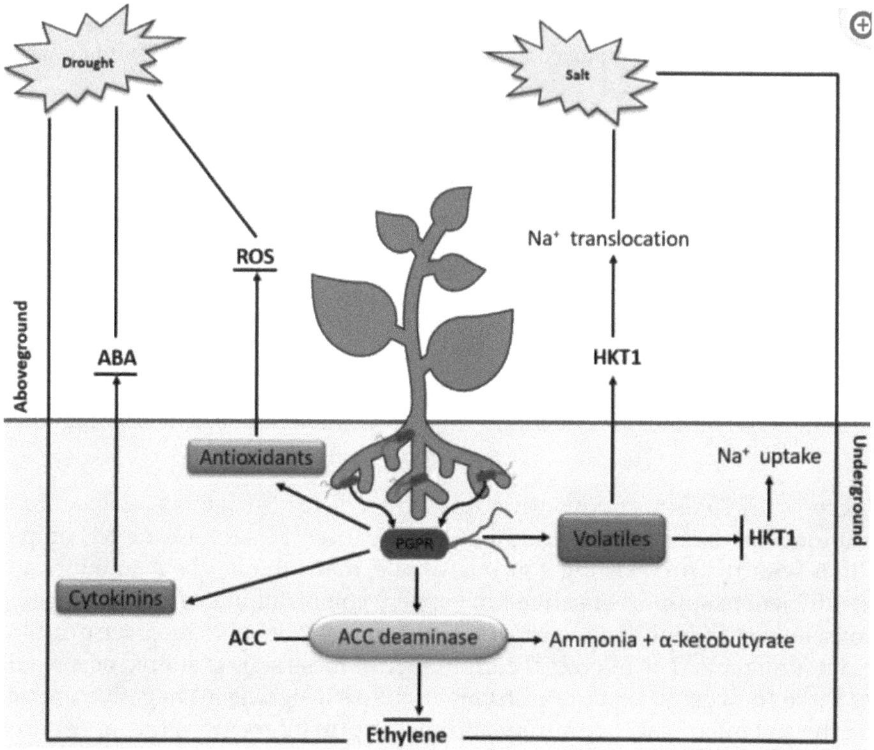

FIGURE 7.6
Outline of the systemic tolerance established by PGPR against the effects of salinity and drought. Arrows denote substances released by the PGPR during stress, such as volatiles, cytokinin, antioxidants and ACC deaminase. ROS destruction is brought on by cytokinins, whereas ABA buildup is caused by antioxidants. The ACC deaminase breaks down ACC and prevents the synthesis of ET. At high salt circumstances, the volatiles released by the PGPR cause HKT1 expression to be downregulated in the roots but upregulated in the shoot tissue. This causes Na^+ to be circulated throughout the entire plant (Pandey et al., 2017; Bhat et al., 2020).

7.10.1 Improved Abiotic Stress Tolerance via PGPR

By PGPR-induced priming of the plant immunity and defenses, a process known as PGPR-plant interactions enhances resistance to abiotic stressors. When compared to unprimed plants, the plant responds to stress in this state more quickly and/or robustly, improving its stress tolerance (Figure 7.4). This state of readiness, known as the "primed state," has been associated with the effective activation of defensive responses, which leads to improved stress resistance (Chen et al., 2018). Many processes, including systemic acquired resistance (SAR) and induced systemic resistance, can be used to describe the increased resistance (ISR). The SAR defense response system, which provides resistance to repeated stress encounters, is activated in distal areas of the plant in response to localized infection, preparing the plant to fight itself in the event of an attack. On the other hand, ISR is mediated by microbes that affect plant growth, such as PGPR. These microbes cause resistance in the plant by colonizing the root system (Kamilov et al., 2015; Canarini et al., 2019). Whereas ISR uses the ET and JA response pathways to produce resistance, SAR is SA-reliant. It has been extensively reported that SAR and ISR have become important forms of priming, however, it is still unclear exactly how these mechanisms work. The accumulation of inactive proteins involved in signal amplification, such as MAPKs, the activation of transcription factors that enhance the transcription of defense-related genes in response to perceived stress, and epigenetic changes involving DNA modifications, histone modifications or chromatin alterations are just a few of the many hypotheses that have been put forth despite the fact that the mechanisms of priming are not fully understood. Since it improves defense responses without compromising a plant's overall fitness and because the resulting stress tolerance or resistance cannot be reversed by microorganisms, plant priming has been viewed as a viable technique for the regulation of stress. This provides long-term resistance (Lee et al., 2012).

7.11 Biosafety of PGPRs

It is thought that PGPRs are good candidates for sustainable agriculture. The fact that these bacteria should not have any negative impacts on the environment or people is a crucial aspect of PGPRs and other biofertilizer agents. Biosafety levels were developed to classify the usable microorganisms in a range of biosafety classes, based on the various categories of risk posed by them. According to the World Health Organization guidelines on the usage of microorganisms and the guidelines on biosafety in microbiological and biomedical laboratories, both of which were published by the U.S. Department of Health and Human Services in 1999. Based on the infectious agents' pathogenicity to human health, route of transmission and known therapies, they were divided into four risk groups (BSL-1-4). These levels have to be closely followed while managing these bacteria. The microbial strains selected for biofertilizer creation should preferably belong to the low-risk group of non-pathogenic BSL-1 microorganisms (Zhang et al., 2021).

7.12 Potential Marker Genes for PGPR Monitoring at Field Level

Genes that code for the enzymes needed to produce antibiotics, hydrogen cyanide, destroy ethylene and promote auxin are potential markers for PGPR activities. Several strains of pseudomonas that produce 2,4-DAPG contain a large number of HCN genes. Using a global collection of isolates, it has been demonstrated that PCR primers targeting a conserved sequence in the *hcnAB* genes are specifically sensitive to the presence of HCN-producing pseudomonas. Similar primers can be used to stop the synthesis of 2,4-DAPG. These primers have typically been used to test soil-cultivated isolates, but they can also be utilized with soil with the caution that gene products may be generated (Lee et al., 2012; Chen et al., 2018). To be sure that the primers have not amplified any other genes, the PCR results should be subjected to further analysis, including DNA melting, size checking (using a 570-bp fragment) and sequencing. These primers have the benefit that the forward primer can be used in conjunction with a GC clamp for PCR product analysis by DGGE to evaluate the genotypic diversity of local DAPG-producing *Pseudomonas* isolates (Canarini et al., 2019). Aminocyclopropane carboxylic acid deaminase is encoded by the *accA* gene, which is another target gene for quantification. The *accA* gene is found in many Gram-positive bacteria, rhizobia and fungi, as well as a large variety of Gram-negative bacteria, including PGPR pseudomonas. Furthermore, care must be taken when interpreting changes in *accA* gene copy counts because some bacteria have highly similar sequences that instead code for enzymes with different functions, including serine deamination (Bai et al., 2015; Stanley & Vander Heijden, 2017).

7.13 Guidelines and Precautions for Using PGPRs as Biofertilizers

The main precautions and recommendations necessary for employing PGPRs as biofertilizers are as follows:(1) it is crucial that the given biofertilizer, which will be applied to fields, is of high quality, comes from a reputable producer, and contains 107 live cells per gramme as an inoculum, (2) the biofertilizer should only be used for the crops listed on the product's commercially accessible packaging since it shows specificity, (3) the name of the crop for which the culture bag is intended should be written on the bag, (4) excess culture should be inoculated during inoculation, and any leftover or residual culture should be placed right away in the field's grooves so that the inoculum bacteria can contact with other rhizosphere microbes and start colonizing them, (5) because biofertilizers are microbial products, they should be stored in cool, shaded areas, preferably at room temperature (25–28°C), before being applied to fields to ensure a longer shelf-life, (6) direct contact between the biofertilizers and agrochemicals (herbicides, weedicides and insecticides) should be rigorously avoided during storage or application, (7) in general, 10 kg of seeds can be successfully treated with 200 g of biofertilizer, (8) soil additions like lime or rock phosphate are typically recommended in unfavorable soil conditions, especially if the soil is extremely acidic.

7.14 Roadmap for the Commercialization of Biofertilizers Based on PGPR

By providing environmentally benign substitutes for synthetic agrochemicals like chemical fertilizers and pesticides, the use of PGPRs as biofertilizers for enhancing plant growth and crop output, enhancing soil fertility and biocontrolling phytopathogens promote sustainable agriculture (Sheth et al., 2016; Archana et al., 2020). The following pathway typically guides the development and commercialization of PGPR-based biofertilizers (Figure 7.7).

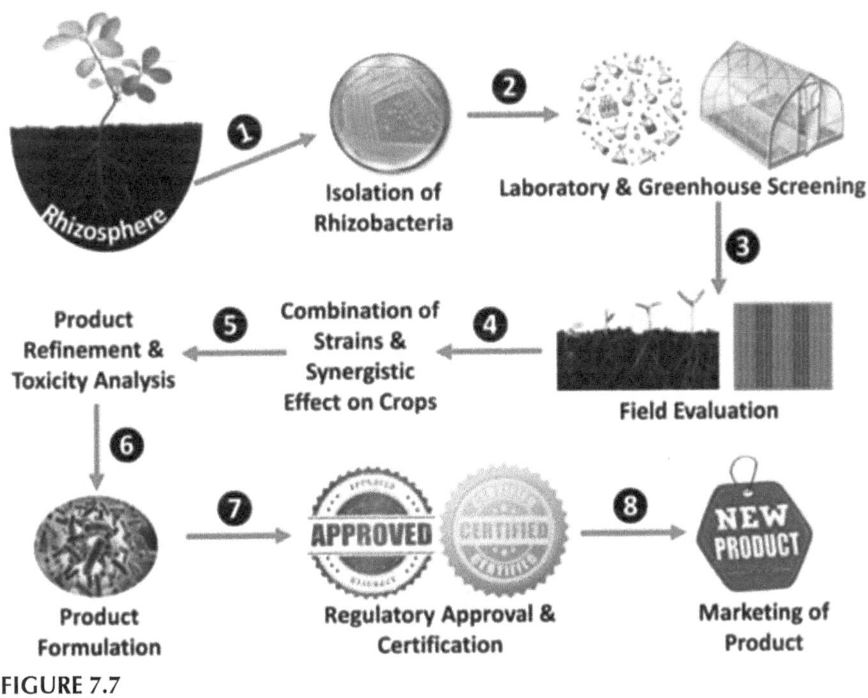

FIGURE 7.7
A plan for commercializing biofertilizers based on PGPR.

7.15 Conclusions and Future Perspectives

Worldwide demand for both food and medically vital crops has increased in recent years as a result of climate change and the population boom. Many techniques, including crop breeding, genetic engineering and greater fertilizer use, have been used to meet these demands. Each of these methods has linked benefits and drawbacks. Nevertheless, the main issues with these methods are that they need a lot of time, and increasing fertilizer use has been demonstrated to have a detrimental effect on soil health. Hence, microorganisms like PGPRs are used to promote sustainable agricultural practices since they do more than only boost soil fertility. Similar findings have been confirmed for medicinal plants, which are protected from abiotic stressors by PGPRs via a variety of metabolic processes. Also, the utilization of PGPRs in the growth of medicinally significant plants revealed enhanced production of metabolites that have been employed in conventional medical procedures. In addition to helping plants grow and develop, PGPRs also directly lessen the impact of plant diseases by thwarting the establishment of hostile pathogens inside plants. They act antagonistically by producing hydrolytic enzymes that break down fungal cells, such as protease, chitinase, lipases and glucanases. They also create antibiotics and siderophores and can control the level of ethylene under pathogenic stress by using the ACC deaminase enzyme. They also create phytohormones, which control the growth of the roots and shoots and facilitate nutrient intake. Moreover, they function as biostimulants, which include phytostimulators, biopesticides and biofertilizers. Therefore, it can be inferred that PGPRs through a variety of processes, reduce the abiotic stress conditions and create an environment that is conducive to plant growth. In addition, it was shown that medicinally significant plants not only reduced stress but also increased the metabolite levels of economically significant substances. Due to their dual advantages, they are a good substitute to encourage sustainable agricultural methods for growing medicinal plants in challenging environments and enhancing soil health. This chapter details potential mechanisms that various PGPRs may employ to optimize the growth of various medicinal plants. Yet, there is still a knowledge gap about the precise biochemical pathways and mechanisms that the PGPRs control to produce these effects. Future research should, in general, concentrate on PGPR-based metabolite engineering, investigation of advantageous PGPR strains, and identification of target genes to facilitate plant growth under stress conditions.

References

Archana, T., Rajendran, L., Manoranjitham, S. K., Santhana Krishnan, V. P., Paramasivan, M., and Karthikeyan, G. Culture-dependent analysis of seed bacterial endophyte, Pseudomonas spp. EGN 1 against the stem rot disease (Sclerotium rolfsii Sacc.) in groundnut. *Egypt. J. Biol. Pest Cont.* **2020**. 30, 1–13.

Arif, I., Batool, M., and Schenk, P. M. Plant microbiome engineering: expected benefits for improved crop growth and resilience. *Trends Biotechnol.* **2020**. 38, 1385–1396. doi: 10.1016/j.tibtech.2020.04.015

Backer, R., Rokem, J. S., Ilangumaran, G., Lamont, J., Praslickova, D., Ricci, E., Subramanian, S., and Smith, D. L. Plant growth-promoting rhizobacteria: Context, mechanisms of action, and roadmap to commercialization of biostimulants for sustainable agriculture. *Front. Plant Sci.* **2018**. 9, 1473.

Bai, Y., Müller, D. B., Srinivas, G., Garrido-Oter, R., Potthoff, E., Rott, M., et al. Functional overlap of the *Arabidopsis* leaf and root microbiota. *Nature* **2015**. 528, 364–369. doi: 10.1038/nature16192

Barea, J. M. Future challenges and perspectives for applying microbial biotechnology in sustainable agriculture based on a better understanding of plant-microbiome interactions. *J. Soil Sci. Plant Nutr.* **2015**. 15, 261–282.

Barea, J. M., Pozo, M. J., Azcón, R., and Azcón-Aguilar, C. Microbial interactions in the rhizosphere. In: F. J. de Bruijn (ed). *Molecular Microbial Ecology of the Rhizosphere*, vol 1. **2013**. Wiley Blackwell, Hoboken, New Jersey, USA, pp: 29–44.

Barnawal, D., Bharti, N., Pandey, S. S., Pandey, A., Chanotiya, C. S., and Kalra, A. Plant growth-promoting rhizobacteria enhance wheat salt and drought stress tolerance by altering endogenous phytohormone levels and TaCTR1/TaDREB2 expression. *Physiol. Plant.* **2017**. 161, 502–514. doi: 10.1111/ppl.12614

Barret, M., Tan, H., Egan, F., Morrissey, J. P., Reen, J., and O'Gara, F. Exploiting new systems-based strategies to elucidate plant–bacterial interactions in the rhizosphere. In: F. J. de Bruijn (ed). *Molecular Microbial Ecology of the Rhizosphere*, vol 1. **2013**. Wiley Blackwell, Hoboken, New Jersey, USA, pp: 57–68.

Bashan, Y., de-Bashan, L. E., Prabhu, S. R., and Hernandez, J. P. Advances in plant growth-promoting bacterial inoculant technology: Formulations and practical perspectives (1998–2013). *Plant Soil* **2014**. 378, 1–33.

Basu, S., Priyadarshini, P., Prasad, R., and Kumar, G. Effects of microbial signaling in plant growth and development. In: R. Prasad and S-H. Zhang (eds), *Beneficial Microorganisms in Agriculture*. **2022**. Springer, Singapore, pp: 329–348. doi: 10.1007/978-981-19-0733-3_14

Behm, J. E., Geurts, R., and Kiers, E. T. Parasponia: A novel system for studying mutualism stability. *Trends Plant Sci.* **2014**. 19, 757–763. doi: 10.1016/j.tplants.2014.08.007

Berendsen, R. L., Vismans, G., Yu, K., Song, Y., de Jonge, R., Burgman, W. P., et al. Disease-induced assemblage of a plant-beneficial bacterial consortium. *ISME J.* **2018**. 12, 1496–1507. doi: 10.1038/s41396-018-0093-1

Bhatt, P., Verma, A., Verma, S., Anwar, M., Prasher, P., Mudila, H., et al. Understanding phytomicrobiome: a potential reservoir for better crop management. *Sustainability*. **2020**. 12, 5446. doi: 10.3390/su12135446

Bona, E., Massa, N., Novello, G., Boatti, L., Cesaro, P., Todeschini, V., et al. Metaproteomic characterization of the *Vitis vinifera* rhizosphere. *FEMS Microbiol. Ecol.* **2019**. 95 , fiy204. doi: 10.1093/femsec/fiy204

Canarini, A., Kaiser, C., Merchant, A., Richter, A., and Wanek, W. Root exudation of primary metabolites: mechanisms and their roles in plant responses to environmental stimuli. *Front. Plant Sci.* **2019**. 10, 157. doi: 10.3389/fpls.2019.00157

Carrion, V. J., Perez-Jaramillo, J., Cordovez, V., Tracanna, V., De Hollander, M., Ruiz-Buck, D., et al. Pathogen-induced activation of disease-suppressive functions in the endophytic root microbiome. *Science.* **2019**. 366, 606–612. doi: 10.1126/science.aaw9285

Carvalhais, L. C., Dennis, P. G., Tyson, G. W., and Schenk, P. M. Rhizosphere metatranscriptomics: challenges and opportunities. In: F. J. de Bruijn (ed). *Molecular Microbial Ecology of the Rhizosphere.* vol 2. **2013**. Wiley Blackwell, Hoboken, New Jersey, USA, pp: 1137–1144.

Chen, Y., Wang, J., Yang, N., Wen, Z., Sun, X., Chai, Y., et al. Wheat microbiome bacteria can reduce virulence of a plant pathogenic fungus by altering histone acetylation. *Nat. Commun.* **2018**. 9, 3429. doi: 10.1038/s41467-018-05683-7

Chouhan, G. K., Verma, J. P., Jaiswal, D. K., Mukherjee, A., Singh, S., de Araujo Pereira, A. P., Liu, H., Abd Allah, E. F., and Singh, B. K. Phytomicrobiome for promoting sustainable agriculture and food security: Opportunities, challenges, and solutions. *Microbiol Res.* **2021** Jul. 248, 126763.

de Bruijn, F. J. Biological nitrogen fixation. In: B. Lugtenberg (ed). *Principles of Plant-Microbe Interactions.* **2015**. Springer International Publishing Switzerland, Heidelberg, pp: 215–224.

Du Jardin, P. Plant biostimulants: Definition, concept, main categories and regulation. *Sci. Hortic.* **2015**. 196, 3–14.

Foo, J. L., Ling, H., Lee, Y. S., and Chang, M. W. Microbiome engineering: current applications and its future. *Biotechnol J.* **2017**. 12, 1600099.

Glick, B. R. Plant growth-promoting bacteria: mechanisms and applications. *Scientifica* **2012**. 963401. doi: 10.6064/2012/963401

Hart, M. M., Antunes, P. M., Chaudhary, V. B., and Abbott, L. K. (2018). Fungal inoculants in the field: is the reward greater than the risk? *Funct. Ecol.* 32, 126–135. doi: 10.1111/1365-2435.12976

Hassan, S., and Mathesius, U. The role of flavonoids in root–rhizosphere signalling: opportunities and challenges for improving plant–microbe interactions. *J. Exp. Bot.* **2012**. 63, 3429–3444. doi: 10.1093/jxb/err430

Ismail, M. A., Amin, M. A., Eid, A. M., Hassan, S. E. D., Mahgoub, H. A., Lashin, I., et al. Comparative study between exogenously applied plant growth hormones versus metabolites of microbial endophytes as plant growth-promoting for *Phaseolus vulgaris* L. *Cells* **2021**. 10, 1059. doi: 10.3390/cells10051059

Jat, S. L., Suby, S. B., Parihar, C. M., Gambhir, G., Kumar, N., and Rakshit, S. Microbiome for sustainable agriculture: a review with special reference to the corn production system. *Arch Microbiol.* **2021** Aug. 203(6), 2771–2793. doi: 10.1007/s00203-021-02320-8. Epub 2021 Apr 21.

Kamilova, F., Okon, Y., Deweert, S., and Horal, K. Commercialization of microbes: Manufacturing, inoculation, best practice for objective field testing, and registration. In: B. Lugtenberg (ed). *Principles of Plant-Microbe Interactions: Microbes for Sustainable Agriculture.* **2015**. Springer, Cham, Switzerland, pp. 319–327.

Khan, M. A., Asaf, S., Khan, A. L., Adhikari, A., Jan, R., Ali, S., et al. Plant growth-promoting endophytic bacteria augment growth and salinity tolerance in rice plants. *Plant Biol.* **2020**. 22, 850–862. doi: 10.1111/plb.13124

Kloepper, J. W., Leong, J., Teintze, M., and Schroth, M. N. Enhanced plant growth by siderophores produced by plant growth promoting rhizobacteria. *Nature* **1980**. 286, 885–886.

Kumar, A., Singh, S., Mukherjee, A., Rastogi, R. P., and Verma, J. P. Salt-tolerant plant growth-promoting *Bacillus pumilus* strain JPVS11 to enhance plant growth attributes of rice and improve soil health under salinity stress. *Microbiol. Res.* **2021**. 242, 126616. doi: 10.1016/j.micres.2020.126616

Kumar, M., Kour, D., Yadav, A. N., Saxena, R., Rai, P. K., Jyoti, A., et al. Biodiversity of methylotrophic microbial communities and their potential role in mitigation of abiotic stresses in plants. *Biologia* **2019**. 74, 287–308. doi: 10.2478/s11756-019-00190-6

Kumawat, K. C., Razdan, N., and Saharan, K. Rhizospheric microbiome: Bio-based emerging strategies for sustainable agriculture development and future perspectives. *Microbiol Res.* **2022** Jan. 254, 126901. doi: 10.1016/j.micres.2021.126901. Epub 2021 Oct 23.

Lee, B., Lee, S., and Ryu, C. M. Foliar aphid feeding recruits rhizosphere bacteria and primes plant immunity against pathogenic and non-pathogenic bacteria in pepper. *Annals Bot.* **2012**. 110, 281–290. doi: 10.1093/aob/mcs055

Levy, A., Gonzalez, I. S., Mittelviefhaus, M., Clingenpeel, S., Paredes, S. H., Miao, J., et al. Genomic features of bacterial adaptation to plants. *Nat. Genet.* **2018**. 50, 138–150. doi: 10.1038/s41588-017-0012-9

Li, G., Zhao, H., Liu, Z., Wang, H., Xu, B., and Guo, X. The wisdom of honeybee defenses against environmental stresses. *Front. Microbiol.* **2018**. 9, 722. doi: 10.3389/fmicb.2018.00722

Liu, C. W., and Murray, J. D. The role of flavonoids in nodulation host-range specificity: an update. *Plants* **2016**. 5, 33. doi: 10.3390/plants5030033

Liu, K., McInroy, J. A., Hu, C. H., and Kloepper, J. W. Mixtures of plant-growth-promoting rhizobacteria enhance biological control of multiple plant diseases and plant-growth promotion in the presence of pathogens. *Plant Dis.* **2018**. 102, 67–72. doi: 10.1094/PDIS-04-17-0478-RE

Lyu, D., Backer, R., Subramanian, S., and Smith, D. L. Phytomicrobiome coordination signals hold potential for climate change-resilient agriculture. *Front. Plant Sci.* **2020**. 11, 634. doi: 10.3389/fpls.2020.00634

Lyu, D., Msimbira, L., Nazari, M., Antar, M., Pagé, A., Shah, A., Monjezi, N., Zajonc, J., Tanney, C., and Backer, R. The coevolution of plants and microbes underpins sustainable agriculture. *Microorganisms* **2021**. 9, 1036.

Majeed, A., Muhammad, Z., Islam, S., Ullah, Z., and Ullah, R. Cyanobacterial application as bio-fertilizers in rice fields: role in growth promotion and crop productivity. *PSM Microbiol.* **2017**. 2, 47–50.

Malusà, E., Pinzari, F., and Canfora, L. Efficacy of biofertilizers: Challenges to improve crop production. In: D. P. Singh, H. B. Singh, and R. Prabha, Eds.*Microbial Inoculants in Sustainable Agricultural Productivity*. **2016**. Springer, New Delhi, India, pp: 17–40.

Meena, K. K., Sorty, A. M., Bitla, U. M., Choudhary, K., Gupta, P., Pareek, A., et al. Abiotic stress responses and microbe-mediated mitigation in plants: the omics strategies. *Front. Plant Sci.* **2017**. 8, 172. doi: 10.3389/fpls.2017.00172

Mercado-Blanco, J. Life of microbes inside the plant. In: B. Lugtenberg (ed). *Principles of Plant-Microbe Interactions*. **2015**. Springer International Publishing Switzerland, Heidelberg, pp: 25–32.

Olivares, J., Bedmar, E. J., and Sanjuan, J. Biological nitrogen fixation in the context of global change. *Mol. Plant-Microbe Interact*. **2013**. 26, 486–494.

Pandey, P., Irulappan, V., Bagavathiannan, M. V., and Senthil-Kumar, M. Impact of combined abiotic and biotic stresses on plant growth and avenues for crop improvement by exploiting physio-morphological traits. *Front. Plant Sci.* **2017**. 8, 537. doi: 10.3389/fpls.2017.00537

Qiu, Z., Egidi, E., Liu, H., Kaur, S., and Singh, B. K. New frontiers in agriculture productivity: Optimised microbial inoculants and in situ microbiome engineering. *Biotechnol Adv*. **2019** Nov 1. 37(6), 107371.

Rana, K. L., Kour, D., Kaur, T., Devi, R., Yadav, A. N., Yadav, N., Dhaliwal, H. S., and Saxena, A. K. Endophytic microbes: biodiversity, plant growth-promoting mechanisms and potential applications for agricultural sustainability. *Antonie Van Leeuwenhoek*. **2020** Aug. 113(8), 1075–1107.

Saad, M. M., Eida, A. A., and Hirt, H. Tailoring plant-associated microbial inoculants in agriculture: a roadmap for successful application. *J Exp Bot*. **2020** Jun 26. 71(13), 3878–3901.

Schreiter, S., Eltlbany, N., and Smalla, K. Microbial communities in the rhizosphere analyzed by cultivation-independent DNA-based methods. In: B. Lugtenberg (ed). *Principles of Plant-Microbe Interactions*. **2015**. Springer International Publishing Switzerland, Heidelberg, pp: 289–298.

Shameer, S., and Prasad, T. N. V. K. V. Plant growth promoting rhizobacteria for sustainable agricultural practices with special reference to biotic and abiotic stresses. *Plant Growth Regul.* **2018.** 84, 603–615. doi: 10.1007/s10725-017-0365-1

Sheth, R. U., Cabral, V., Chen, S. P., and Wang, H. H. Manipulating bacterial communities by in situ microbiome engineering. *Trends Genet*. **2016**. 32, 189–200. doi: 10.1016/j.tig.2016.01.005

Singhal, P., Jan, A. T., Azam, M., and Haq, Q. M. R. Plant abiotic stress: a prospective strategy of exploiting promoters as alternative to overcome the escalating burden. *Front. Life Sci.* **2016**. 9, 52–63. doi: 10.1080/21553769.2015.1077478

Smith, D. L., Praslickova, D., and Ilangumaran, G. Inter-organismal signaling and management of the phytomicrobiome. *Front. Plant Sci.* **2015**. 6, 722. doi: 10.3389/fpls.2015.00722

Spence, C., and Bais, H. Probiotics for plants: rhizospheric microbiome and plant fitness. In: F. J. de Bruijn (ed). *Molecular Microbial Ecology of the Rhizosphere*. vol 2. **2013**. Wiley Blackwell, Hoboken, New Jersey, USA, pp: 713–721.

Stanley, C. E., and van der Heijden, M. G. Microbiome-on-a-chip: new frontiers in plant–microbiota research. *Trends Microbiol*. **2017**. 25, 610–613. doi: 10.1016/j.tim.2017.05.001

Thakur, N., Nigam, M., Mann, N. A., Gupta, S., Hussain, C. M., Shukla, S. K., Shah, A. A., Casini, R., Elansary, H. O., and Khan, S. A. Host-mediated gene engineering and microbiome-based technology optimization for sustainable agriculture and environment. *Funct Integr Genomics*. **2023** Feb 8. 23(1), 57. doi: 10.1007/s10142-023-00982-9.

Thomas, L., and Singh, I. Microbial biofertilizers: Types and applications. In: B. Giri, R. Prasad, Q. S. Wu, and A. Varma (Eds). *Biofertilizers for Sustainable Agriculture and Environment*. **2019**. Springer, Cham, Switzerland, pp. 1–19.

Wang, C., Liu, D., and Bai, E. Decreasing soil microbial diversity is associated with decreasing microbial biomass under nitrogen addition. *Soil Biol. Biochem.* **2018**. 120, 120–133. doi: 10.1016/j.soilbio.2018.02.003

Wang, Q., Dodd, I. C., Belimov, A. A., and Jiang, F. Rhizosphere bacteria containing 1-aminocyclopropane-1-carboxylate deaminase increase growth and photosynthesis of pea plants under salt stress by limiting Na+ accumulation. *Funct. Plant Biol.* **2016**. 43, 161–172. doi: 10.1071/FP15200

Woo, S. L., Ruocco, M., Vinale, F., Nigro, M., Marra, R., Lombardi, N., et al. Trichoderma-based products and their widespread use in agriculture. *Open Mycol. J.* **2014**. 8, 71–126. doi: 10.2174/1874437001408010071

Wu, D., Zhao, M., Shen, S., Fu, Y., Sasaki, T., Yamamoto, Y., et al. Al-induced secretion of organic acid, gene expression and root elongation in soybean roots. *Acta Physiol. Plant.* **2013**. 35, 223–232. doi: 10.1007/s11738-012-1067-y

Xiang, G., Ma, W., Gao, S., Jin, Z., Yue, Q., and Yao, Y. Transcriptomic and phosphoproteomic profiling and metabolite analyses reveal the mechanism of NaHCO3-induced organic acid secretion in grapevine roots. *BMC Plant Biol.* **2019**. 19, 383. doi: 10.1186/s12870-019-1990-9

Xu, K., Lee, Y. S., Li, J., and Li, C. Resistance mechanisms and reprogramming of microorganisms for efficient biorefinery under multiple environmental stresses. *Synthetic Systems Biotechnol.* **2019**. 4, 92–98. doi: 10.1016/j.synbio.2019.02.003

Yang, J. L., Fan, W., and Zheng, S. J. Mechanisms and regulation of aluminum-induced secretion of organic acid anions from plant roots. *J. Zhejiang University-Science B* **2019**. 20, 513–527. doi: 10.1631/jzus. B1900188

Zhang, J., Cook, J., Nearing, J. T., Zhang, J., Raudonis, R., Glick, B. R., Langille, M. G. I., and Cheng, Z. Harnessing the plant microbiome to promote the growth of agricultural crops. *Microbiol Res.* **2021** Apr. 245, 126690.

Zhang, R., Vivanco, J. M., and Shen, Q. The unseen rhizosphere root–soil–microbe interactions for crop production. *Curr. Opi. Microbiol.* **2017**. 37, 8–14. doi: 10.1016/j.mib.2017.03.008

8

Plant Growth-Promoting Rhizobacteria for Nutrient Mobilization and Enhancing Yield and Quality of Vegetable Crops

Sherly J., Raja Shankar and P. Sivasakthivelan

8.1 Introduction

Vegetables are the most important component of a balanced diet and act as a protective food. A high vegetable diet has been associated with a reduced risk of human cardiovascular disease (Mullie and Clarys, 2011). Consumers in the past few decades have become increasingly concerned about the nutrient density and quality of vegetables they are consuming, which led to an investigation on the impact of nutrition on the quality of vegetable produces under several production scenarios, as the application of nutrients significantly impacts on fruit color, texture, disease susceptibility, juice composition and the development of physiological disorders, which has direct effect on shelf-life (Singh et al., 2013). Further, the nutrient levels that produce maximum yield may not always support the highest fruit quality and maximum quality retention, and poor management of fertilizers will increase physiological disorders due to deficiencies of some nutrients or increase of other nutrients that may lead to toxicity and subsequent poor shelf-life. However, owing to health benefits and high productivity, the area under vegetable production is constrained by more chemical fertilizer application. Majumdar and Gupta (2000) estimated that about 50% of applied fertilizers leach down into the soil, and its side effects are reflected on human health in the form of diseases such as methmoglobinemia in children. Hence, augmentation of crop yields from the existing cultivable areas under environmentally safe scenarios is an ideal approach (Kumar, 2013).

Soil bacteria that colonize in roots of plants following their inoculation onto seed and that enhance plant growth are defined as plant-promoting rhizobacteria. Upon colonizing, these growth-promoting rhizobacteria grow in and near the plant tissues, which stimulate plant growth by various mechanisms by nutrient mobilizers, plant growth, phytohormone producers,

DOI: 10.1201/9781003625674-8

biocontrol agents of insect pests and diseases. PGPR are well known to produce numerous microbial metabolites, which result in a reduction soil pH that plays a role in nutrient mobilization in the crop ecosystem. However, PGPR's mode of action and mechanism vary with the associated host plant and growing environments. Moreover, many other factors influence PGPR action, namely biotic factors (plant genotype, growth stages, host defense mechanisms and the presence of other microbial populations and soil abiotic factors and environmental conditions). The effect of PGPR on crops growing in different agronomic regions may range from neutralto positive to deleterious.

8.1.1 Salinity

Soil salinity itself is considered a major abiotic threat to crop production, which has affected almost one billion ha of land globally (Fageria et al., 2012; Ranjbar et al., 2016) and caused production loss worth 27.3 billion US$ (Munns and Tester, 2008; Qadir et al., 2014). Under saline conditions, the cellular ion balance in plant tissue is disturbed by high salt concentrations, which lead to the production of reactive oxygen species (ROS) and accumulation of Na+ and Cl^- ions (Grover et al., 2011). Destruction of cellular structure and biomolecules happens due to the excess level of ROS (O, superoxide and hydrogen peroxide), such as degradation of chlorophyll and peroxidize lipids, consequently photosynthetic activity damages cell membranes and eventually induces cell death (Apel and Hirt, 2004). Na+ and Cl^- ions interfere with enzyme function and stomatal opening and closing, resulting in osmotic stress and reduced photosynthesis (Kang et al., 2014). Furthermore, high concentrations of Cl^- ions inhibit nitrate reductase activity, which leads to nutrient imbalance (Nadeem et al., 2014), elevated ethylene levels, and premature senescence and defoliation (Grover et al., 2011).

Hence, physical removal of salt or chemical application is cost expensive and environmental hazards, however, microbial inoculants could improve plant growth and yield under saline conditions (Lugtenburg et al., 2013). There are two types of plant growth-promoting bacteria: exosymbiotic and endosymbiotic bacteria (Gerhardt et al., 2017), where the former are associated with the outside of roots in the rhizosphere (PGPR: plant growth-promoting rhizobacteria) and the latter resides within roots (PGPEB: plant growth-promoting endophytic bacteria) and the stress-tolerant plants and bacteria co-evolved survival strategies to adapt to high salt environments (Zhao et al., 2016). The colonization of either allochthonus or autochthonous or both PGPR starts very quickly on the surface of seeds and roots in response to chemically variable photosynthates released by plants in and around root/soil surfaces. Consequently, on colonization and ramifications, the PGPR starts facilitating growth by supplying major and minor nutrients to plants (Khan et al., 2009) and moderating plant hormone levels through producing ACC deaminase enzyme or indirectly by producing exopolysaccharides and biofilm, osmoprotectants, antioxidant enzymes and volatile organic compounds (VOCs) to survive and stimulate plant growth (Glick, 2012) needs to be explored.

8.2 Nutrient Mobilization

The global energy crisis has increased the cost of chemical fertilizers, however the exorbitant application of fertilizer has increased health and environmental issues. It is estimated that the manufacturing of 1-kg fertilizer of N requires six times more energy than that needed for producing either P or K fertilizers. The energy required for 1-kg fertilizer of N, P and K (11.2, 1.1 and 1 KWh, respectively) is estimated (Da Silva et al., 1978), which clearly indicates that N fertilizer demands greater energy despite its abundant availability in the air. Similarly, a spoonful of soil contains more organisms, such as 1 billion bacteria, 10–20 million actinomycetes, 1 million fungi, 10,020 million protozoa, 100,000 algae, and some worms and insects (Cosico, 1994). Hence effective utilization of these microorganisms on nutrient mobilization is cost effective and environmentally safe.

8.2.1 Nitrogen Fixation

Adding of N to the soil nutrient pool is called N fixation, which is mediated by symbiotic prokaryotic microorganisms (*Rhizobium, Mesorhizobium, Bradyrhizobium, Azorhizobium, Allorhizobium* and *Sinorhizobium*) and asymbiotic/free-living organisms (*Azoarcus, Azospirillum, Burkholderia, Gluconacetobacter, Pseudomonas, Azotobacter, Arthrobacter, Acinetobacter, Bacillus, Enterobacter, Erwinia, Flavobacterium, Klebsiella* and *Acetobacter*) by a highly conserved and energetically expensive enzyme called nitrogenase. Followed by N fixation, a sequential oxidation of N that occurs as ammonia (nitrification) into nitrate by two metabolically distinct groups of bacteria such as ammonia-oxidizing bacteria (*Nitrosomonus, Nitrosospira, Nitrosovibrio, Nitrosolobus* and *Nitrosococcus*) transforming ammonia to nitrite, and nitrifying bacteria (*Nitrobacter, Nitrococcus, Nitrospira* and *Nitrospina*) transforming nitrite to nitrate. Alternatively, 15–70% of ammonium derived from organic matter mineralization is reported to be eliminated through nitrification and denitrification processes (Betrand et al., 2015) by lowered oxygen concentration due to microorganism respiration in rhizosphere.

8.2.2 Phosphate Solubilization

Next to N, phosphorous is the most important nutrient for biological processes, is present at levels of 400–1200 mg/kg of soil, where only a very small amount (1 mg or less) of P is in soluble forms, while the rest is insoluble and, hence, not available for plant uptake (Khan et al., 2009). However, phosphate-solubilizing microorganisms (*Achromobacter xylosoxidans, Bacillus polymyxa, Pseudomonas putida, Acetobacter diazotrophicus, Agrobacterium radio-bacter, Bradyrhizobium mediterranium, Enterobacter aerogenes, Pantoea agglomerans, Gluconacetobacter diazotrophicus* and *Rhizobium meliloti*), which hydrolyses the insoluble form of P to soluble form through solubilization (inorganic P) and mineralization (organic P).

8.2.3 Potassium Solubilization Bacteria

Potassium solubilization is done by a wide range of saprophytic bacteria, fungal strains and actinomycetes (Bakhshandeh et al., 2017; Gundala et al., 2013). There are strong evidences that soil bacteria are capable of transforming soil K to the forms available to plant effectively (Meena et al., 2016).

8.2.4 Plant Growth Regulator

Plant growth regulators are naturally occurring organic substances that exert, at low concentrations, a major influence on plant growth and upregulation of physiological processes. Phytohormones such as auxin, gibberellin, ethylene, cytokinin and abscisic acid are reported to be produced by plants, by microorganisms, and even by algae. Indole acetic acid (IAA) is synthesized by its physiological precursor tryptophan (Trp) (Spaepen et al., 2007a) through five Trp-dependent pathways that occurs in bacteria (Di et al., 2016). Similarly, gibberellins are synthesized not only by plants and fungi but also by bacteria (Morrone et al., 2009) such as *Acetobacter diazotrophicus, Azospirillum lipoferum, A., Bacillus pumilus, B. cereus, B. macrolides, Herbaspirillum seropedicae, Acinetobacter calcoaceticus, Burkholderia cepacia* and *Promicromonospora* sp. have been reported as gibberellin producers (Kang et al., 2012). Cytokinin-producing bacteria such as *Azotobacter, Azospirillum, Agrobacterium, Pseudomonas, Paenibacillus, Bacillus, Achromobacter* and *Klebsiella* involved in plant growth regulation (Maheswari et al., 2015; Wong et al., 2015) have been reported. The other growth-promoting hormones synthesized during stress conditions is involved with an enzyme 1-aminocyclopropane-1-carboxylate (ACC) deaminase synthesized by a wide range of rhizospheric bacteria (Glick et al., 2007) decreases the deleterious ethylene amounts and balances ABA levels in stressed plants.

8.3 PGPR on Vegetable Production and Quality

Vegetables require a considerably higher quantity of plant nutrients for better growth and optimum yield (Solaiman and Rahbbani, 2006).

Adesemoye and Ugoji (2006) assessed the efficacy of *Pseudomonas* sp. against three crops like okra, tomato and African spinach (*Amaranthus* sp.). The *Pseudomonas* isolates due to their great plant growth-promoting abilities facilitated the growth of each plant. In tomato, Ahirwar et al. (2015) reported that *Pseudomonas fluorescence* (SS5) inoculated tomato had a significant increase in root and shoot weight, length, fruit yield per plant and total fruit yield. The tomato yield was increased by 57% when *P. fluorescence* (SS5) inoculated plants were grown in field as compared to pot culture (19%). Additionally, the total dry matter accumulation in field-grown tomato had increased substantially by 28%. Ordookhani et al. (2010) found that tomato seeds inoculated with *Pseudomonas, Azotobacter* and *Azospirillum* had enhanced levels of fruit lycopene content (80.42 mg/kg fresh fruit) and antioxidant properties.

In brinjal, Fu et al. (2010) reported that *Pseudomonas* sp. inoculated plants had higher shoot Ca^{2+} but no increase in shoot Na^+ concentration compared to non-inoculated plants grown under salinity stress, which exhibited significantly decreased K^+ concentration and increased Na^+ concentration. In pepper, Silva et al. (2014a) found that the Rhizobium strains on pepper decreased organic acid content and no significant change in phenolics and capsaicinoids of leaves and no improvement in antioxidant activity of fruits. Silva et al. (2014b) found that inoculation of rhizobium strains caused a significant decrease in sterols and triterpenes and a substantial increase in fatty acids. In pumpkin, Habibi et al. (2013) found that the maximum seed yield (118.5 g m^{-2}), oil yield (43.31 g m^{-2}) and fruit yield (6056.2 g m^{-1}) were obtained with treatment that pumpkin seeds inoculated with free-living nitrogen-fixing bacteria (NO) and phosphate solubilizing bacteria (PO) + 50% organic fertilizer. The maximum oil percentage (48.15%) was obtained at pumpkin seeds inoculated with PSB + 50% chemical fertilizer. It is concluded that the application of biofertilizers with 50% chemical and organic fertilizers (CO), reduced the need for C and produced higher oil, seed and fruit yield. Similarly, Habibi et al. (2013) found that the biofertilizer treatments had significant effects on chlorophyll, photosynthesis rate, fruit diameter, fruit fresh weight and fruit and seed yield of summer squash. Maximum seed yield (97.97 gr/m^2) was obtained from 50% chemical fertilizer + nitroxin + phosphate solubilizing bacteria and the highest fruit yield (6272.2 and 5790/5 gr/ha), photosynthesis rate (15.56 and 14.05 µmol m^{-2} s^{-1} CO_2) and chlorophyll (53.06 and 50.08 SPAD values).

In watermelon, Kharat et al. (2021) reported that RDF 100% through FYM + biofertilizer (Azotobacter) was found with high fruit weight (1.81 kg), pulp weight per fruit (1.47 kg), fruit length (25.67 cm) and diameter (15.27 cm), early to first (78 days) and total harvest of fruit (97.93 days). However, the highest yield was obtained with RDF 100% through FYM + biofertilizer (Azotobacter) through high number of fruits per vine (1.80), yield per vine (3.258 kg) and yield per hectare (57.9175 t/ha) as compared to other organic sources and control. In potato, Pishchik et al. (1998) reported the inoculation effect of nitrogen fixing *Klebsiella* on yield of non-legumes such as potato showed a significant increase in potato yield and N content was obtained after inoculation with *K. mobilis* strains CIAM880 and CIAM853, when low doses of nitrogenous fertilizer were used. Naqqash et al. (2016) observed that inoculation of nitrogen fixing bacteria, namely, *Azospirillum*, *Enterobacter* and *Rhizobium* under axenic conditions resulted in differential growth responses of potato. Of these, *Azospirillum* sp. TN10 showed the highest increase in fresh and dry weight of potato over control plants.

In cauliflower, Bhuysal to Bashyal (2011) found that the maximum stem height, stem diameter, highest curd height, curd diameter, fresh curd weight and curd yield were recorded at 120 kg nitrogen and 2 kg biofertilizer ha^{-1}. Cauliflower curd yield obtained at 120 kg nitrogen ha^{-1} did not significantly differ with the curd yield recorded at 60 kg nitrogen and 2 kg biofertilizer ha^{-1}. The earliest days to curd initiation and maturity were recorded at 30 kg nitrogen and 2 kg biofertilizer ha^{-1}. The highest vitamin C content of cauliflower curds and the most attractive curd color were recorded at 60 kg nitrogen and 2 kg biofertilizer ha^{-1}, while the appearance and overall acceptability were recorded at 120 kg nitrogen and 2 kg biofertilizer ha^{-1}. In okra, Bamboriya et al. (2018) found that application of vermin-compost @ 6 t ha^{-1} significantly increased yield $plant^{-1}$ (311.46 g), ha^{-1} (141.48 q) and $picking^{-1}$ (6.11 kg) of okra. However, biofertilizers combined use of Azospirillum + PSB recorded significantly higher yield (326.01 g $plant^{-1}$, 134.91 q ha^{-1} and 5.83 kg $picking^{-1}$). The highest protein (1.97%) content and the lowest crude fiber content (1.54%) of fruit recorded under vermin-compost, which was on a par with poultry manure. In cabbage, Rana et al. (2019) reported that the minimum number of days to 50 per cent head maturity (67.50) was recorded in biofertilizers + boric acid 0.5%. The significantly maximum polar diameter (13.75 cm) and maximum equatorial diameter (13.36 cm) was recorded in biofertilizers + ferrous sulphate 0.5%. Maximum TSS (5.72 °Brix) was observed in biofertilizers + boric acid 0.5%. However, maximum vitamin C content (59.97 mg/100 g) was recorded in biofertilizers + zinc sulphate 0.5%. In onion, Amitkumar et al. (2019) revealed that combined application of PSB (25%) + rhizobium (25%) + trichoderma (50%) increases the maximum height of plants, the number of leaves per plant, bulb diameters (6.8 cm), neck thickness of bulb, length of bulb (6.63 cm), number of scales (8.7), fresh weight of bulb (61.93 g), total soluble solid (T.S.S.) except weight of bulb and yield (45.67 t/ha) in Trichoderma (100%) under Lucknow conditions.

In pea, an effective microorganism culture consists of co-existing beneficial microorganisms, the main ones being the species of photosynthetic bacteria, viz. *Rhodopseudomonas plastris* and *Rhodobacter sphacrodes*; lactobacilli, viz. *Lactobacillus plantarum*, *L. casei* and *Streptococcus lactis*; yeasts (*Saccharomyces* spp.) and Actinomycetes (*Streptomyces* spp.) had improved crop growth and yield by increasing photosynthesis, producing bioactive substances such as hormones and enzymes, controlling soil diseases and accelerating decomposition of lignin materials in the soil (Hussain et al. 2002). Javaid (2006) found that foliar application of effective microorganisms enhanced nodulation such as nodule number (217%) and nodule biomass (167%), grain yield (126%) and green manure (145%) in pea. Rather et al. (2010) reported that co-inoculation of all the three biofertilizers, i.e., Rhizobium, Azotobacter and PSB produced significantly higher growth characteristics as compared to absolute control and when inoculated them individually. Rhizobium + azotobacter + PSB gave highest growth in terms of plant height (45.26 cm), number of leaves/plant (13.33), number of branches/plant (4.20), number of nodules/plant (38.46), fresh weight and dry weight of nodules (562.34 and 122.62 mg, respectively). The yield attributes like pod length (9.95 cm), number of pods/plant (16.60), number of seeds/pod (6.98) and 1000 grain weight (112.1 g) and yield of grain (19.8 q/ha) and straw of pea (25.52 q/ha) increased by co-inoculation of bio-fertilizers. In carrot and lettuce, Florex-Felix et al. (2013) assessed the impact of *R. leguminosarum* strain PEPV16 registered a significant increase in macro and micronutrients of both crops. Also, the rhizobial inoculation enhanced the N and P uptake by lettuce and carrot plants. The P uptake in lettuce shoots was increased by 15%, while 40% increase in P concentration was recorded in carrot roots. The increase in Fe content of both crops was attributed to the production of siderophores by *R. leguminosarum* strain PEPV16.

8.4 Augmentation of Yield of Vegetables Using PGPR Under Soil Salinity

A normal soil constitutes 600 million bacteria belonging to about 20,000 species, which is reduced to 1 million with 5000–8000 species in stress conditions. Plant growth-promoting rhizobacteria (PGPR) are in symbiotic association with the plant system, which helps in combating abiotic stress and increases the overall productivity and yield. On the other hand, the majority of vegetable crops have very low salinity threshold that are 2.5 dS m^{-1} (Snapp et al. 1991). Accordingly, the area of soils with restrictions for vegetable crop production is therefore greater than the area that is defined as "salinized", since a saline soil is generally defined as showing an EC value of the saturation extract (ECe) in the root zone that exceeds 4 dS m^{-1} (approximately 40 mM NaCl) at 25°C and having an exchangeable Na^+ level of 15% (Shrivastava & Kumar, 2014). However, a wider range of saline tolerance has been recorded based on crop performance and yield. Machado and Serralheiro (2017) reported asparagus as the most salt-tolerant vegetable crop. According to Grieve et al. (2012), the salinity tolerance of different vegetables based on threshold salinity (ds m^{-1}) in the following order

TABLE 8.1

Vegetable Inoculation with Plant Growth-Promoting Rhizobacteria

Crop	Inoculant of PGPR	References
Tomato	*Pseudomonas fluorescens* and *Trichoderma viridi; P. fluorescens CHA0; P. marginalis, P. putida* and *P. syringae; B. subtilis, Serratia* and *Arthrobacter* sp.	Shiva et al. (2013), Ardebili et al. (2011)
Brinjal	*Azospirillum, Azotobacter*	Patel et al. (2011)
Chilli	*B. subtilis, Bacillus strains; Serratia and Arthrobacter* sp.	Ashwini and Srividya (2014); Amaresan et al. (2012)
Okra	*B. subtilis* and *B. firmus*	Abbasi et al. (2013)
Potato	*B. cereus, B. subtilis, Azotobacter*	Singh (2013)
Broccoli	*Bacillus cereus, Brevibacillus reuszeri* and *Rhizobium rubi*	Yildrim et al. (2011)
Cabbage	*B. megaterium, Pantoea agglomerans* and *B. subtilis*	Turan et al. (2014)
Spinach	*B. cereus, B. licheniformis, B. megaterium, B. subtilis, Bacillus strains, P. putida, Paenibacillus polymyxa*	Cakmakci et al. (2007)
Radish	*B. subtilis* and *P. fluorescens*	Mohamed and Gomaa (2012)
Cucumber	*B. subtilis (BA-142), B. megaterium (GC subgroup A. MFD-2), Acinetobacter baumannii (CD-1)* and *P. agglomerans (FF)*	Dursun et al. (2010)
Onion	*Spirulina platensis* + cow dung *Scenedesmus subspicatus* + humic acid	Dinesh kumar et al. (2020), Gemin et al. (2019)
Lettuce	*Chlorella vulgaris* *Scenedesmus quadricauda*	Ergun et al. (2020), Puglisi et al. (2020)

where sugar beet (7.1), peas (3.4), tomato and cucumber (2.5), spinach (2.0), cabbage (1.8), potato (1.7), cauliflower (1.5), broccoli (1.3), radish and onion (1.2) and carrot (1.0). Based on Musikovic and Ilin (2012), the majority of vegetables tend to grow under brackish and moderately brackish conditions and none of the vegetables survive economically at saline and highly saline conditions. The data given in Table 8.2 indicated that the moderately saline soil with <7 dsm^{-1} could result in a 50% yield reduction invariably in all vegetables (Mungal, 1993).

There are PGPR, which improves the salt tolerance of plants through the following mechanism: (1) inducing the antioxidant system; (2) maintaining water balance within the plant, releasing bound phosphorus and potassium from the soil, chelating iron, and fixing atmospheric N (Santoyo et al., 2016); (3) selectively absorbing K^+ and excluding Na^+ to maintain a high K^+/Na^+ ratio (Islam et al., 2015); (4) using PGPR release of exopolysaccharides (EPSs) (Timmusk et al., 2014); as the formation of protective biofilms reduces the toxicity of Na^+; (5) maintaining plant hormone levels (Etesami et al., 2014); and increasing osmotic regulatory substances (Munns, 2002). If these traits are suitably exploited, microorganisms can play a significant role in alleviating salt effects on crop plants (Shrivasata and Kumar, 2015).

TABLE 8.2

Classification of Waters with Respect to Total Salt Concentration and Salt Tolerance of Selected Vegetables

Water Designation	EC (dsm^{-1})	Crop	Threshold EC (dsm^{-1})	*50% Yield Reduction at dsm^{-1}	Ranking
Fresh water	<0.6				
Slightly brackish	0.6–1.5	Bean	1.0	3	Sensitive
		Carrot	1.0	6	Sensitive
		Musk melon	1.0	6	Moderately Sensitive
		Egg plant	1.1	7	Moderately Sensitive
		Onion	1.2	5	Sensitive
		Radish	1.2	5	Moderately Sensitive
		Lettuce	1.3	5	Moderately Sensitive
		Pepper	1.5	7	Moderately Sensitive
Brackish	1.5–3.0	Garlic	1.7	7	Moderately Sensitive
		Potato	1.7	7	Moderately Sensitive
		Cabbage	1.8	10	Moderately Sensitive
		Celery	1.8	10	Moderately Sensitive
		Spinach	2.0	10	Moderately Sensitive
		Cucumber	2.5	7	Moderately Sensitive
		Tomato	2.5	7	Moderately Sensitive
Moderately saline	3–3.8	Pea	3.4	7	Moderately Sensitive
		Beet root	4.0		Moderately Tolerant
		Asparagus	4.1		Tolerant
Saline	8–15.0	–	–		–
Highly saline	15–45.0	–	–		–

Source: Maksimovic and Ilin, 2012; *Mungal, 1993.

Halophilic bacteria have the potential to remove sodium ions from soil and increase metabolic and enzymatic activities in plants. Arora et al. (2016) used halophilic bacteria for the remediation of saline and sodic soils. The bacterial inoculation improved soil properties by decreasing soil pH from 9.4 to 8.6, and increasing microbial biomass C from 82 to 137 µg/g. Similarly, in a pot experiment irrigated with saline water (5% NaCl), inoculation of halophilic bacterial consortium increased fresh weight, dry weight, shoot length and root length of maize plants by 194.5, 98.97, 15.37 and 7.4 cm, respectively, compared to the uninoculated control plants. A low-cost microbial bio-formulation "CSR-BIO," a consortium of *Bacillus pumilus, Bacillus thuringenesis* and *Trichoderma harzianum*, is rapidly becoming popular with the farmers in many states (Damodaran et al., 2013). This bio-formulation acts as a soil conditioner and nutrient mobilizer and has been found to increase productivity in sodic and normal soils by 22–43%. Niu et al. (2018) reported that salt tolerant plant beneficial microbes improved crop productivity in arid and semiarid regions. The diazotrophic salt tolerant bacterial strains of *Pseudomonas, Agrobacterium, Streptomyces, Klebsiella* and *Ochromobacter* isolated from the roots of a halophytic plant *Arthrocnemum indicum* showed salinity tolerance ranging from 4–8% NaCl and improved the productivity of peanut (Sharma et al., 2016). A salt-tolerant bacteria strain *Staphyllococus xylosus* ST-1 caused 25% greater plant growth at 100 mM (Afrasayab et al., 2010), *Bacillus mojavensis* VKAK-1 changing plant water relation (Kapadia et al., 2022). A nitrogen-fixing salt-tolerant bacterium *Swaminathan halotolerant* PAS51T isolated from wild rice associated with the mangrove ecosystem (Loganathan and Nair, 2004). A salt-tolerant rhizobium (L-19 and L-304) produced higher nodulation, yield, and nitrogen fixation in lentils (Islam et al., 2013).

Unfortunately, the application of ST-PGPR in alleviating salt stress of vegetable crops is very scarce. However, there are 129 isolates of rhizospheric bacteria from turmeric, and two isolates of *Bacillus amyloquefaciens* exhibited plant growth-promoting and antimicrobial activity against fungal pathogens (Passari et al., 2018). Among the eight *Steptomyces* sp. stains isolated, A39 produced the highest amount of IAA and solubilizes a high amount of phosphate even at higher salt concentration (Palaniyandi et al., 2014) and *Streptomyces variabilis* (4NC), *S. fradiae* (8PK), and *Streptomyces strain* C-2012 have the potential for plant growth promotion and help plants to alleviate salt stress (Vurukonda et al., 2018; Olanrewaju et al., 2019; Toiba et al., 2019). The microbes such as *Bacillus, Klebseilla, Agrobacterium, Pseudomonas* and *Ochrobacterium* sp. showed salt tolerance levels up to 10% NaCl (Sharma et al., 2016; Zheng et al., 2018; Chen et al., 2019). Kapadia et al. (2022) identified microbial isolates at Dandi, Navsari, 12 isolates exhibited PGP traits at a high salt concentration of 15–20% NaCl, where AD9 belongs to *Bacillus halotolerans* had higher ammonia production and P solibilization at 15%, AD32 (*Bacillus* sp. clone ADCNO) had higher IAA production at 20% NaCl. AD2 (*Streptomyces* sp. clone ADCNB) and AD26 (*Achromobacter* sp. clone ADCNI) showed ACC deaminase activity and AD32 (*Bacillus* sp. clone ADCNL) gave a high siderophore activity.

High salinity reduces the uptake of available P by plant roots due to sorption processes to the soil colloid, however P-solubilizing bacteria could survive, solubilize fixed and applied P in soil (Banerjee et al., 2010). In soybean, seeds inoculated with *Bacillus megaterium* A12 as the efficient solubilizer of P, which significantly increased plant growth at 30 and 90 mM NaCl concentrations. The PGPR regulates plant hormones such as IAA to plant root architecture, where *Pseudomonas* sp. PDmznCd2003 is not affected by soil salinity stress at 4–16 dSm^{-1}. In tomato, Mayak et al. (2004) isolated an *Achromobacter piechaudii* bacterium from the Arava region of Southern Israel, which significantly increased the fresh and dry weight of tomato seedlings at high salt concentration, similar to *P. stutzeri* (Tank and Sarf, 2010). Inoculation of tomato seedlings with *A. chroococcum* 764 under both moderate and severe (50 and 100 mM NaCl, respectively) salt stress increased fresh fruit weight and fruit number as compared to the control (Van Oosten et al., 2018). In egg plant, *Xanthobacter autotropicus* BM13, *E. areogenes* BM10 and *Bacillus brevis* FK2 alleviated the negative effect of salt and improved growth and yield (Abd

El-Azeem et al., 2012). In cucumber, a ST-PGPR strain *Rhodopseudomonas palustris* G5 increased shoot weight, root length, fresh weight, total chlorophyll content and soluble sugar content in cucumber seedlings grown under 3% NaCl condition (Ge and Zhang, 2019). In potato, Tahir et al. (2019) conferred that a consortium of salt-tolerant *Bacillus* strains enhanced potato tuber yield by the production of auxin, and antioxidant enzymes and regulating the uptake of Na+, K+ and Ca^{+2} in normal and salt-affected soils. In lettuce, Kohler et al. (2009) found that the salinity decreased lettuce growth, regardless of the biological treatment and of the salt stress level. The plants inoculated with *P. mendocina* had significantly greater shoot biomass than the control plants. At the highest salinity level (4g NaCl/kg soil), the water content was greater in leaves of plants treated with *P. mendocina* or *G. mosseae.* Salt stress decreased sugar accumulation and increased foliar proline concentration, particularly in plants inoculated with the PGPR. The PGPR strain induced a higher increase in these antioxidant enzymes in response to severe salinity. In peas, Barnawal et al. (2014) assessed the tolerance to salinity in pea plants by reducing stress ethylene levels through ACC deaminase-containing rhizobacteria *Arthrobacter protophhormiae* (SA3) and promoting plant growth through improved colonization of beneficial microbes like *Rhizobium leguminosarum* (R) and *Glomus mosseae* (G). This tripartite collaboration increased plant weight by 53%, reduced prolein content, lipid peroxidation and increased pigment content under 200-mM salt condition. It is also detected that decreased ACC oxidase (ACO) activity induced by SA3 and reduced ACC synthase (ACS) activity in AMF.

8.5 Conclusion and Future Research

There is a rising demand for the cultivation of vegetable crops in saline-affected soil by considering compatible, ecologically sound and environmental friendly methods. The larger total saline soil as compared to the total vegetable growing area clearly indicated the inevitability of saline soil exploitation to supply adequate vegetables 400 g/person. Considering the fact, the presence of ACC deaminase-containing bacteria, higher colonization of AMF and nodulation of rhizobia could be achieved, which is restricted by high ethylene concentration under stress. A combination of rhizobacteria *Arthrobacter protophhormiae* (SA3) + *Rhizobium leguminosarum* (R) + *Glomus mosseae* (G) could be attempted for all vegetables to induce stress tolerance of vegetables for saline (8-15 EC) and highly saline soil (15-45EC). In the future, extensive research needs to be emphasized in this area, particularly on the field performance of potential microorganisms as a source of biofertlizers in highly stressed soil. Further, CRISPR technology is being applied for elucidating gene functions and fine-tuning plant response pathways for tolerance to drought, salinity and other stresses in vegetables. The interaction of CRISPR-mediated varieties for stress with salt-tolerant PGPR under saline and highly saline conditions needs to be unearthed. Sequencing technology and quantitative analysis of differential gene expression in response to various signals and stresses such as induced proteins 26i and 42i associated with membrane proteins and high root colonization potential in saline soil (Qurashi and Sabri, 2012), Protein 41i a survival protein (SurE), essential for the survival of osmotic stress (2.5M NaCl) in bacteria (Visick et al., 1998), and GST gene over-expression involved in ISR for protecting cells from oxidative damages (Kandasamy et al., 2009) be explored extensively.

References

Abbasi M. W., Zaki M. J. and Anis M. 2013. Application of Bacillus species cultured on different low cost organic substrates for the control of root-knot nematode infection on okra (*Abelmoschus esculentus* Moench), *Pak. J. Bot.* 45: 1079–1084.

Abd El-Azeem S. A. M., Elwan M. W. M., SungJ. K. and Ok Y. S. 2012. Alleviation of salt stress in eggplant (*Solanum melongena* L.) by plant-growth-promoting rhizobacteria, *Commun. Soil Sci. Plan.* 43: 1303–1315. doi: 10.1080/00103624.2012.666305

Adesemoye, A. O., & Ugoji, E. O. (2006). Evaluating Pseudomonas species for biological control of root diseases in okra (Abelmoschus esculentus), tomato (Solanum lycopersicum), and African spinach (Amaranthus sp.). *African Journal of Biotechnology*, 5(10), 866–870.

Afrasayab S., Faisal M. and Hasnain S. 2010. Comparative study of wild and transformed salt bacterial strains on *Triticum aestivum* grown under salt stress, *Braz. J. Microbiol.* 41: 946–955.

Ahirwar N. K., Gupta G., Singh V., Rawlley R. K. and Ramana S. 2015. Influence on growth and fruit yield of tomato (*Lycopersicon esculentum* Mill.) plants by inoculation with *Pseudomonas fluorescence* (SS5): Possible role of plant growth promotion, *Int. J. Curr. Microbiol. Appl. Sci.* 4: 720–730.

Amaresan N., Jayakumar V., Kumar K. and Thajuddin N. 2012. Isolation and characterization of plant growth promoting endophytic bacteria and their effect on tomato (*Lycopersicon esculentum*) and chilli (*Capsicum annuum*) seedling growth, Ann. *Microbiol.* 62: 805–810.

Apel K. and Hirt H. 2004. Reactive oxygen species: Metabolism, oxidative stress, and signal transduction, *Annu. Rev. Plant Biol.* 55: 373–399

Ardebili Z. O., Ardebili N. O. and Hamdi S. M. M. 2011. Physiological effects of Pseudomonas fluorescens CHA0 on tomato (*Lycopersicon esculentum* Mill.) plants and its possible impact on *Fusarium oxysporum* f. sp. Lycopersici, *Aust. J. Crop Sci.* 5: 1631–1638.

Arora S., Singh Y. P., Vanza M. and Sahni, D. 2016. Bioremediation of saline and sodic soils through halophilic bacteria to enhance agricultural production, *J. Soil Water Conserv.* 15: 302–305. doi: 10.5958/2455-7145.2016. 00027.8

Ashwini N. and Srividya, S. 2014. Potentiality of Bacillus subtilis as biocontrol agent for management of anthracnose disease of chilli caused by *Colletotrichum gloeosporioides* OGC1, *3 Biotech* 4: 127–136.

Bakhshandeh E., Pirdashti H. and Lendeh K. S. 2017. Phosphate and potassium-solubilizing bacteria effect on the growth of rice, *Ecol. Eng.* 103: 164–169.

Bamboriya J. S., Naga S. R., Sharma S. R., Choudhary M. R. and Bamboriya S. D. 2018. Productivity, quality and profitability of Okra (*Abelmoschus esculentus*) as influenced by organic manures and biofertilizers, *Int. J. Bio-resou. Stress Manag.* 9(4): 506–509.

Banerjee G., Scott Graig J. S. and Walton J. D. 2010. Improving enzymes for biomass conversion: A basic research perspective, *Bioenergy Res.*, 3: 82–92.

Barnawal D., Bharti N., Maji D., Chanotiya C. S. and Kalra A. 2014. ACC deaminase-containing *Arthrobacter protophormiae* induces NaCl stress tolerance through reduced ACC oxidase activity and ethylene production resulting in improved nodulation and mycorrhization in *Pisum sativum, J. Plant Physiol.* 171: 884–894.

Bashyal L. N. 2011. Response of cauliflower to nitrogen fixing biofertilizer and graded levels of nitrogen, *J. Agric. Environ.* 12: 41–49.

Bertrand J. C., Bonin P., Caumette P., Gattuso J. P., Grégori G., Guyoneaud R., Le Roux X., Matheron R. and Poly F. 2015. Biogeochemical cycles. In: Jean-Claude Bertrand (ed) *Environmental microbiology: fundamentals and applications: Microbial ecology*, Springer Science + Business Media, Dordrecht. 511–617.

Cakmakc I. R., Erat E., Erdogan U. and Dönmez M. F. 2007. The influence of plant growth–promoting rhizobacteria on growth and enzyme activities in wheat and spinach plants, *J. Plant Nutr. Soil Sci.* 170: 288–295.

Chen Y., Jin P., Cui Z., Xu T., Zhao R. and Zheng Z. 2019. Identification and characterization of Janthinobacteriumsvalbardensis F19, a novel low-C/N-tolerant denitrifying bacterium, *Appl. Sci.* 9: 1937. doi: 10.3390/app9091937

Cosico W. C.1994. Are microbial inoculants necessary for composting?, *Philippine Agric.* 77(1): 67–76.

Damodaran T., Rai R. B., Jha S. K., Sharma D. K., Mishra V. K. and Dhama K. 2013. Impact of social factors in adopon of CSRBIO- A cost effective, eco-friendly bio-growth enhancer for sustainable crop production, *South Asian J. Exper. Bio.* 3: 158–165.

Da Silva, J. A. T., Dung, L. N. T., & Tai, H. H. (1978). *Energy consumption in fertilizer production*. In: Proceedings of the International Fertilizer Society, London: International Fertilizer Society.

Di D. W., Zhang C., Luo P., An C. W. and Guo G. Q. 2016. The biosynthesis of auxin: How many paths truly lead to IAA?, *Plant Growth Regul.* 78: 275–285.

Dineshkumar R., Subramanian J., Arumugam A., Ahamed Rasheeq A. and Sampathkumar P. 2020. Exploring the microalgae biofertilizer effect on onion cultivation by field experiment, *Waste Biomas. Valorizat.* 11: 77–87. doi: 10.1007/s12649-018-0466-8

Dursun A., Ekinci M. and Dönmez M. F. 2010. Effects of foliar application of plant growth promoting bacterium on chemical contents, yield and growth of tomato (*Lycopersicon esculentum* L.) and cucumber (*Cucumis Sativus* L.), *Pak. J. Bot.* 42: 3349–3356.

Ergun O., Dasgan H. and Isık, O. 2020. Effects of microalgae *Chlorella vulgaris* on hydroponically grown lettuce, *Acta Horticult.* 2: 169–176. doi: 10.17660/ActaHortic.2020.1273.23

Etesami H., Hosseini H. M., Alikhani H. A. and Mohammadi L. 2014. Bacterial Biosynthesis of 1-Aminocyclopropane-1-Carboxylate (ACC) Deaminase and Indole-3-Acetic Acid (IAA) as endophytic preferential selection traits by rice plant seedlings, *J. Plant Growth Regul.* 33: 654–670.

Fageria N. K., Stone L. F. and Santes A. B. D. 2012. Breeding for salinity tolerance, Plant Breed. *Abiotic Stress Toler.* 103–122.

Flores-Félix D., Menéndez E., Rivera L .P., Marcos-García M., Martínez-Hidalgo P., Mateos P. F., Martínez-Molina E., Velázquez M. E., García-Fraile P. and Rivas R. 2013. Use of *Rhizobium leguminosarum* as a potential biofertilizer for *Lactuca sativa* and *Daucus carota* crops, *J. Plant Nutr. Soil Sci.* 176: 876–882.

Fu Q., Liu C., Ding Y., Lin Y. and Guo B. 2010. Ameliorative effects of inoculation with the plant growth-promoting rhizobacterium *Pseudomonas sp.* DW1 on growth of eggplant (*Solanum melongena* L.) seedlings under salt stress, Agric. *Water Manag.* 97: 1994–2000.

Ge H. and Zhang F. 2019. Growth-promoting ability of *Rhodopseudomonas palustris* G5 and its effect on induced resistance in cucumber against salt stress, *J. Plant Growth Regul.* 38: 180–188. doi: 10.1007/s00344-018-9825-8

Gemin L. G., Mógor Á. F., De Oliveira Amatussi J. and Mógor G. 2019. Microalgae associated to humic acid as a novel biostimulant improving onion growth and yield, *Sci. Horticult.* 256: 108560. doi: 10.1016/j.scienta.2019.108560

Gerhardt K. E., Gerwing P. D. and Greenberg, B. M. 2017. Opinion: Taking phytoremediation from proven technology to accepted practice, *Plant Sci.* 256: 170–185.

Glick B. R. 2012. Plant growth promoting bacteria Mechanims and applications, *Scientifica*, 963401:1-15.

Glick B. R., Cheng Z., Czarny J. and Duan J. 2007. Promotion of plant growth by ACC deaminase- producing soil bacteria, *Eur. J. Plant Pathol.* 119: 329–339.

Grieve C. M., Grattan S. R. and Maas E. V. 2012. Plant salt tolerance. In: Wallender WW and Tanji KK (eds) *ASCE manual and reports on engineering practice no. 71 Agricultural Salinity Assessment and Management, 2nd Edn,* ASCE, Reston, VA. 405–459.

Grover M., Ali S. Z., Sandhya V., Rasul A. and Venkateswarlu B. 2011. Role of microorganisms in adaptation of agriculture crops to abiotic stresses, *World J. Microbiol. Biotechnol.* 27: 1231–1240.

Gundala P. B., Chinthala P. and Sreenivasulu B. 2013. A new facultative alkaliphilic, potassium solubilizing, Bacillus Sp.SVUNM9 isolated from mica cores of Nellore District, Andhra Pradesh, India. Research and Reviews, *J. Microbiol. Biotechnol.* 2: 1–7.

Habibi A., Heidari G. R., Sohrabi Y. and Mohamadi K. H. 2013. Effect of biofertilizers and chemical fertilizers on yield and yield components of pumpkin (*Cucurbita pepo* L. Convar. pepo Var. styriaca), *Iran. J. Med Aromatic Plant.* 28(4): 604–614.

Hussain T., Anjum A. D. and Tahir J. 2002. Technology of beneficial microorganisms, *Nat. Farming Environ.* 3: 1–14.

Islam M. Z., Sattar M. A., Ashrafuzzaman M., Zulkerami B. and Shamsuddoha A. T. M. 2013. Evaliuating some salinity tolerant rhyzobacterial strains to Lentil production under salinity stress, *Int. J. Agric. Biol.*, 15: 499–504.

Islam F.,Yasmeen T., Arif M.S., Ali S., Ali B., Hameed S. and Zhou W. 2015. Plant growth promoting bacteria confer salt tolerance in *Vigna radiata* by up-regulating antioxidant defense and biological soil fertility, *Plant Growth Regul.* 80: 23–36.

Javaid A. 2006. Foliar application of effective microorganisms on pea as an alternative fertilizer, *Agron. Sustain. Dev.* 26: 257–262.

Kang S. M., Khan A. L., Hamayun M., Hussain J., Joo G. J., You Y. H., Kim J. G. and Lee I. J. 2012. Gibberellin- producing *Promicromonospora* sp. SE188 improves *Solanum lycopersicum* plant growth and influences endogenous plant hormones, *J. Microbiol.* 50: 902–909.

Kang S. M., Khan A. L., Waqas M., You Y. H., Kim J. H., Kim J. G., Hamayun M. and Lee I. J. 2014. Plant growth-promoting rhizobacteria reduce adverse effects of salinity and osmotic stress by regulating phytohormones and antioxidants in *Cucumis sativus*, *J. Plant Interact.* 9: 673–682.

Kandasamy S., Logananthan K., Muthuraj R., Duraisamy S., Seetharaman S., Thiruvengadam R. and Ramasamy S. 2009. Understanding the molecular basis of plant growth promotional effect of Pseudomonas fluorescens on the rice through protein profiling, *Proteome Sci.* 7: 47.

Kapadia C., Patel N., Rana A., Vaidya H., Alfarraj S., Ansari M. J., Gafur A., Poczal P. and Sayyed R. Z. 2022. Evaluation of plant growth promoting and salinity Ameliorating potential of Halophilic Bacteria isolated from saline soil, Front. *Plant Sci.* 13: 946217.

Kharat B. M., Shinde S. J. and Jadhav S. D. Effect of different organic sources on yield and yield attributing character of watermelon (*Citrullus lanatus* Thunb.), *J. Pharmacogn. Phytochem.* 10(1): 827–832.

Khan A. A., Jilani G., Akhtar M. S., Naqvi S. M. S. and Rasheed M. 2009. Phosphorus solubilizing bacteria: Occurrence, mechanisms and their role in crop production, *J. Agric. Biol. Sci.* 1: 48–58.

Kohler J., Hernández J. A., Caravaca F. and Roldán A. 2009. Induction of antioxidant enzymes is involved in the greater effectiveness of a PGPR versus AM fungi with respect to increasing the tolerance of lettuce to severe salt stress. *Environ. Exp. Bot.* 65: 245–252.

Kumar A., Meena M. L., Shivran B. C., Pal H. and Meena B. L. 2019. Impact of bio-fertilizer on growth, yield and quality of onion (*Allium cepa* L.) Cv. PUSA RED, *Plant Arch.* 19(1): 772–776.

Kumar S. 2013. The role of biopesticides in sustainably feeding the nine billion global populations, *J. Biofertilizer Biopestic.* 4: e114.

Loganathan P. and Nair S. 2004. *Swaminathania salitolerants* gen. nov.sp.nov., a solt tolerant nitrogen fixing and phosphate solubilizing bacyterium from wild rice (*Proterasia corctata*Tateoka), *Int. J. Sys. Evol. Microbiol.* 54: 1185–1190.

Lugtenburg B. J., Maltanova N., Kamilova F. and Berg G. 2013. Plant growth promotion by microbes, *Mol. Microb. Ecol. Rhyzosphere.* 2: 561–573.

Machado R. M. A. and Serralheiro R. P. 2017. Soil salinity: Effect on vegetable crop growth. Management practices to prevent and mitigate soil salinization, *Horticulturae* 3: 30.

Maheshwari D. K., Dheeman S. and Agarwal M. 2015. Phytohormone-producing PGPR for sustainable agriculture. In: Maheshwari DK (ed) *Bacterial metabolites in sustainable agroecosystem, sustainable development and biodiversity*, Springer International Publishing, Switzerland. 12: 159–182.

Majumdar D. and Gupta N. 2000. Nitrate pollution of groundwater and associated human health disorders, *Indian J. Environ. Health.* 42: 28–39.

Maksimovic I. and Ilin Z. 2012. Effects of salinity on vegetable growth and nutrients uptake. In: Shui Lee T (ed) *Irrigation systems and practices in challenging environments*, ISBN 978-953-51-0420-9, Chapter-9, InTech: 169–190. www.intechopen.com/books/

Mangal J. L. 1993. Vegetable crops. In: Manchanda HR, Karwasra SPS and Sharma HC (eds) *Salinity research at H. A. U.*, CCS Haryana Agricultural University, Hisar.

Mayak S., Tirosh T. and Glick, B. R. 2004. Plant growth-promoting bacteria confer resistance in tomato plants to salt stress. *Plant Physiol. Biochem.* 42: 565–572. doi: 10.1016/j.plaphy.2004.05.009

Meena V. S., Maurya B. R., Verma J. P. and Meena R. S. 2016. *Potassium solubilizing microorganisms for sustainable agriculture*, Springer India, New Delhi. 293–313.

Mohamed H. I. and Gomaa E. Z. 2012. Effect of plant growth promoting Bacillus subtilis and Pseudomonas fluorescens on growth and pigment composition of radish plants (*Raphanus sativus*) under NaCl stress, *Photosynthetica*. 50: 263–272.

Morrone D., Chambers J., Lowry L., Kim G., Anterola A., Bender K. and Peters R. J.2009. Gibberellin biosynthesis in bacteria: Separate ent-copalyldiphosphate and ent-kaurene synthases in *Brady rhizobium japonicum*, *FEBS Lett.* 583: 475–480.

Mullie P. and Clarys P. 2011. Association between cardio- vascular disease risk factor knowledge and lifestyle, *Food Nutr. Sci* . 2(10): 1048–1053.

Munns, R. 2002. Comparative physiology of salt and water stress, *Plant Cell Environ.* 25: 239–250.

Munns R. and Tester M. 2008. Mechanism of salinity tolerance, *Annu. Rev. Plant Biol.* 59: 651–681.

Nadeem S. M., Ahmad M., Zahir Z. A., Javaid A. and Ashraf M. 2014. The role of mycorrhizae and plant growth promoting rhizobacteria (PGPR) in improving crop productivity under stressful environments, *Biotechnol. Adv.* 32: 429–448.

Naqqash T., Hameed S., Imran A., Hanif M. K., Majeed A. and Van Elsas J. D. 2016. Differential response of potato toward inoculation with taxonomically diverse plant growth promoting rhizobacteria, *Front Plant Sci.* 7: 144.

Niu X., Song L., Xiao Y. and Ge, W. 2018. Drought-tolerant plant growth-promoting rhizobacteria associated with foxtail millet in a semi-arid agroecosystem and their potential in alleviating drought stress, *Front. Microbiol.* 8: 2580. doi: 10.3389/fmicb.2017.02580

Olanrewaju O. S., Ayangbenro A. S., Glick B. R. and Babalola O. O. 2019. Plant health: Feedback effect of root exudates-rhizobiome interactions, *Appl. Microbiol. Biotechnol.* 103: 1155–1166. doi: 10.1007/s00253-018-9556-6

Ordookhani K., Khavazi K., Moezzi A. and Rejali F. 2010. Influence of PGPR and AMF on antioxidant activity, lycopene and potassium contents in tomato, *Afr. J. Agric. Res.* 5: 1108–1116.

Palaniyandi S. A., Damodharan K., Yang S. H. and Suh J. W. 2014. *Streptomyces sp.* strain PGPA39 alleviates salt stress and promotes growth of "Micro Tom" tomato plants, *J. Appl. Microbiol.* 117: 766–773. doi: 10.1111/jam.12563

Passari A. K., Lalsiamthari P. C., Zothanpuia V. V., Mishra V. K. and Yadav M. K. 2018. Biocontrol of Fusarium wilt of Capsicum annuum by rhizospheric bacteria isolated from turmeric endowed with plant growth promotion and disease suppression potential, *Eur. J. Plant Pathol.* 150: 831–846. doi: 10.1007/s10658-017-1325-3

Patel B. N., Solanki M. P., Patel S. R. and Desai J. R. 2011. Effect of bio-fertilizers growth, physiological parameters, yield and quality of brinjal cv. Surati Ravaiya, *Indian J. Hortic.* 68: 370–374.

Pishchik V. N., Chernyaeva I. I., Kozhemaykov A. P., Vorobyov N. I., Lazarev A. M. and Kozlov L. P. 1998. Effect of inoculation with nitrogen- xing*Klebsiella*on potato yield. In: Malik KA et al (eds) *Nitrogen xation with non-legumes*, Kluwer Academic Publishers, Great Britain. 223–235.

Puglisi I., La Bella E., Rovetto E. I., Lo Piero A. R. and Baglieri A. 2020. Biostimulant effect and biochemical response in lettuce seedlings treated with A *Scenedesmus quadricauda* extract, *Plants (Basel Switzerland)* 9: 123. doi: 10.3390/plants9010123

Qadir M., Quillerou E., Nnagia V., Nurtaza G., Singh M., Thomas R. J., Drechsel P. and Nable A. D. 2014. Economics of salt induced land degradation and restoration, *Nat. Resour. Forum.* 38: 282–295.

Qurashi A. W. and Sabri A. N. 2012. Bacterial exopolysacchirdes and biofilm formulation stimulate chickpea growth and soil aggregation under saline stress, *Braz. J. Microbiol.* 43: 1183–1191.

Rajnjar F. and Jalali M. 2016. The combination of geostatics and geochemical simulation for the site specific management of soil salinity and sodicity, *Comput. Electron. Agric.* 121: 301–312.

Rana S., Thakur K. S., Bhardwaj R. K. and Kansal S. 2019. Effect of biofertilizers and micronutrients on growth and quality attributes of cabbage (*Brassica oleracea* var. *capitata* L.). *International Journal of Chemical Studies*, 7(2). 1325–1328.

Rather S. A., Hussain M. A. and Sharma N. L. Effect of bio-fertilizers on growth, yield and economics of field pea (*Pisum sativum* L.). *Inter. J. Agric. Sci.* 6(1): 65–66.

Santoyo G., Moreno-Hagelsieb G., del Carmen Orozco-Mosqueda, M. and Glick B. R. 2016. Plant growth-promoting bacterial endophytes, *Microbiol. Res.* 183: 92–99.

Sharma S., Kulkarni J. and Jha B. 2016. Halotolerant rhizobacteria promote growth and enhance salinity tolerance in peanut. *Front. Microbiol.* 7: 1600. doi: 10.3389/fmicb.2016.01600

Sharma S., Sayyed R., Sonawane M., Trivedi M. and Thivakaran G. 2016. *Neurospora* sp. SR8, a novel phosphate olubilizer from rhizosphere of soil of Sorghum in Kachh, Gujarat, India, *Indian J. Exp. Biol.* 54: 644–649.

Shiva N., Gomathi G., Karthika S., Ramya S., Senathipathi B., Senthil P., Surendar K. K. and Kumar S. R. 2013. Physiological effects of *Pseudomonas fluorescens* on tomato, *Int. J. Hortic.* 18: 104–108.

Shrivastava P. and Kumar R. 2014. Soil salinity: A serious environmental issue and plant growth promoting bacteria as one of the tools for its alleviation. I*Saudi Journal of Biological Sciences*, 21(1), 1–7.

Shrivasata P. and Kumar R. 2015. Soil salinity: A serious environmental issue and plant growth promoting bacteria as one of the tools for its alleviation, *Saudi J. Biologi. Sci.* 22: 123–131. doi: 10.1016/j.sjbs.2014.12.001

Silva L. R., Azevedo J., Pereira M. J., Carro L., Velazquez E., Peix A., Valentão P. and Andrade P. B. 2014a. Inoculation of the nonlegume *Capsicum annuum* (1.) with Rhizobium strains. 1. Effect on bioactive compounds, antioxidant activity, and fruit ripeness, *J. Agric. Food Chem.* 62: 557–564.

Silva L. R., Azevedo J., Pereira M. J., Carro L., Velazquez E., Peix A., Valentão P. and Andrade P. B. 2014b. Inoculation of the nonlegume *Capsicum annuum* l. with Rhizobium strains. 2. Changes in sterols, triterpenes, fatty acids, and volatile compounds, *J. Agric. Food Chem.* 62: 565–573.

Singh S. K., Singh R. S. and Awasthi O. P. 2013. Influence of pre-harvest and post-harvest treatments on shelf life and quality attributes of ber fruits, *Indian J. Hortic.* 70(4): 610–613.

Singh U. N. 2013. Effect of bio-fertilizers on yield and economic traits of potato at two fertility levels, Hort. *Flora Res. Spectr.* 2: 262–264.

Snapp S. S., Shennan C. and Bruggen A. V. 1991. Effects of salinity on severity of infection by Phytophthora par-asitica Dast., ion concentrations and growth of tomato, *Lycopersicon esculentum*, Mill, *New Phytol.* 119: 275–284.

Solaiman, A. H. M., & Rabbani, M. G. (2006). Nutrient management for sustainable vegetable production. In: *Proceedings of the 4th International Vegetable Science Conference*, Department of Horticulture, Bangladesh Agricultural University, Mymensingh, Bangladesh

Spaepen S., Vanderleyden J. and Remans R. 2007a. Indole-3-acetic acid in microbial and microorganism- plant signaling, FEMS Microbiol. *Rev.* 31: 425–448.

Tahir M., Ahmad I., Shahid M., Shah G. M., Farooq A. B. U . and Akram M. 2019. Regulation of antioxidant production, ion uptake and productivity in potato (*Solanum tuberosum* L.) plant inoculated with growth promoting salt tolerant Bacillus strains. *Ecotox. Environ. Safe.* 178: 33–42. doi: 10.1016/j.ecoenv.2019.04.027

Tank, N. and Saraf, M. 2010. Salinity-resistant plant growth promoting rhizobacteria ameliorates sodium chloride stress on tomato plants. *J. Plant Interact.* 5: 51–58. doi: 10.1080/17429140903125848

Timmusk S., Abd El-Daim I.A., Copolovici L., Tanilas T., Kännaste A., Behers L., Nevo E., Seisenbaeva G., Stenström E. and Niinemets U. 2014. Drought-tolerance of wheat improved by rhizosphere bacteria from harsh environments: Enhanced biomass production and reduced emissions of stress volatiles, *PLoS One* 9: e96086.

Tolba S. T. M., Ibrahim M., Amer E. A. M. and Ahmed D. A. M. First insights into salt tolerance improvement of Stevia by plant growth-promoting *Streptomyces species, Arch. Microbiol.* 201: 1295–1306. doi: 10.1007/s00203-019-01696-y

Turan M., Ekinci M., Yildirim E., Güneş A., Karagöz K., Kotan R. and Dursun A. 2014. Plant growth-promoting rhizobacteria improved growth, nutrient, and hormone content of cabbage (*Brassica oleracea*) seedlings, *Turk. J. Agric. For.* 38: 327–333.

Van Oosten M. J., Di Stasio E., Cirillo V., Silletti S., Ventorino V. and Pepe, O. 2018. Root inoculation with *Azotobacter chroococcum* 76A enhances tomato plants adaptation to salt stress under low N conditions, *BMC Plant Biol.* 18: 205. doi: 10.1186/s12870-018-1411-5

Visick J. E., Ichikawa J. K. and Clarke S. 1998. Mutations in the Escherishia Coli SurE gene increase isoasparrtyl accumulations in a strain lacking the pcm repair methyltransferase but suppress survival phenotypes. *FEMS Microbiol. Lett.* 167: 19–25.

Vurukonda S. S. K. P., Giovanardi D. and Stefani E. 2018. Plant growth promoting and biocontrol activity of streptomyces spp. As endophytes, *Int. J. Mol. Sci.* 19: 952. doi: 10.3390/ijms19040952

Wong W. S., Tan S. N., Ge L., Chen X. and Yong J. W. H. 2015. The importance of phytohormones and microbes in biofertilizers. In: Maheshwari DK (ed) *Bacterial metabolites in sustainable agro- ecosystem, sustainable development and biodiversity*, Springer International Publishing, Switzerland. 12: 105–158.

Yildirim E., Karlidag H., Turan M., Dursun A. and Goktepe, F. 2011. Growth nutrient uptake, and yield promotion of broccoli by plant growth promoting rhizobacteria with manure, Hortic. *Sci.* 46: 932–936.

Zhao S., Zhou N., Zhao Z. Y., Zhang K., Wu G. H. and Tian C. Y. 2016. Isolation of endophytic plant growth-promoting bacteria associated with the Halophyte Salicornia europaea and evaluation of their promoting activity under salt stress, Curr. *Microbiol.* 73: 574–581.

Zheng B. X., Ibrahim M., Zhang D. P., Bi Q. F., Li H. Z. and Zhou G. W. 2018. Identification and characterization of inorganic-phosphate-solubilizing bacteria from agricultural fields with a rapid isolation method, *AMB Express* 8: 47. doi: 10.1186/s13568-018-0575-6

9

Probiotic Fruit and Vegetable Products: Biotechnological Tools for Functional Enrichment

Anika, Priyanka Thakur, Atul Dhiman, Vikas Kumar and Satish Kumar

9.1 Introduction

In India, the green revolution and subsequent efforts to increase the food production through various scientific interventions have led to an increase in food sufficiency. Yet, the country continues to struggle with the issue of inefficient food storage techniques, which result in significant food waste leading to both economic and nutritional losses. Such food losses exacerbate the current malnutrition and nutrition crisis in the country. This is where the significance of fruits and vegetables comes into play as they are among the richest sources of various macro and micronutrients along with various health-promoting substances. Fruits and vegetables, which are perishable commodities, have high nutritive value and contribute significantly to human well-being. The World Health Organization (WHO) supports the global movement to encourage people to eat more fruits and vegetables for their healthy well-being to reduce the pressure on the healthcare system. Fruits and vegetables serve as important sources of many vital compounds including antioxidants, and are cholesterol-free, low in salt and rich in many minerals like calcium, zinc, phosphorus, iron, potassium and vitamins. However, they are available as a seasonal surplus, resulting in a glut in the market during one season and scarcity during another. Hence, processing of fruits and vegetables is a great alternative to make them available throughout the year. The value-added products from fruits and vegetables have also gained significant market share in recent years due to increased health awareness among the people and their desire for a healthy and disease-free life. Various fruit and vegetable products like-unfermented fruit beverages (juice, nectar, squash, RTS, crush, syrup, etc.), fermented fruit beverages (wine, champaigne, cider, etc.), jam, jelly marmalade, leather, preserve, chutney, sauces, ketchup, pickles, etc., are being produced nowadays at a large scale in the food processing industry.

DOI: 10.1201/9781003625674-9

These processed products are wholesome, safe, nutritious, highly acceptable by the consumers and available throughout the year. In addition to this, it also helps in generating new income sources for farmers and reduction in imported processed products and helps to increase export earnings.

However, there is a growing demand among the consumers for the food products with the improved immunological response especially after the COVID-19 outbreak all over the world. Hence, the concept of functional foods has become more prevalent among both the processing industries and the consumers. By enhancing or inhibiting immunological responses, functional foods can improve the immune system, boost host defense mechanisms against infection, and reduce allergies and inflammation. Prebiotics and probiotics are the two food components whose production is constantly increasing with a vision to combat various health issues that persist for a long time and cause degenerative diseases. Because of their ability to promote health, probiotics and prebiotics are currently the most well-known functional foods (Ashaolu 2020). The present chapter has been conceptualized to provide an insight into the historic developments in the field of probiotics, their production innovations, product development, biotechnological advances for enhanced bioactivity and the researchable issues identified. It is anticipated that this chapter shall cater to the needs of the research community by providing a detailed update on the topic and the future research areas to explore, while for students, it shall be ready to refer to material to update their scientific know-how on the science of the probiotics.

9.2 The Fundamental Concept of Probiotics, Prebiotics and Their Role in Overall Health Management

"Probiotics" are living microbes (typically beneficial bacteria, yeast) that, when taken in adequate levels, boost the host's health. "Probiotics" include a wide range of microbes. The genera *Lactobacillus* and *Bifidobacterium*, which make up the majority of the normal gut microbiota in a variety of mammalian species, are two of the most commonly utilized probiotics. Some other genera utilized as probiotics are *Enterococci, Leuconostoc* and yeast like *Saccharomyces*. But, due to the dairy products' optimal culture environment for probiotic viability, most of the probiotics are dairy-based and hence, the dairy industry alone accounts for about 78% of yearly probiotic sales globally. Dairy-based probiotics have immense health potential, however, issues like lactose intolerance and other disorders, such as hypersensitivity to protein and the fat present in milk-based products limit their consumption in specific population groups. This has resulted in popularization of the fruit and vegetable-based probiotic food products with an upsurge in their production, which is further supported by the trends like vegetarianism and veganism (Ashaolu 2020; Lillo-Pérez et al. 2021). Examples of probiotic fruit and vegetable products include fermented vegetables, such as kimchi and sauerkraut, and fermented fruit beverages, such as kefir and kombucha. Apart from this, several findings elucidate numerous health benefits of probiotics in humans, which can be inferred to as the proficiency of probiotics in lowering blood pressure and hypertension, improve cholesterol levels in the blood, aid in managing diabetes, lower the risk of degenerative diseases, lessen symptoms of gastrointestinal disorders, and prevent various allergies (Lillo-Pérez et al. 2021). The diversity of various probiotic foods, their health potential and potential improvement strategies have been listed in Table 9.1.

According to International Scientific Association of Probiotics and Prebiotics (ISAPP), 2016, prebiotic is "a substrate that is selectively exploited by host microorganisms delivering a health benefit". Prebiotics can be used in place of probiotics or as an additional supplement to sustain the healthy microbes in the gut. The most well-known prebiotics are naturally occurring carbohydrates found in fruits and vegetables like banana, greens, asparagus, berries, garlic, onion and chicory, while some other potential prebiotics are peptides, proteins and certain lipids. The major health benefits of prebiotics are laxation, blood glucose level regulation, prevention of colorectal cancer, prevention of IBD (irritable bowel syndrome), improvement in immune responses, etc. (Sanders et al. 2018; Ashaolu 2020). Including fruits and vegetables in the regular diet on a daily basis has shown a positive impact on health as they provide the necessary accessory substances to sustain normal health, especially prebiotics. Further, the fermented fruit and vegetable-based food products have both the pre- and probiotics in abundance, which have additional health benefits.

9.3 The Emerging Concept of Parabiotics and Postbiotics

Recent studies have proposed that the viability of bacteria is not an obligatory requirement in functional foods to realize the desired health effects, which are often related to probiotics. Thus, giving rise to the emerging concept of paraprobiotics and post-biotics, which exhibit the inessentiality of bacterial viability for the attainment of health-promoting benefits, thus, broadening the potential opportunity for improvement in the field of functional foods (Aguilar-Toalá et al. 2018). Parabiotics are "inactivated and dead microorganisms/microbial fragments", released in enough proportions, which aid in improving human health and is also known as ghost or inactivated probiotics. Parabiotics mainly consist of non-living microorganisms that are either whole or in cellular fragment and constitute polysaccharide like

TABLE 9.1

An Exposition on Various Probiotic Food Products Along with Their Health Benefits and the Scope for the Potential Improvement by Utilization of Various Biotechnological Tools

Sr. No	Food Product	Microorganisms nvolved	Key Bioactive Compounds	Proposed Health Benefits	Potential Improvement by Biotechnological Intervention	References
1.	Kefir	Lactic acid bacteria (LAB), (genera *Lactobacillus*, *Lactococcus*, *Leuconostoc*, and *Streptococcus*) Yeasts (primarily *Kluyveromyces*, *Candida*, and *Saccharomyces*)	Carotenoid, flavonoid, anthocyanin	• Anti-inflammatory • Antioxidant • Cancer-fighting • Anti-microbial • Anti- diabetic • Anti-hypertensive • Anti-hypercholesterolemic	GMO's in kefir production can potentially enhance the production of specific compounds, such as vitamins or enzymes	Azizi et al. 2021; Egea et al. 2022
2.	Kimchi	LAB (Genera, *Leuconostoc*, *Weissella*, and *Lactobacillus*, *Pediococcus*, *Lactococcus*, *Enterococcus*, and *Streptococcus*)	Gingerol, capsaicin, carotenoids, allyl compounds, isothiocyanate, and indole-3-carbinol	• Immunomodulatory effects • Antimutagen • Antioxidation • Angiotensin-converting enzyme inhibition	Genetic improvement for potential production of vitamin B_{12}. Addition of specific strains of bacteria can increase the production of vitamins B_{12}	Cha et al. 2023; Panghal et al. 2018; Azizi et al. 2021
3.	Shalgam juice	*Lactiplantibacillus pentosus*, *L. plantarum subsp. plantarum and Saccharomyces cerevisiae*	Anthocyanin	• Favourable health benefit • Prevent heart disorders • Anti-inflammation	Interventions to improve the stability of anthocyanin as some strains of *Lactobacillus* have been shown to improve the stability of anthocyanins	Akman et al. 2021; Panghal et al. 2018
4.	Gilaburu juice	LAB	Polyphenols, phenolics, flavonoids and anthocyanins	• Acid, bile tolerance • Antimicrobial activity • Antibiotic resistance	Use of novel high-pressure processing technology to retain the bioactive content of the juice	Akman et al. 2021

5.	Hardaliye	*L. casei and L. paracasei, L. sanfranciscensis, L. brevis, L. pontis, L. acetotolerans, L. vaccinostercus. L. plantarum, L. delbrueckii, and L. rahamnosus*	Anthocyanin, flavonoids, polyphenols	• Decreases plasma levels: homocysteine, dienoconjugate, malondialdehyde • Anti-inflammatory activity • Anti-oxidant	Gene-editing technique to increase the expression of specific enzyme/protein involved in production of bioactive compounds of fruit used in hardaliye production	Azizi et al. 2021; Behera and Panda 2020
6.	Orange juice	*L. plantarum*	Flavonoid, anthocyanins	• Anti-microbial • Anti-oxidant • Anti-inflammatory	Increasing the viability of the probiotic bacteria in orange juice Use of biotechnological intervention to creat probiotic orange juice is still a developing field, and more research is needed to fully understand the safety, efficacy, and consumer acceptance of these products	Behera and Panda 2020
7.	Mango juice	*L. acidophilus, L. delbruekii, L. plantarum, L. plantarum* and *L. casei*	Phenolic, flavonoids, carotenoids	• Anti-oxidant • Anti-inflammatory	Enzyme technology can be used to reduce sugar content of mango juice, while maintaining its taste and flavor. Genetic engineering to increase the vitamin C content of mango juice	Behera and Panda 2020; Panghal et al. 2018
8.	Jiang-gua (fermented cucumber)	*Weissella (W.) cibaria, W. hellenica, L. plantarum, L. lactis,* and *Enterococcus casseliflavus*	Carotenoid, phenolic acid, flavonoid (quercetin), chlorogenic acid	• Anti-oxidant • Anti-inflammatory • Anti-tumor • Immunomodulatory properties	Enzyme technology can be used to increase fiber content in jiang gua, while genetic engineering can be used to increase the production of certain flavor compounds	Behera and Panda 2020

(continued)

TABLE 9.1 (Continued)

An Exposition on Various Probiotic Food Products Along with Their Health Benefits and the Scope for the Potential Improvement by Utilization of Various Biotechnological Tools

Sr. No	Food Product	Microorganisms nvolved	Key Bioactive Compounds	Proposed Health Benefits	Potential Improvement by Biotechnological Intervention	References
9.	Tursu	*L. plantarum*, *P. pentosaceus* and *L. mesenteroides*	Polyphenols. flavonoids, carotenoids	• Anti-oxidant • Anti-inflammatory • Preventing cardiovascular disease, diabetes	Improvement of flavor, nutritional properties and texture by genetic manipulations	Behera and Panda 2020
10.	Pak-Gard-Dong	LAB	Secondary metabolite, organic acids, polyphenols, and flavonoids	• Boost immune system • Anti-inflammatory	Possible synergistic health effect may be explored	Panghal et al. 2018
11.	Black current juice	*L. plantarum*	Vitamin C, anthocyanin, phenolic compounds	• Improving cognitive function • Reducing risk of chronic disease (cancer, heart disease) • Anti-oxidant activity	Enzyme technology can be used to reduce the sugar content of blackcurrent juice Genetic manipulations to increase anthocyanin content of juice, which is a potent antioxidant	Panghal et al. 2018
12.	Peach juice	*L. delbrueckii*	Vitamin C, vitamin A, Carotenoid (beta-carotene. lycopene), phenolic acid, flavonoids	• Healthy beverage for lactose-allergic consumers • Prevent chronic disease (heart disease, cancer, and diabetes) • Anti-microbial • Anti-inflammatory	Genetic engineering can be used to increase the antioxidant production	Panghal et al. 2018
13.	Peach pickle	*L. brevis, L. lactis, Weissella cibaria, W. paramesenteroides, W. minor, Leuconostoc mesenteroides, and Enterococcus faecalis*	Organic acid, phenolic compound (flavonoids, phenolic acid)	• Anti-microbial • Anti-oxidant • Anti-inflammatory • Preventing degenerative disease	Higher retention of the antioxidants in peach pickle by enhancing the phenolic compounds	Panghal et al. 2018

14.	Olive drink	*L. pentosus, L. plantarum, Pediococcus cerevisiae, L. mesenteroides, L. brevis*	Organic acids, phenols, antioxidants	• Anti-oxidant • Anti-atherogenic	Added bacteria can ferment sugars in drinks, producing additional health-promoting metabolites and reducing the sugar content	Azizi et al. 2021
15.	Cashew apple juice	*L. casei*	High ascorbic acid, phenol, antioxidants	• Vitamin • Anti-oxidant • Polyphenol, all of which have a number of beneficial effects on human health.	Enzyme technology to increase bioactive compounds and production of enzymes that degrade undesirable compounds in the juice, i.e., tannins	Pereira et al. 2014
16.	Pomegranate juice	*L. bulgaricus and L. plantarum*	Antioxidants such as vitamin E, ascorbic acid, anthocyanins, and vitamin C	• Antiviral • Antibacterial • Cancer-fighting • Antimutagenic	Efforts directed for the improved bioavailabilty and absorption of anthocyanin in the gut	Panghal et al. 2018
17.	Kanji	*L. plantarum, L. acidophilus, L. casei*	Beta carotene, tocopherol, ascorbic acid	• Improving lactose metabolism • Preventing gastrointestinal infections • Boosting immunity against chronic conditions	Proper fermentation method can improve the stability of bioactive compounds	Panghal et al. 2018; Kaur et al. 2023
18.	Moringa and beetroot beverage	*L. plantarum, Enterococcus hirae*	Phenolic acid, nitrates, betaine	• Nutritious beverage • Free radical scavenging efficiency	Beverage can help to increase the bioavailability and absorption of nutrients in the gut	Panghal et al. 2018
19.	Yan-tsai-shin (fermented broccoli)	*L. plantarum, Enterococcus sulphurous, W. paramesenteroides, W. minor, Leuconostoc mesenteroides and W. cibaria*	Phenolic compounds (quercetin, kaempferol, isorhamnetin)	• Anti-cancer properties • Anti-oxidant • Anti-inflammatory	Enzymes to improve the texture and nutritional value, adding amylase in fermentation break down carbohydrate into simple sugars making them digestable	Panghal et al. 2018

(continued)

TABLE 9.1 (Continued)

An Exposition on Various Probiotic Food Products Along with Their Health Benefits and the Scope for the Potential Improvement by Utilization of Various Biotechnological Tools

Sr. No	Food Product	Microorganisms nvolved	Key Bioactive Compounds	Proposed Health Benefits	Potential Improvement by Biotechnological Intervention	References
20.	Sauerkraut	*L. mesenteroides, L. plantarum, L. brevis, L. rhamnosus and L. plantarum*	Vitamins C and B, phenolic acid	• Prevent cirrhosis and diarrhea	During fermentation, controlling of temperature and pH can optimize growth of probiotic strains while inhibiting the growth of harmful bacteria	Azizi et al. 2021; Zabat et al. 2018
21.	Dhamuoi	*Leuconostoc mesenteroides and L. plantarum*	Carotenoids, polyphenols (catechins, epicatechins, quercetin)	• Anti-oxidant • Reduces macular degeneration • Prevent DNA damage • Anti-diabetic • Prevent Alzheimer's disease	GM technique is used to increase the level of carotenoids	Rasika et al. 2021
22.	Gundruk	*Pediococcus and L. plantarum, L. fermentum, L. casei*	Polyphenol, flavonoids, carotenoids, vitamin C, vitamin K	• Prevent heart disease • Prevent diabetes • Reduce risk of cancer	Biofilm technology helps to improve the survival and stability of probiotic bacteria during processing and storage	Manas Ranjan Swain et al. 2014
23.	Paocai	*L. pentosus, L. plantarum, Leuconostoc mesenteroides, L. brevis, L. lactis, and L. fermentum*	Vitamin C. K, organic acid, flavonoids, carotenoid, polyphenol	• Anti-oxidant activity • Reduces risk of food • orne disease • Reduce inflammation • Reduce constipation	Genetic engineering techniques can be used to modify the genes that confer resistance to environmental stresses, such as temperature or pH fluctuations	Rasika et al. 2021

24.	Sayurasin	*L. mesenteroides, L. confuses, L. plantarum and P. pentosaceus*	Vitamins & minerals (B, iron)	• Anti-oxidant • Anti-inflammatory • Anti-microbial	Genetic engineering can be used to modify genes of LAB strains, enhance the production of desirable compounds, such as vitamins or antioxidants	Rasika et al. 2021
25.	Kombucha	*Torulopsis species, Saccharomyces ludwigii, S. cerevisiae, S. bisporus, and Zygo saccharomyces*	Vitamins C and B2, polyphenols like catechins	• Improved digestive health • Increase energy and immunity • Reduced risk of chronic disease • Detoxification	Nutrient fortification optimization for the fermentation	Pamungkas et al. 2022
26.	Jiang-gua (cucumber)	*Weissella cibaria W. hellenica, L. plantarum, Leuconostoc lactis, Enterococcus casseliflavus*	Vitamin C, flavonoids (quercetin, kaempferol), triterpenoids (cucurbitacin B, lupeol)	• Anti-bacterial • Anti-oxidant • Anti-inflammatory • Improve blood vessel function • Anti- tumorous • Anti-diabetic	GM technique to increase the level of vitamin C and other antioxidants	Kumar et al. 2022
27.	Inziangsang (mustard leaves)	*L. brevis, L. plantarum, Pediococcus acidilactici*	Phenolic compound, vitamin B12	• Improve immune function • Improve digestion • Anti-inflammatory • Anti-microbial • Anti-oxidant	Adding protease enzyme during fermentation can help break down protein, making them more easily digestible	Manas Ranjan Swain et al. 2014
28.	Nozawana-Zuke (turnip, mustard)	*L. curvatus*	Vitamin C, vitamin K, polyphenols	• Anti-oxidant • Anti-inflammatory • Anti-cancerous	New packaging technologies like vacuum packaging or modified atmosphere packaging can be used to prevent oxidation and spoilage	Manas Ranjan Swain and Ray 2017

(continued)

TABLE 9.1 (Continued)

An Exposition on Various Probiotic Food Products Along with Their Health Benefits and the Scope for the Potential Improvement by Utilization of Various Biotechnological Tools

Sr. No	Food Product	Microorganisms nvolved	Key Bioactive Compounds	Proposed Health Benefits	Potential Improvement by Biotechnological Intervention	References
29.	Sinki and Sunki (radish and turnip)	*L. plantarum, L. brevis, L. fermentum, L. fallax and P. pentosaceus and Bacillus coagulans*	Phenolic compound, carotenoids	• Anti-oxidant • Reduces risk of chronic disease	Addition of prebiotics, such as insulin or fructooligosaccharides, can help to promote the growth of beneficial bacteria in the gut	Manas Ranjan Swain et al. 2014
30.	Fruit bars	*L. casei, L. plantarum, L. acidophilus, Bifidobacterium bifidum*	Polyphenols, flavonoids, carotenoids, fiber	• Anti-inflammatory • Anti-oxidant • Neuroprotective effect • Maintain cholesterol level • Improve blood sugar level	Biotechnology can be used to develop innovative packaging materials and technologies to extend shelf-life	Rasika et al. 2021
31.	Vegetable soups	*Lactobacillus phallushillus, L. rhamnosus, Bifidobacterium lactis*	Carotenoids, phenolic acid, flavonoids, vitamin C, vitamin A, folate	• Support immune system • Lower the risk of chronic disease • Promotes healthy digestion	Biotechnological interventions, such as fermentation, addition of probiotic cultures, and prebiotic ingredients can be used to create a nutritious and flavorful probiotic vegetable soup with enhanced health benefits	Kumar et al. 2022
32.	Goyang	*L. plantarum, L. brevis, Lactococcus lactus, Enterococcus faecium, Pediococcus pentosaceus*	Flavonoids (apigenin, luteolin, quercetin), vitamins, coumarins, phthalides,	• Anti- microbial • Anti inflammatory • Anti-oxidant • Anti-tumor • Reduces risk of chronic disease	Gene modification to enhance the production of antioxidants, vitamins and minerals	Tamang and Tamang 2007

33.	Sayur Asin	*Leuconostoc mesenteroides, Lactobacillus confuses, L. curvatus, Bacillus, Staphylococcus, and Corynebacterium, Candida sake, Candida guilliermondii*	Vitamin C, beta-carotene, polyphenols, glucosinolates, allicin	• Boost metabolism • Anti-inflammatory • Support healthy bones, skin, and immune function. • Anti-oxidant	Gene modification to enhance the nutritional content and increase the level of vitamins and minerals Enzyme treatment can be used to improve the texture and flavor by breaking down proteins (proteolytic enzymes)	Ryan, Cooper, and Tauer 2013
34.	Tempoyak	*L. brevis, L. mali, L. fermentum, L. durianis, Leuconostoc mesenteriodes*	Polyphenols, carotenoids (beta carotene, lutein), flavonoids (quercetin, kaemferol)	• Anti-microbial • Anti-mutagenic • Immunostimulatory	Genetic modification can be used to improve the characteristics of bacteria used in fermentation, leads to the production of higher levels of beneficial compounds, such as anti-oxidants	Rajagukguk and Arnold 2021

exopolysaccharide, cell-surface-associated proteins, derivatives of peptidoglycan upon break, etc. (Nataraj et al. 2020). Examples of parabiotic fruit and vegetable products include grape seed extract and pomegranate extract, which contain polyphenols that can act as prebiotics and promote the growth of beneficial bacteria in the gut. Parabiotics have an advantage over probiotics in terms of their ability to prevent the risk of microbial translocation, infection or enhanced inflammatory responses mainly in people with imbalanced immune system. Major effects of paraprobiotics on human health are immunological modulation, antagonistic effects against pathogens, anti-tumor effect, etc. (Aguilar-Toalá et al. 2018). According to ISAPP, post-biotic is termed as the formulation of non-living microbes and their constituents that enhance the health benefits. These are intentionally disabled microorganisms that either include several metabolites or other cell constituents that have a positive effect on health (Sanders et al. 2018). Post-biotics can be categorized on the basis of their elemental structure, i.e., carbohydrates (e.g., teichoic acid), proteins (e.g., lactocepin), organic acids (e.g., propionic acid) etc. or their physiological functions, which include anti-inflammatory, anti-oxidant, anti-hypertensive, anti-obesogenic, immunomodulation, etc. Major bacterial strains, which are used as a source of post-biotics are *Lactobacillus* and *Bifidobacterium, Streptococcus, Faecalibacterium.* Post-biotics have emerged as novel functional food ingredients and have gained significant importance in the recent past while their positive impacts on health is gradually becoming more and more relevant, as supported by recent research findings. The diversity of the pre-, pro-, para- and post-biotics is illustrated in Figure 9.1 along with their health benefits.

9.4 Diversity and Potential of the Fruit and Vegetable-Based Probiotic Food Products

Fruit and vegetable-based probiotics are being potentially developed as an alternative to the dairy-based probiotics due to the concern of lactose intolerance and are high in demand due to the growing health concerns among the consumers especially after the occurrence of COVID-19. Fruits and vegetables are a suitable substrate as they do not contain lactose, cholesterol or other dairy allergens while they are designated as the foods rich in various health-promoting phytochemicals. Probiotic foods have been prepared using products made from fruits and vegetables, such as beverages, purées, fermented vegetables, table olives and minimally processed fruits etc. As an instance, the fruit and vegetable juices are considered as rich sources of the bioactive compounds, which can help in building up the phytochemical

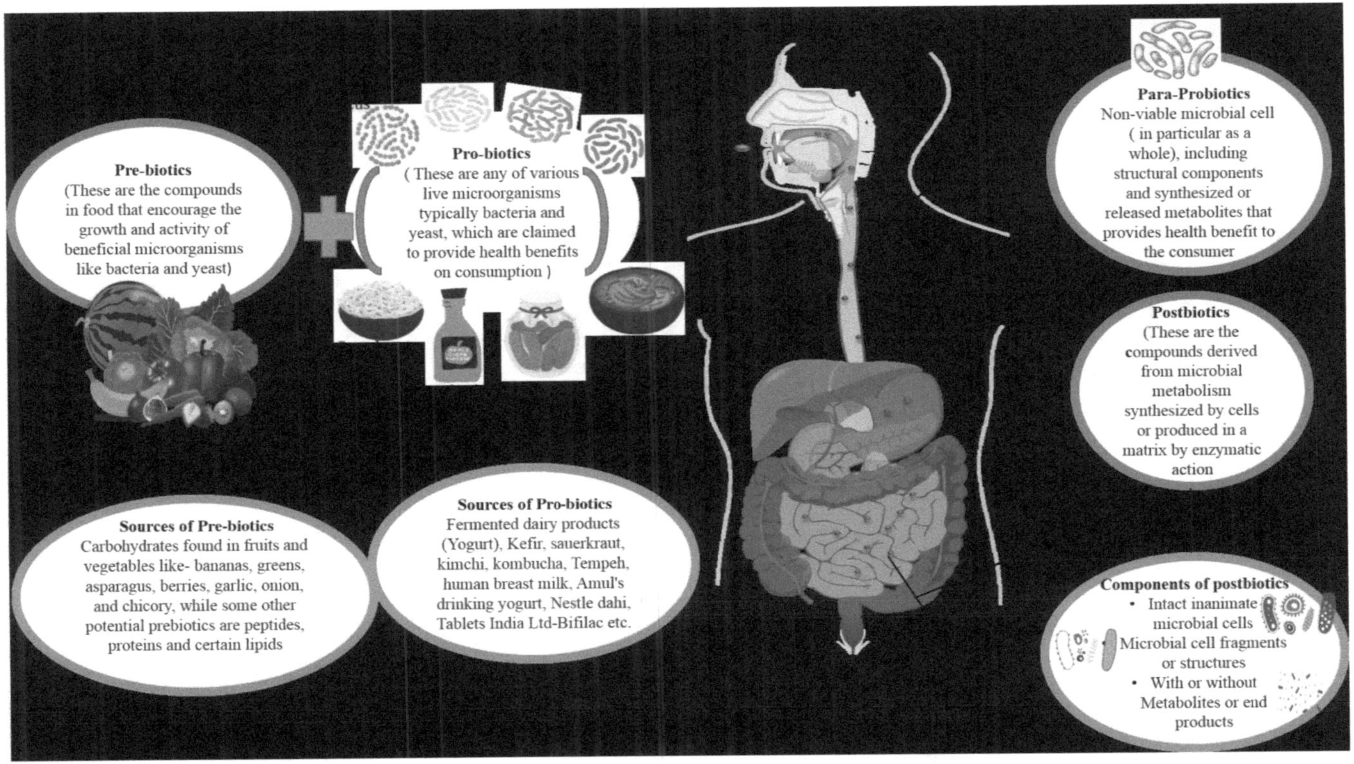

FIGURE 9.1
Illustration of pre-, pro-, para- and post-biotics along with their health benefits to consumers.

profile of the body. Apart from fresh juices, fermented beverages from fruit and vegetable juices are also becoming popular, though they have been under commercial consumption for ages. Lactic acid fermented juices have improved sensory, nutritional, and shelf-life qualities compared to the fresh juices.

In food processing, fermentation serves a variety of purposes, including preserving food by producing several metabolites, such as different organic acids and various potential bacteriocins, increasing the safety requirement of food by hindering various toxic pathogens, and enriching the nutrition content and sensory attributes of the product (Profir et al. 2015). Probiotics originating from fruits and vegetables have become more popular in recent times as the demand for plant-based products has grown more abruptly. The desire of customers for healthier substitutes to dairy probiotics has increased as a result of the growing understanding of the allergenicity and high amounts of cholesterol and fat in milk. This has forced the food community to develop appropriate fruit and vegetable-based probiotic drinks in order to satisfy the need for beverages made without dairy and with vegetarian ingredients (Mojikon et al. 2022). Therefore, in order to meet the need for beverages prepared without dairy and with vegetarian ingredients, it is required to develop the usage of fruits and vegetables to generate probiotic food products. Fruit jam is another perfect example of functional food to provide probiotics because it is simple to make, an excellent source of sugar, and consumed by the majority of rural residents. *U. kirkiana*, an underutilized fruit, was used to create a probiotic jam for the southern African population that lacks resources and thus can be an excellent technology for the nutritional security in such regions (Chawafambira et al. 2020). Fruits such as pineapple, mango, grape, strawberry, olive and orange are used to create probiotic fruit-based products. Turkey's native drink, *Hurduliye*, is made from grapes and is formulated by LAB (*L. casei* and *L. paracasei*) fermentation using crushed mustard seeds, and benzoic acids as additional supplements (Behera and Panda 2020). Cha et al. (2023) investigated whether *kimchi* (fermented cabbage probiotic food) contains various bioactive substances, i.e., indole-3-carbinol, capsaicin, carotenoids, allyl compounds and gingerol, which has been reported to improve the microbiome in the digestive tract. There is currently no evidence to suggest that fruit and vegetable-based probiotic foods can directly prevent COVID-19. However, consuming a healthy diet that includes a variety of fruits and vegetables can help support overall immune system function, which is important for fighting infections, including COVID-19. Probiotic foods such as *kefir* and fermented vegetables can also help in supporting gut health and may indirectly support immune function. More scientific and systematic study is required to comprehend the effect of those microorganisms on gut health in the immune system to validate their use as potential therapeutic agents and support their commercial recommendation. It is important to note that a healthy diet and lifestyle can support overall health and immunity (Akar Şahingöz and Yalçın 2022). Together with the usage of probiotics, the general population's consumption of fermented foods like kefir has increased, which further supports the claims for various therapeutic applications of such products (Egea et al. 2022). The fruit of the wild bush known as *gilaburu*, are used to prepare "gilaburu juice" made by pressing the fruits once fermentation is complete, which serve as a good source of different bioactive compounds and are rich in various prebiotic substances and have been reported to support the growth of probiotic microbes in the gut. Some traditionally fermented foods contain large amounts of LAB due to the prolonged fermentation process, which can have additional health benefits (Akman et al. 2021). Non-dairy food matrices (fruits and vegetables) have demonstrated the potential for providing probiotics with high viability and may be an excellent alternative to give benefits without milk proteins, lactose, saturated fat and cholesterol (Min et al. 2019).

9.5 Challenges in Production, Processing and Storage of Fruit and Vegetable-Based Probiotics

The development of fruit and vegetable-based probiotic products is an emerging as well as challenging field in the food processing industry. The various problems faced during production, processing and storage of probiotic foods based on fruits and vegetables is attributed to the fact that the raw materials used can serve as a ground for the microbial contamination and can have variable compositions that can affect the viability and activity of probiotic microorganisms. Various problems may arise during the production, processing and storage of fruit and vegetable-based probiotic foods. The most important among all is the microbial contamination. Fruits and vegetables and their products are rich in nutrients, which can provide an excellent ground for the microorganisms to breed and spoil them. Foods can be contaminated with harmful bacteria, viruses and parasites, which can compete with probiotic microorganisms and compromise their viability and activity. This risk can however be reduced by implementing good agricultural and hygiene practices during production and processing. Variability of composition of the fruits and vegetables in terms of their nutrient content, acidity and water activity, can also affect the viability and activity of probiotic microorganisms. This variability can be addressed by adjusting the processing conditions to optimize the growth and survival of the probiotics. The limited shelf-life of fruits and vegetables due to very high moisture content can promote the growth of spoilage microorganisms and cause the degradation of nutrients and flavors. This can be mitigated by applying appropriate preservation methods, such as refrigeration, freezing or drying. Additionally, some of the fruits and vegetables have strong flavors and odors that can affect the acceptance of the probiotic foods if there are no flavors produced by the microorganisms. This can be addressed by using flavor-masking techniques or by combining fruits and vegetables with other ingredients to improve the taste and quality of the final product. Further various regulatory compliance requirements for the probiotic food during production, processing, labeling, and marketing in many countries pose a challenge for the producers. This can be addressed by staying up-to-date with the latest regulations and by working with regulatory authorities to ensure compliance.

9.6 Conclusion

Fruit and vegetable products are excellent source of prebiotics and play an important role in health management by providing nutrients, bioactive compounds and microbial components that can modulate the gut microbiota and enhance digestive, metabolic and immune functions. Incorporating a variety of fruits and vegetables into the diet can help to maintain a healthy lifestyle. The prospect of probiotic food development is wide, as probiotics can be added to many different types of food products, including dairy products, fermented vegetable beverages, such as kefir and kombucha, meat product, cereals and beverages. With the advancement in science and technology in the field of functional foods, it is likely that the new sources of pre-, pro- and parabiotics will be discovered in fruits and vegetables. Novel biotechnological tools are seen as the potential technologies for the improvement in the production, processing and marketing of such products. It has been realized that the post-biotics market is growing at a rapid pace and technological interventions are desired on how effectively we can grow probiotics in the laboratory on petri plates and extract the post- and parabiotic components from them, process them and supply the purest possible form to the consumers. Although, the science of pre- and probiotics is well established, yet the concepts of parabiotics and post-biotics are relatively new and require extensive scientific and systematically designed studies with medically proven and well-documented health benefits to increase and justify the applications of these functional food ingredients in our regular diet for our well-being. Future work may be directed on increasing the viability of the probiotics in the food and gut, effective technological tools for the extraction of the para- and post-biotic compounds and novel food product development with the most efficient delivery of the biomolecules to the target organs for addressing specific health problems for the general well-being of the masses.

References

Aguilar-Toalá, J. E., R. Garcia-Varela, H. S. Garcia, V. Mata-Haro, A. F. González-Córdova, B. Vallejo-Cordoba, and A. Hernández-Mendoza. 2018. "Postbiotics: An Evolving Term within the Functional Foods Field." *Trends in Food Science and Technology* 75: 105–14. https://doi.org/10.1016/j.tifs.2018.03.009.

Akar Şahingöz, Semra, and Eren Yalçın. 2022. "The Covid-19 Pandemic and Fermented Products." *Journal of Tourism and Gastronomy Studies*, no. June. https://doi.org/10.21325/jotags.2022.1020.

Akman, Perihan Kubra, Gorkem Ozulku, Fatih Tornuk, and Hasan Yetim. 2021. "Potential Probiotic Lactic Acid Bacteria Isolated from Fermented Gilaburu and Shalgam Beverages." *Lwt* 149(April): 111705. https://doi.org/10.1016/j.lwt.2021.111705.

Ashaolu, Tolulope Joshua. 2020. "Immune Boosting Functional Foods and Their Mechanisms: A Critical Evaluation of Probiotics and Prebiotics." *Biomedicine and Pharmacotherapy* 130: 110625. https://doi.org/10.1016/j.biopha.2020.110625.

Azizi, Nor Farahin, Muganti Rajah Kumar, Swee Keong Yeap, Janna Ong Abdullah, Melati Khalid, Abdul Rahman Omar, Mohd Azuraidi Osman, Sharifah Alawieyah Syed Mortadza, and Noorjahan Banu Alitheen. 2021. "Kefir and Its Biological Activities." *Foods* 10 (6): 1–26. https://doi.org/10.3390/foods10061210.

Behera, Sudhanshu S., and Sandeep K. Panda. 2020. "Ethnic and Industrial Probiotic Foods and Beverages: Efficacy and Acceptance." *Current Opinion in Food Science* 32: 29–36. https://doi.org/10.1016/j.cofs.2020.01.006.

Cha, Jeongmin, Yeon Bee Kim, Seong Eun Park, Se Hee Lee, Seong Woon Roh, Hong Seok Son, and Tae Woong Whon. 2023. "Does Kimchi Deserve the Status of a Probiotic Food?" *Critical Reviews in Food Science and Nutrition*: 1–14. https://doi.org/10.1080/10408398.2023.2170319.

Chawafambira, A., M. M. Sedibe, A. Mpofu, and M. C. Achilonu. 2020. "The Potential of Uapaca Kirkiana Fruit Jam for the Delivery of Lactobacillus Rhamnosus Yoba as a Probiotic Food." *African Journal of Food, Agriculture, Nutrition and Development* 20 (4): 16161–77. https://doi.org/10.18697/ajfand.92.19355.

Egea, Mariana Buranelo, Daiane Costa dos Santos, Josemar Gonçalves de Oliveira Filho, Joana da Costa Ores, Katiuchia Pereira Takeuchi, and Ailton Cesar Lemes. 2022. "A Review of Nondairy Kefir Products: Their Characteristics and Potential Human Health Benefits." *Critical Reviews in Food Science and Nutrition* 62 (6): 1536–52. https://doi.org/10.1080/10408398.2020.1844140.

Kaur, Preetinder, Ruchika Zalpouri, Ritika Modi, Param Pal Sahota, Tarsem Singh Dhillon, and Amrit Kaur. 2023. "Development and Standardization of Processing Technique for Ready-to-Use Lab Fermented Kanji Mix Using Refractance Window Dried Black Carrot Powder." *Scientific Reports* 13 (1): 1–12. https://doi.org/10.1038/s41598-023-27450-5.

Kumar, Sachin, Gurdeep Rattu, Swati Mitharwal, Abhishek Chandra, Sourabh Kumar, Aman Kaushik, Vijendra Mishra, and Prabhat K. Nema. 2022. "Trends in Non-Dairy-Based Probiotic Food Products: Advances and Challenges." *Journal of Food Processing and Preservation* 46 (9). https://doi.org/10.1111/jfpp.16578.

Lillo-Pérez, Siegried, María Guerra-Valle, Patricio Orellana-Palma, and Guillermo Petzold. 2021. "Probiotics in Fruit and Vegetable Matrices: Opportunities for Nondairy Consumers." *Lwt* 151 (April). https://doi.org/10.1016/j.lwt.2021.112106.

Min, Min, Craig R. Bunt, Susan L. Mason, and Malik A. Hussain. 2019. "Non-Dairy Probiotic Food Products: An Emerging Group of Functional Foods." *Critical Reviews in Food Science and Nutrition* 59 (16): 2626–41. https://doi.org/10.1080/10408398.2018.1462760.

Mojikon, Floyd Darren, Melisa Elsie Kasimin, Arnold Marshall Molujin, Jualang Azlan Gansau, and Roslina Jawan. 2022. "Probiotication of Nutritious Fruit and Vegetable Juices: An Alternative to Dairy-Based Probiotic Functional Products." *Nutrients* 14 (17). https://doi.org/10.3390/nu14173457.

Nataraj, Basavaprabhu H., Syed Azmal Ali, Pradip V. Behare, and Hariom Yadav. 2020. "Postbiotics-Parabiotics: The New Horizons in Microbial Biotherapy and Functional Foods." *Microbial Cell Factories* 19 (1): 1–23. https://doi.org/10.1186/s12934-020-01426-w.

Pamungkas, Barolym Tri, Ayu Safitri, Firman Rezaldi, Muhammad Andry, Lucky Dita Agustiansyah, M. Fariz Fadillah, Fajar Hidayanto, and Hari Hariadi. 2022. "Antifungal *Trycophyton Rubrum* and *Trycophyton Mentagrophytes* in Liquid Bath Soap Fermented Probiotic Kombucha Flower Telang (*Clitoria Ternatea* L) as a Pharmaceutical Biotechnology Product." *BIOTIK: Jurnal Ilmiah Biologi Teknologi Dan Kependidikan* 10 (2): 179. https://doi.org/10.22373/biotik.v10i2.15160.

Panghal, Anil, Sandeep Janghu, Kiran Virkar, Yogesh Gat, Vikas Kumar, and Navnidhi Chhikara. 2018. "Potential Non-Dairy Probiotic Products – A Healthy Approach." *Food Bioscience* 21: 80–89. https://doi.org/10.1016/j.fbio.2017.12.003.

Pereira, Ana Lúcia Fernandes, Francisca Diva Lima Almeida, Micael Andrade Lima, José Maria Correia da Costa, and Sueli Rodrigues. 2014. "Spray-Drying of Probiotic Cashew Apple Juice." *Food and Bioprocess Technology* 7 (9): 2492–99. https://doi.org/10.1007/s11947-013-1236-z.

Profir, Alina Georgiana, Camelia Viorica Neagu, and Camelia Vizireanu. 2015. "Impact of Nutrients on the Probiotic Survival and Sensory Properties of Vegetables Juice." *Romanian Biotechnological Letters* 20 (6): 11041–48.

Rajagukguk, Yolanda Victoria, and Marcellus Arnold. 2021. "Tempoyak: Fermented Durian Paste of Malay Ethnic and Its Functional Properties." *International Journal of Gastronomy and Food Science* 23 (December 2020): 100297. https://doi.org/10.1016/j.ijgfs.2020.100297.

Rasika, D. M. D., Janak K. Vidanarachchi, Selma F. Luiz, Denise Rosane Perdomo Azeredo, Adriano G. Cruz, and Chaminda Senaka Ranadheera. 2021. "Probiotic Delivery through Non-Dairy Plant-Based Food Matrices." *Agriculture (Switzerland)* 11 (7): 1–23. https://doi.org/10.3390/agriculture11070599.

Ryan, Cooper, and Tauer. 2013. "済無No Title No Title No Title." *Paper Knowledge. Toward a Media History of Documents*, no. January 2013: 12–26.

Sanders, M. E., D. Merenstein, C. A. Merrifield, and R. Hutkins. 2018. "Probiotics for Human Use." *Nutrition Bulletin* 43 (3): 212–25. https://doi.org/10.1111/nbu.12334.

Swain, Manas Ranjan, and Ramesh C. Ray. 2017. "Nutritional Values and Bioactive Compounds in Lactic Acid Fermented Vegetables and Fruits." *Lactic Acid*

Fermentation of Fruits and Vegetables, no. April 2015: 37–52. https://doi.org/10.1201/9781315370378.

Swain, Manas Ranjan, Marimuthu Anandharaj, Ramesh Chandra Ray, and Rizwana Parveen Rani. 2014. "Fermented Fruits and Vegetables of Asia: A Potential Source of Probiotics." *Biotechnology Research International* 2014: 1–19. https://doi.org/10.1155/2014/250424.

Tamang, Buddhiman, and Jyoti Prakash Tamang. 2007. "Role of Lactic Acid Bacteria and Their Functional Properties in Goyang, a Fermented Leafy Vegetable Product of the Sherpas." *Journal of Hill Research* 20 (2): 53–61.

Zabat, Michelle A., William H. Sano, Jenna I. Wurster, Damien J. Cabral, and Peter Belenky. 2018. "Microbial Community Analysis of Sauerkraut Fermentation Reveals a Stable and Rapidly Established Community." *Foods* 7 (5). https://doi.org/10.3390/foods7050077.

10

Fruit Wine From Non-Grape Fruit

Ankit Kumar, Rajat Chandel, Satish Kumar, Rakesh Gehlot, Ramandeep Kaur and Vikas Kumar

10.1 Introduction

Fruits from various indigenous plants are edible with a short harvesting span, causing surplus output. Although many of them are consumed instantly, however, during peak harvest times, significant amounts are wasted due to high temperature, fluctuating humidity, insufficient storage facilities, inappropriate handling, problematic transit and various microbial diseases. The food industries use a range of processing and preservation technologies to enhance the shelf stability of these perishable goods to ensure their year-round availability and safer delivery to areas outside the growing region (Barrett & Lloyd 2012; Joshi & Kumar 2011). Thus, using ripe fruits or their juices for winemaking is an appealing way to use up extra and overripe fruit in an efficient way.

Winemaking, one of humankind's oldest technologies ever known, is currently proven as the most economically successful biotechnological process concerning food industries (Moreno-Arribas & Polo 2005). Since the beginning of civilization, winemaking has evolved along with agricultural and human development (Chambers & Pretorius 2010). Earlier, in Neolithic cultures in the Middle East and China, the biomolecular archaeological evidence of plant additives in fermented beverages first appeared. During this time, the first domesticated flora and fauna served as the foundation for complex societies and long-term habitation (McGovern et al. 2009). Earlier, in ancient China, fermented drinks were frequently made from millet, rice and fruits. Yet, a variety of natural materials, including herbs and tree resins were used in Egypt in the past along with wine to formulate herbal therapeutic beverages and serve as a source of polyphenols and other bioactive substances (McGovern et al. 2009). Most of these bioactive substances are released throughout the winemaking process into an aqueous ethanolic solution, increasing their biological availability for human consumption (Shahidi 2009). Major polyphenol resveratrol found in grape skin and red

 DOI: 10.1201/9781003625674-10

wine reportedly showed cardioprotective, anti-inflammatory, anti-cancer, anti-diabetes and neuroprotective properties due to its ROS scavenging, metal chelating and enzyme modulating activities (Srikanta et al. 2016). Other useful phenolic compounds present in red wines are gallic acid, caffeic acid, quercetin, catechin and anthocyanins (Hill & Ganeshananthan 2008; Hiscott 2015).

Wine is always linked with grapes technically and commercially from age-old times. Due to its importance as a model plant for winemaking, almost all the molecular and genetic research was concentrated on grapes (Pretorius & Høj 2005). Besides grapes, nature has blessed us with substrates/raw materials that can be utilized for wine industries. Today's market offers a vast range of fruits that vary in size, color, flavor and nutritional value (Voguel 2003). According to the definition, wine is a fermented beverage created from grapes only, otherwise, wine is given the prefix of the fruit from which it is derived. It is a diversified commodities class and is considered the by-product of must's yeast fermentation. Clinical studies suggest that moderate red wine consumption is providing several health benefits, i.e., lowering mortality owing to cardiovascular morbidity as it contains antioxidant polyphenolics in abundance (Gresele et al. 2011; Halpern 2008). These phenolic acids can scavenge free radicals and reactive oxygen species, making them beneficial in preventing and treating diseases caused by them (Shahidi 2009). Moreover, wine polyphenols influence oral and intestinal flora (Requena et al. 2010)). Therefore, this chapter provides an overview of non-grape fruit wines, their production, flavor profile and their use in human nutrition.

10.2 Types of Fruit Wines

A variety of fruit wines have been formulated throughout the world, however, there is not any particular system of wine classification. Wines are particularly classified on broad attributes based on geographic origin, variety and type of fruit, the process of fermentation and vinification; and sometimes bear generic names. The most common parameter of classification is based on their alcohol content and wines are classified as table (10–11%) and fortified/dessert (16–23%) wines. Fortified wines possess brandy to enhance their alcohol content. Most of the table wines produced around the world are still wines that possess no carbon dioxide or sparkle in them. Sparkling wines possess a considerable amount of carbon dioxide, which is sparged artificially. For instance, Champagne – which originates in the Champagne region of France is a delicate sparkling wine, which is secondary fermented in tanks for carbonation (Jeandet et al. 1997). Another classification is based on the sugar content of wines. The wines that possess nearly 8% sugar content are considered Sweet, while the wines that possess negligible sugar are termed dry wines. For example, table wines with low alcohol content and no sugar are dry wines, whereas dessert/fortified wines are sweet wines that possess a considerable amount of sugar (Joshi et al. 2012). Wines are also classified based on the type of fruit, i.e., Cider made from apples, Boukha from figs, Perry from pear, etc. Another category is Natural wines, which are produced using a wine base, herbs, spices, fruit juice essences, aromatics, etc. These type of wines are known as vermouth containing 15–21% alcohol content (Marrelli 2021).

10.3 Non-Grape Fruits Used for Winemaking

From prehistoric periods, only grapes were considered for the production of fruit wines. Nowadays, a variety of fruits cultivated worldwide are used for the preparation of wines. Several non-grape fruits have been reported from various parts of the world, which are utilized in the production of wine, i.e., apple, cherry, berry, wild apricot, plum, kiwifruit, peach, strawberry, etc. The most popular non-grape fruit wine is cider made from apples throughout the apple-growing countries in the world. On the other hand, in Asian countries, i.e., Japan, China, Korea and Taiwan, wines made from plums are very popular. A variety of red wines are prepared from blends of various fruits, i.e., blueberries, currants, blackberries, etc. Low sugar content is a major concern in other fruit wines; therefore, additional sugar can be added must to accelerate the fermentation process conferred by yeast. Also, some fruits, i.e., strawberries, cherries, raspberries and pineapples, etc., are highly acidic, which may turn into sour-tasting wine. To overcome the fruit's tart acidity, sugar and water are added to counter the fruit's tart acidity (Joshi 2009).

10.4 Various Tropical, Sub-Tropical and Temperate Fruits Are Used For Winemaking

10.4.1 Apple

Apple (*Malus domestica*) is the most commercially significant, widely consumed and marketable fruit worldwide. It has different organoleptic and nutritional properties. Many western countries call it by various names, i.e., cidre (France), sidre (Italy), sidra (Spain), and apfel wein in Germany and Switzerland (Swami et al. 2014). Apple wine has grown to be the second-largest fruit wine sector as a result of rising demand brought on by the global overproduction of apples (Fan et al. 2006). Many sugars, i.e., glucose, fructose, sucrose, etc., are present in apple juice in various concentrations (Joshi et al. 2013; Joshi & Kumar 2017).

From juice extraction to aging, there are several phases involved in the complicated process of making apple wine. The quality of the ingredients and the production methods impart an impact on the final product's quality. In winemaking, each phase is essential to the end product's flavor, aroma and color. Juice extraction is the initial step in the production of apple wine. The taste, aroma and color of the finished product are significantly influenced by the juice's quality, which is influenced by apple variety, level of ripeness and processing methods. To extract the juice, apples are cleaned, sorted and crushed. After that, the juice is pressed to separate the liquid from the sediments (Viana et al. 2020). Following juice extraction, must is produced, and fermentation can be started. The fermentation process and the quality of the finished product are influenced by the temperature, pH and nutritional profile of the must. Although the fermentation period can vary, the yeast normally completes the process in about two weeks (Oliveira et al. 2021). Clarification is carried out after the fermentation process is finished. The wine is racked, which includes moving the liquid from one container to another to clear off any sediment. After clarification, fining is carried out. Fining agents like bentonite can be used for clarification because they bind to the particles and assist them sink to the bottom of the container. The wine is next filtered to remove any impurities and produce a stable, clear product (Ballesteros et al. 2021). Aging is the last stage in the production of apple wine. Usually, oak barrels are used to mature the wine, which can enhance the flavor and aroma. Depending on the intended features, the aging process might take anywhere from six months to a year to produce a mature product. The wine goes through chemical changes throughout this period that affect the flavor, fragrance and color of the finished product (Ramalhosa et al. 2015).

Besides, there are some issues with apple winemaking. First, the fruit wine industry has a persistent issue with incomplete/slow fermentations of apple juice, which results in spontaneous tank capacity loss due to prolonged processing times. It may also generate potential off-taste in finished products owing to the higher concentration of sugar residues, primarily fructose (Jamir et al. 2019). Second, selecting a better strain of fructophilic yeast is very important as apple juice contains a lot of fructose. Several volatile components are present in apple wine, which evolved throughout the fermentation process. To measure these volatile chemicals in wine, older techniques, i.e., liquid–liquid extraction, direct injection, static headspace sampling and solid phase extraction were used, which were laborious, time-consuming and occasionally required huge amounts of samples and solvents. Therefore, to determine flavor volatiles with an alcohol concentration below 12%, the headspace solid-phase microextraction-gas chromatography/mass spectrometry technique (HS-SPME-GC/MS) was developed (Hernández et al. 2013; Wang et al. 2004).

10.4.2 Pear

Pear (*Pyrus communis*) wine, commonly known as perry, is a fermented alcoholic beverage prepared from pears. It is a widely consumed beverage with a long history that dates back thousands of years. Pear wine has a distinct flavor and aroma in addition to numerous possible health advantages, i.e., inflammatory and antioxidant qualities along with a promising nutritional profile (Chandel et al. 2023). Pear wine contains antioxidants that can help shield cells from free radical damage, which may lower the chance of developing chronic illnesses, i.e., cancer and heart disease (Li et al. 2017). Pear has also been reported to enhance gut health by encouraging the development of good bacteria in the digestive tract. According to studies, pear wine includes prebiotic substances that can specifically promote the growth of gut-healthy bacteria like *Bifidobacterium* and *Lactobacillus* (Zhang et al. 2020). Pear wine is low in calories and rich in vitamins and minerals. Five ounces of pear wine constitute a normal serving, which has 120–130 calories, 1 gram of fat, and 12–13 grams of carbohydrates. Moreover, it has potassium, magnesium and vitamin C in trace amounts (U.S.D.A. 2022).

10.4.3 Apricot

Apricot wine is a tasty and aromatic alcoholic beverage produced by fermenting apricots. To produce apricot wine, it is essential to choose the right apricot cultivar. Goldcot, Blenheim and Tilton are the cultivars of apricot that are most frequently utilized in the making of apricot wine. These cultivars are chosen for their distinctive flavor and fragrance profiles, which influence the wine's final flavor (Kaur et al. 2020). Due to the inclusion of numerous bioactive substances, such as polyphenols, flavonoids and carotenoids, apricot wine provides several health benefits as other wines. These substances have anti-inflammatory and anti-oxidant qualities that may help lower the chance of developing chronic illnesses like diabetes, cardiovascular disease and cancer (Kaur et al. 2020). An average of 120 calories can be found in a 5-ounce serving. Apricot wine, however, contains a lot of sugar and alcohol, therefore it should only be drunk occasionally (Adaramola & Benjamin 2018). For the production of apricot wine, the apricots are picked, then cleaned, and pitted. The juice is then extracted after the apricots have been crushed. The juice is mixed with sugar and yeast, and the mixture is left to ferment for a few weeks. Alcohol is produced during the fermentation process from the juice's sugar. After fermentation is finished, the apricot wine is stored for many months in a cool, dark environment to improve its flavor and aroma (Kaur et al. 2020).

10.4.4 Medlar

Little self-fertile trees known as medlars (*Mespilus germanica*) are native to southwest Asia and south eastern Europe. These are attractive trees that appear exceptionally hardy, bearing fruit even during the harsh summers, when other apple species struggle to survive. This fruit has an appearance somewhere between a little apple to a rose hip. As they cannot ripen before harvesting, they must be picked hard and allowed to "blet," which is a process of softening that resembles partial rotting. These fruits are kept in cool, dark storage in sawdust or bran until they soften and take on a fragrant flavor. Moreover, the fruits can be used to create a sherry-like country wine. In a nutshell, the bletted fruits are combined with sugar, water, and yeast and allowed to ferment, producing wine (Khan 2018).

10.4.5 Plum

In Germany, the Pacific Coastal states, Japan, Korea and China, plum (*Prunus domestica*) wine is highly preferred (Amerine et al. 1972). Umeshu is a Japanese alcoholic beverage made from plums. Plum wine is known as meijiu in China, maesil ju in Korea and is sold under the brand names Mae Hwa Su, Mae Chui Soon and Seol Joong Mae. Wumeijiu, a "smoked plum liquor," is a Japanese-style plum liquor that is produced in Taiwan by combining the liquors of prunus species and oolong tea. Plum wine is a fantastic source of vitamins, minerals and anti-oxidants, including vitamin C and vitamin K, which supports a healthy immune system and strong bones, respectively (Joshi et al. 2014). Moreover, it has potassium and magnesium, both of which are important for maintaining heart health and maintaining healthy muscles and nerves (Koh et al. 2016). Plum wine aids in digestion, immunity and lowering the risk of chronic illnesses, including cancer and heart disease when consumed in moderation (Zhang et al. 2021).

Plum wine is made by choosing fresh, ripe plums that are free of any deformity and thoroughly washing them to remove dirt or debris. The plums are then steeped for several months in a mixture of sugar and alcohol. The plums are taken out and the liquid is filtered. To get the desired flavor, the strained liquid is then put in bottles and matured for a few more months (Wakita & Imura 2018). The flavor and aroma of the finished plum wine can vary depending on the quality of the plums used in its preparation (Qiu et al. 2020). The flavor and aroma of the wine can also be affected by the temperature at which the plums are fermented. A stronger and fruitier flavor can result from a higher fermentation temperature, whilst a softer flavor can be developed at lower temperatures (Shepherd & Waterhouse 2020). Moreover, an extended aging period can produce a flavor that is more sophisticated and complex (Wakita & Imura 2018).

10.4.6 Blueberry

Many *Vaccinium* species, including highbush (*V. corymbosum*), lowbush (*V. angustifolium*), rabbiteye (*V. ashei*), large cranberries and blueberries are significant commercial crops. Blueberry beverages derived from fruits possess substantial levels of bioactive compounds having anti-diabetic properties (Basu et al. 2010; Stull et al. 2010). Johnson et al. (2011) revealed that highbush blueberries are a rich source of anti-oxidants and phenolic compounds along with alpha-amylase as well asglucosidase inhibitory ability, which is crucial for the control of type 2 diabetes. Blueberry wine also possesses a variety of phenolic compounds, i.e., kaempferol, vanillic acid, ferulic acid, gallic acid, caffeic acid and protocatechuic acid. In terms of overall polyphenolic content, blueberry wines contained an average of 1623.37 mg/L ellagic acid equivalent, 20.82 mg/L cyanidin equivalents, and 21.21 mM/L Trolox equivalents. The findings revealed that blueberry wines have potential health benefits and, as a result, could help the wine economy's finances (Johnson & Gonzalez de Mejia 2012).

10.4.7 Cherry

Due to high perishability, cherry fruits (*Prunus cerasus*) may not be acceptable for fresh consumption if they get damaged during transportation. Therefore, to prevent post-harvest losses, the fruits are frequently processed into juice as well as wine. Often tart cherries are used to make cherry fruit wine that gives enough acid. There are various types of cherry wine produced by Michigan winemakers in the United States, including spicy variants and cherry-grape blends. A fortified fruit wine called Cherry Kijafa is made in Denmark from cherries and natural flavors. It is typically shipped to North America and Europe and contains 16% alcohol by volume on average. To produce cherry wine, cultivars including Montmorency, Balaton and Northstar are frequently used (Hao et al. 2019). Besides, Shandong cherry wine has gained popularity in China due to its distinctive flavor, Recently, various *S. cerevisiae* strains (BM44, RC212, RA17, D21, D254 and GRE) were used to ferment tart cherries of the "Early Richmond" type to examine their impact on the synthesis of volatiles (Lin et al. 2018). Resultantly, acetic acid and 3-methylbutanol were the two substances that all of the made wines were high in, followed by 2-methylpropanol and ethyl lactate. The polyphenols (neochlorogenic and chlorogenic acids) and anthocyanins (cyanidin 3-rutinoside and cyanidin 3-glucosylrutinoside) in the cherry wines were all present in higher concentrations (Lin et al. 2018).

10.4.8 Elderberry

Sambucus nigra, a common species of elderberry, is found in North Africa, Asia, Europe and the United States. The species is a significant source of ingestible fruits loaded with numerous phytochemicals and secondary metabolites, i.e., phenolics and anthocyanins, possessing immune-stimulating, anti-inflammatory as well as anti-oxidant properties (Thole et al. 2006; Youdim et al. 2000). The elderberry wine possesses dark red, with a pH of 3.9, and included a moderate amount of ethanol (13.2% v/v). Wine as well as elderberry must be observed as an abundant source of phenolic chemicals (neochlorogenic acid, chlorogenic acid, quercetin-3-rutinoside, kaempferol-3-rutinoside, quercetin-3-glucoside and other cyanidin anthocyanins). The wine's anti-oxidant potential was comparable to that of red wine and was substantially associated with its phenolic content (Schmitzer et al. 2010).

10.4.9 Peach

Peach (*Prunus persica*), sweet drupe fruit belongs to the *Rosaceae* family. The suitable peach cultivar must be chosen as the initial step in the manufacturing of peach wine. Peach wine is made using a variety of peach varieties, including O'Henry, Elberta, and Red Haven. In comparison to white wines with TPC (243.67 mg/LGAE) and TFC (129.67 mg/L CAE), peach wine has greater TPC (402.53 mg/L GAE) and TFC (332.67 mg/L CAE), which results in higher anti-oxidant activity. Caffeic acid, chlorogenic acid and catechin are the three primary phenolic substances discovered in peach wine. These substances have anti-inflammatory and anti-oxidant qualities that may help lower the chance of developing chronic illnesses like diabetes, heart disease and cancer (González-SanJosé et al. 2021).

10.4.10 Raspberry

Raspberry (*Rubus* spp.) is abundantly grown in Korea and can be processed into jam, ice cream, juice and wine (Hager et al. 2008). It has also been utilized as a medicine since prehistoric times (Lim et al. 2012). Red raspberry is a tasty and revitalizing beverage that has grown in popularity in recent years due to its distinct flavor and health advantages. Red raspberry wine is made through several procedures, including choosing the right raspberry cultivar, cleaning the fruit, fermenting it and aging it. Heritage, Encore and Prelude are the most popular raspberry cultivars utilized in the making of red raspberry wine (Mogren et al. 2019).

Black raspberry wines from Korea are an abundant source of phenolics widely consumed as a traditional drink (Ku & Mun 2008). It has little bitter astringent and flavor, caused by its two to three times higher proanthocyanidin content when compared to commercial grape wine (Lim et al. 2012). By using size exclusion chromatography, the immune-stimulating polysaccharide was extracted and identified from Korean black raspberry wine (Hwang & Shin 2011). However, from Korean black raspberry wine, five anti-oxidative active compounds were also extracted and identified, including 4-hydroxybenzoic acid, 4-(2-hydroxyethyl)-phenol, 3,4-dihydroxy benzoic acid, 4,5-trihydroxybenzoic acid, and pyrocatechol (Cho et al. 2011). It is also possible that the phytochemicals present in raspberries possess high anti-oxidant activity, which can protect mammalian cells from biological oxidative stress (Weber & Liu 2002). Raspberries typically include phenolic acids, such as p-coumaric, ferulic, caffeic and ellagic acids. Black raspberry (*Rubus occidentalis*) wine was examined for its levels of polyphenols, anthocyanins and ascorbic acids, as well as their anti-oxidant, anti-inflammatory and anti-proliferative properties. In comparison to fruit, the black raspberry fruit dramatically reduced the growth of HT-29 colon cancer as well as LNCap prostate cancer cells in a dose-dependent manner (Jeong et al. 2010).

10.4.11 Persimmon Wine

Persimmon (*Diospyros kaki*) is one significant underutilized fruit crop. Its color ranges from yellow-orange to deep red. It comes in two variants, i.e., astringent and non-astringent. The astringent cultivars Hachiya, Saijo and Tanenashi are a few well-liked ones. The tannin content in non-astringent persimmons is very low, and therefore, can be consumed while still hard. Fuyu, Jiro and Gosho are a few of the well-known cultivars that are non-astringent (Ding 2016). The persimmon fruits' taste, color and nutritional as well as therapeutic value make them popular among the population. It also possesses a variety of phytochemicals and nutrients, i.e., carbohydrates, vitamins, organic acids, tannins, dietary fiber, triterpenoids, polyphenols and carotenoids (Altuntas et al. 2011). Other phytochemical substances, i.e., polyphenols, flavonoids and carotenoids, are also found in persimmon wine. Catechin, epicatechin and gallic acid are the three main polyphenols found in persimmon wine. Moreover, it has substantial amounts of beta-carotene, resveratrol, and tannins. These components make persimmon wine a powerful anti-oxidant (Kang et al. 2014). Moreover, persimmons have a significant anti-oxidant capacity that may protect cells from oxidative damage. Condensed tannins present in persimmons are primarily responsible for the fruit's anti-oxidant action and may lower the risk of diabetes, hypertension, cardiovascular disease, malignancies, etc. (Hwang et al. 2011). A peculiar quality of persimmon is that it changes and slows down the pace of alcohol metabolism, which lessens the effects of a hangover (Orwa et al. 2020). In addition to the fruits, the leaves also possess therapeutic qualities. Ursolic acid and oleanolic acid are two typical triterpenoids found in its leaves. These are isomeric pentacyclic molecules (Zhou et al. 2010). Both substances show numerous bioactivities, i.e., anti-inflammatory, hepatoprotective, cardiovascular, anti-tumoral, gastroprotective and anti-HIV properties. Sharma and

Mahant (2015) standardized a process for producing persimmon wine with better quality utilizing a variety of yeast strains. The Fuyu cultivar's fruit was used to maximize the ingredients in persimmon pulp for making wine. At a temperature of 27°C, the diluted pulp (1:1) was fermented, and citric acid and DAHP were utilized as acidifying agents and nitrogen sources, respectively (Sharma & Mahant 2015). The wine also had several beneficial qualities, including higher levels of alcohol, total phenols, esters and anti-oxidant activity as well as comparable levels of total/reducing sugars. The wine also received higher ratings for several sensory qualities, including color, flavor, appearance, fragrance, total acidity, volatile acidity, body, sweetness and overall impression. The flavor profile of persimmon wine was distinctive, combining sweet and tart flavors. Its color was golden yellow, and its aroma was delicious with a touch of spice. The cultivar, fermenting procedure, and aging all have an impact on the flavor and aroma of persimmon wine (Ghosh 2015).

10.4.12 Banana

Musa spp. is the world's most valuable fruit crop concerning its economic value in the global market (Aurore et al. 2009). Being a seasonal crop, it possesses a limited shelf-life specifically in tropical regions of the world due to the high humidity and temperature. After reaching maturity and being harvested, ripe bananas degrade quickly. Large amounts of ripe bananas are typically wasted due to improper handling, insufficient storage and subpar transportation facilities, even though they are consumed to a significant level. Consequently, modern as well as traditional processing techniques will be helpful to increase bananas' shelf-life and lowering post-harvest grudges. Banana juice fermentation is regarded as an alluring method of using up extra and overripe bananas. Banana possesses a rich nutritional profile including protein (0.08%), ash (0.35%), vitamin C (9 mg/100 ml) and a pH of 4.45. For the production of banana wine, the juice was stored at 307°C for 14 days after it had attained 18°Brix concentration and then inoculated with *S. cerevisiae* at 3% concentration. During fermentation, parameters, i.e., soluble solids, pH and specific gravity get reduced, however, titratable acidity increases (Akubor et al. 2003). Cavendish bananas are commercially utilized in the production of banana wine, whereas a range of other cultivars, i.e., Malkia and Meru bananas are employed in its production. Banana beer can also be produced; however, it degrades easily during storage, unlike commercially manufactured banana wine, which is a clear, mildly effervescent alcoholic beverage. The final product's sweetness and alcohol content can vary depending on the yeast strain used and the quantity of sugar used. The production of banana wine starts with pre-treating must with pectinase (0.05%) and α-amylase (0.05%) in a sequential manner, which boost the hydrolysis of complex polysaccharides (pectin, starch, etc.), which increases the clarity, decreases viscosity and increases the proportion of extracted juice as well as increases the concentration of total soluble sugars and reducing sugars by 15% and 39%, respectively. However, a meagre difference in soluble solids and alcohol concentrations was observed after 25 days of fermentation despite the fact that the banana juice was pre-treated with enzymes' combination (Cheirsilp & Umsakul 2008). On the other hand, Byaruagaba-Bazirake et al. (2013) used recombinant yeast strains capable of degrading various polysaccharides (xylans, glucans, pectins, starch) rather than commercial enzymes. Due to this, wine yield increased by 5.5% as compared to control yeast. Consequently, instead of using traditional enzyme preparations in the fermentation of banana wine, this recombinant yeast strain might be used (Byaruagaba-Bazirake & Rensburg 2013).

10.4.13 Cacao

Theobroma cacao (Cacao) is famous for its beans, which are used to make chocolate and other related goods. Certain Brazilian states, particularly Bahia, have traditionally based their economies on the production and trade of cocoa beans. The cacao fruit's pulp is a by-product of processing the fruit and a source for the production of beverages and other goods. In an experiment, UFLA CA 1162 yeast strain was used to ferment the cacao fruit pulp, and GC/MS and GC/FID were used to analyze the cacao wine's minor and major constituents. The wines under analysis contained C6 compounds, monoterpenic alcohols, acetate esters, ethyl esters, monoterpenic oxides, volatile phenols, carbonyl compounds, acids, sulfur compounds, sugars, etc. The findings indicated that using cacao fruits to make wine is a practical option that enables the use of crop surpluses and other Cacao fruits that are not being used, formulating a new product for the market (Duarte et al. 2010).

10.4.14 Custard Apple

Custard apple (*Annona squamosa*), also called sugar apple or sweetsop is indigenous to tropical America and is grown in a variety of tropical climates all over the world. Due to the sweet, white, delicate flesh as well as the high nutritional content of mature custard apple fruits, there is significant consumer demand for them (Pinto et al. 2005; Kumar et al. 2016a). Its fruits can also be used to make juice, ice cream, milkshakes, and soft drinks. They are typically eaten fresh as dessert fruits. Custard apples are difficult to consume fresh due to their high susceptibility to spoilage, quick ripening and squashy texture (Okigbo & Obire 2009; Kumar et al. 2016b). Also, early picking can lead to poor fruit quality, whereas fruits left on the tree to ripen are frequently eaten by birds and bats and have the propensity to shatter and degrade when they are overripe (Setyabudi 2022). To reduce post-harvest losses, it is, therefore, more important to prepare mature custard apple fruits into acceptable products, i.e., wine. Using the NCIM 3282 strain of *S. cerevisiae* yeast, Jagtap and Bapat (2015) investigated the custard apple's potential for use in the formulation of wine and evaluated its anti-oxidant potential, TPC, as well as DNA protecting activity. With various anti-oxidant assays, custard apple wine demonstrated high free radical scavenging activity. Major phenol in its wine was protocatechuic acid, gallic acid, caffeic acid, *p*-coumaric acid, and gentisic acid, which were detected using RP-HPLC. Custard apple wine was also found able to prevent DNA damage caused by radiation in pBR322 plasmid DNA (Jagtap & Bapat 2015).

10.4.15 Gabiroba

Gabiroba (*Campomanesia pubescens*) is majorly used by natives of the western as well as southern Brazilian regions. In addition to being consumed raw, gaboiroba fruit is also processed to make jams, ice cream, juices and desserts. According to Duarte et al. (2009), gabiroba fruit pulp can serve as an excellent substrate for winemaking because it has a pH of 4.1 and a sugar content of roughly 14°Brix. The wine produced from the fermentation of gabiroba fruit with UFLA CA 1162 yeast strain possesses C6 compounds, specifically (E)-2-hexanol and (E)-3-hexen-1-ol. The composition of alcohol in Gabiroba wine is 1-butanol, 3-methyl-1pentanol, 1-pentanol-3-methyl-3-butene-1-ol, 1-heptanol, 1-octanol, 2-ethyl-1-hexanol, benzyl alcohol furfurol and 2-phenoxyethanol. Yet, the gabiroba wines also included various ethyl esters, i.e., ethyl-3-methyl butanoate (fruity aroma) and ethyl hexanoate (green apple aroma). Also, the sensory examination of the Gabiroba fruit wine yielded acceptable ratings for the attributes of appearance, scent, and taste of 63%, 60% and 52%, respectively (Duarte et al. 2010).

10.4.16 Guava

Guava (*Psidium guajava*), a member of the *Myrtaceae* family, is one of the preferred fruits in India. The fruit has a well-liked flavor and aroma and is also a good source of pectin, ascorbic acid, sugars, as well as some minerals. The fermentation of guava juice (22°Brix) was optimized concerning osmotolerance, alcohol tolerance, inoculum size, initial pH of the medium, amount of SO_2, amount of diammonium phosphate and incubation temperature using two separate *S. cerevisiae* strains, i.e., NCIM 3287 and NCIM 3095. It was observed that *S. cerevisiae* NCIM 3095 produced guava wine with substantially superior results (Sevda & Rodrigues 2011). These guava wines were subjected to GC/MS and GC/FID analysis, which identified 124 volatile constituents in the wine, including 24 alcohols, 52 esters, 11 ketones, 6 aldehydes, 7 acids, 6 terpenes, 4 phenols and derivatives, 4 sulphur compounds, 4 lactones and 5 other compounds along with various aroma-active volatiles, i.e., (E)-damascenone, ethyl 3-phenylpropionate, ethyl octanoate, 3-methyl butyl acetate, ethyl hexanoate, 2-methyl tetrahydrothiophene-3-one and 4-methoxy-2,4-dimethyl-3(2H)-furanone (Pino & Queris 2011).

10.4.17 Jackfruit

The jackfruit (*Artocarpus heterophyllus*), is found all across the tropics and subtropics and is a member of the *Moraceae* family. Unfortunately, the fruits are extremely vulnerable for spoiling and prone to microbial infections, which cause large post-harvest losses. To reduce post-harvest losses, it is, therefore, more important to prepare mature jackfruit into acceptable products. Jagtap et al. (2011) made wine from jackfruit pulp and assessed flavonoid, TPC and antioxidant potential. The jackfruit wine has high capacities for DMPD (78.05%), DPPH (69.34%) and NO (62.45%). Gallic and protocatechuic acid, two phenolic constituents, were discovered in jackfruit wine. Moreover, the pBR322 plasmid DNA damage from H_2O_2 and UV radiation was also observed to be prevented by the jackfruit wine. Resultantly, if taken in moderation, Jackfruit wine has been shown to have anti-oxidant and DNA damage protection properties, which means it has health benefits and the potential to be a great source for formulating nutraceuticals (Jagtap et al. 2011).

10.4.18 Strawberry

Due to their great nutritional value and high perishability, strawberries (*Fragaria* spp.) are used right away in sweets or other valuable goods like purees, juices, jams, wine, etc. (Sharma et al. 2009). The right strawberry cultivar is chosen to start the process of making strawberry wine. For the creation of strawberry wine, cultivars like Seascape, Chandler, and Sweet Charlie are frequently utilized. Strawberry wine differs from other varieties of wine due to its distinctive chemical and flavor composition. It has a distinctive strawberry flavor and aroma owing to its biochemical profile. Many bioactive substances, i.e., polyphenols, flavonoids and anthocyanins, are present in strawberry wine's chemical composition, which are responsible for both its flavor and health benefits. With an average of 120 calories in each 5-ounce serving, it has comparatively few calories as well. However, strawberry wine is strong in sugar and alcohol, which should be eaten in moderation (Kong et al. 2019).

10.4.19 Jamun

The tropical tree *Syzygium cumini*, native to India, produces ovoid, oblong, shining, crimson berries that are rich in anthocyanins. Jamun is a member of the *Myrtaceae* family. Jamun fruits have therapeutic benefits and are used to treat diabetes due to their beneficial effects on the pancreas gland (Marrelli 2021). Because of its balanced sugar, acid, and tannin contents, the fruit is becoming more and more popular among customers (Das 2009). The berries' high levels of TSS (16°B), which is well balanced by high acidity (1.19%) and tannin (340 mg/100 g), aid in fermentation. Jamun pulp has a significant amount of fermentable sugar that can be used for making red wine and also possesses several health benefits (Patil et al. 2012), including diuretic, anti-scorbutic, astringent, stomatic, carminative and anti-diabetic properties (Patel et al. 2005). The fruit's impact on the pancreas makes it a good nutritional cure for treating diabetes, heart issues, hemorrhoids and several liver issues. The fruit has high anti-oxidant potential in addition to being a very rich source of anthocyanins (Chowdhury & Ray 2007). There are several cultivars of jamun fruit, and each one has distinctive qualities. Allahabad Safeda, Rajarani, Jyoti, Kalipatti and Navsari are some of the well-known cultivars. Because of their high juice content, strong color, and flavor, Kalipatti and Navsari are among the cultivars that are most frequently utilized for wine production (Bchir et al. 2020). Using *S. cerevisiae* to ferment anthocyanin-rich Jamun fruits, a sparkling red wine with a high tannin content and a low alcohol concentration (6%) was produced. Jamun wine also had therapeutic benefits, including the capacity to treat bleeding piles and act as an anti-diabetic. Jamun wine was acknowledged as an alcoholic beverage by the sensory analysis. Yet, it greatly differed from commercial grape wine, especially in taste, flavor and lingering aftertaste, which is possible because Jamun wine has a high tannin concentration (Chowdhury & Ray 2007).

10.4.20 Kiwifruit

Kiwi wine has been made from a sizeable fraction of the kiwifruit (*Actinidia* spp.) produced in New Zealand, California, Italy, Chile, France, Australia and Israel. One of the intriguing things about kiwi wine is that its flavor has nothing to do with the fruit used to make it (Craig 1988). *Actinidia deliciosa*, commonly known as "Hayward" kiwifruit, was recently examined by Towantakavanit et al. (2011) for its fruit maturity value as a component of wine production. In the results, it was observed that pectinase pre-treatment on the pulp of ripe kiwifruit boosted wine production yield from 63.35% to 66.19%. Compared to wine made from ripe fruit (10.3%), wine made from overripe fruits has a higher soluble solid content (13.5%) and potassium content. However, variable fruit ripeness did not impact the acidity, color, TPC and anti-oxidant potential of the wine. In terms of sensory quality, alcohol concentration, TPC, minerals and production yield, the pectinase-treated overripe fruit-based kiwi wine was superior (Towantakavanit et al. 2011).

10.4.21 Lychee

A member of the *Sapindaceae* family, the lychee (*Litchi chinensis*) is a native of Southern China. The improper handling and inadequate storage of fruits result in severe post-harvest losses since the pericarp gets reddish colored before becoming brown (Mahattanatawee et al. 2007). Hence, the excess lychee fruit output can be turned into wine, increasing the grower's income and lowering post-harvest losses (Alves et al. 2011). It is used in China to make alcoholic beverages because it has a good flavor and is a rich source of vitamins and minerals. There are many cultivars of lychee, but Emperor, Bengal, Brewster and Mauritius are the most frequently utilized for making wine. These varieties are renowned for their size, sweetness and juice content (Kadam et al. 2015). Lychee fruits that are at their peak of maturity are cleaned, peeled and then submerged in a 70°B sugar solution for four hours at 50°C. Then the fruits are drained and pulped in a pulper. By diluting the pulps with water, the temperature is brought to 22°C. Then the culture of *S. cerevisiae* (5%) is added to the lychee juice. Following standard procedure, the fermentation is allowed to continue until the TSS reaches 7°B. The wine is then blended with an equal amount of fresh lychee juice, filtered and put into glass bottles, and aged. Then, the bottles are pasteurized for 20 minutes in water at 62.5°C (Vyas & Kochhar 1989). For the creation of lychee wine, the suitability of lychee juice concentrate was examined. In an experiment, three yeast strains (UFLA CA1183, UFLA CA11 and UFLA CA1174) were combined to make lychee wines. The acceptance rate for the UFLA CA1183 and UFLA CA11 yeast-fermented wines was greater than 75%. Yet, UFLA CA1183 strain was observed as most suitable (Alves et al. 2011). The majority of terpenoids derived from litchi juice decreased, even vanished during alcoholic fermentation, while terpenol oxides, ethers and acetates increased, according to research on the evolution of volatile components during the making of litchi wine and aroma profiles of litchi wines, which were determined using HS-SPME-GC/MS. However, the OAVs in young litchi wines with fruity and floral characteristics were highest for ethyl hexanoate, isoamyl acetate, ethyl butanoate, ethyl octanoate and cis-rose oxide. It was observed that the lower temperature improved key aroma retention and also increased the shelf-life of young litchi wines (Alves et al. 2011).

10.4.22 Mango

Mango (*Mangifera indica*) is one of the most expensive tropical desert fruits. It tastes excellent and has a deep, creamy, aromatic flavor. Almost 80% of the crop worldwide is grown in 85 countries of Asia and other eastern regions. India, China, Mexico and Pakistan are the principal nations that produce mangoes. There are many different kinds of mangoes, but Alphonso, Kesar and Totapuri are some of the most widely used cultivars for making mango wine (Bisht & Barman 2013). Reddy and Reddy (2011) investigated the effects of various fermentation conditions (pH, temperature, SO_2 content, aeration) on yeast growth, duration, fermentation rate and developed volatile composition. They performed the fermentation using yeast strain, *S. cerevisiae* 101.

Gas chromatography was used to analyze the composition of the main volatile substances with low boiling points under various fermentation temperatures and aeration conditions. It was observed that the levels of volatile chemicals and yeast growth were both significantly influenced by temperature. As the fermentation temperature was raised to 25°C, it was noticed that the final alcohol concentrations were decreased (Reddy & Reddy 2011). Mango wine typically has an alcohol concentration of 10% to 14% (Mukherjee et al. 2014).

Similarly to this, Reddy and Reddy (2009) evaluated the manner of alcohol production during the fermentation of two mango cultivars, Totapuri and Banginapalli. Totapuri wine has more volatiles than Banginapalli wine. Pectinase therapy boosted mango juice yield and improved wine sensory quality during fermentation by increasing iso-amyl alcohol, n-propanol, 2-phenyl ethanol, methanol, and n-butanol synthesis (Reddy & Reddy 2009). Exogenous-glucosidase was also used to increase the aroma of mango wine, as evidenced by higher amounts of fatty acids and ethyl esters, without appreciably altering the pH, sugars and organic acid levels. The addition of glucosidase was found to be beneficial in intensifying, diversifying and balancing the mango wine aroma profile by accelerating the production of terpenols, benzene derivatives and acetate esters (Li et al. 2013).

Varakumar et al. (2011) examined the carotenoid content of wine prepared from seven different mango cultivars. 15.3–26.5% degradation for β-carotene was observed in all tested samples. Yet, it was discovered that Totapuri wine significantly degraded concerning its β-carotene content. Wine made from the Alphonso, Sindhura, and Banginapalli cultivars demonstrated the highest anti-oxidant activity, whereas the Alphonso, Banginapalli and Sindhura wines displayed the highest inhibitory effects on the oxidation of low-density lipoprotein (Varakumar et al. 2011). According to a recent study by Kondapalli et al. (2014), γ-irradiation increased TPC and TFC in a dose-dependent manner, and decreased microbial loads, improving the wine's quality. Also, the radiation dose greatly increased the anti-oxidant activities of mango wines. Moreover, this mango wine proved effective at preventing DNA damage caused by UV, H_2O_2, and gamma radiation (500 Gy) (Kondapalli et al. 2014). Varakumar et al. (2011) investigated the interactions between three *S. cerevisiae* strains separately and in conjunction with *Oenococcus oeni* during the process of malolactic acid fermentation in mango wine (sequentially and simultaneously). The wine generated with simultaneous inoculation had higher levels of aliphatic alcohol and ethyl acetate. Additionally, the amounts of 1-propanol did not differ significantly, but there were considerable variations in the amounts of amyl alcohol and isobutanol. Compared to the control, the mango wine that was simultaneously inoculated with *O. oeni* had higher sensory scores in flavor, fruity scent and overall acceptance regardless of the yeast strain used (Varakumar et al. 2011).

10.4.23 Citrus

Citrus fruits typically have a short shelf-life and experience post-harvest losses (Panesar et al. 2009). In addition to using citrus fruits to produce juice, efforts have been undertaken to generate alcoholic beverages through fermentation. Oranges, lemons, limes, grapefruits and tangerines are just a few examples of citrus fruits. Many studies have been conducted to determine the potential of citrus fruits for wine production. When the physicochemical qualities of wines made from different citrus fruits (mandarin, kinnow, galgal and orange) were studied, it was found that kinnow wine had the highest fermentation rates, followed by orange and galgal wine (Joshi et al. 2012). Oranges and lemons are the two kinds that are most frequently used to make wine. Citrus cultivars including Valencia oranges, Blood oranges and Eureka lemons are some of the most widely used for making wine (Manzocco et al. 2017). The strong acidity of citrus fruits is one of the issues with citrus wine manufacturing. Due to this, obtaining the desired alcohol concentration during fermentation may be difficult.

The flavor and aroma of the wine can also be impacted by the presence of limonin and other bitter chemicals in citrus fruits (Manzocco et al. 2017). Concerning flavor profile, Blood orange wine was examined by GC/FID and GC/MS for a total of 64 volatiles, i.e., higher alcohols, terpenes, esters, acids, lactones, volatile phenols, acetal compounds, aldehyde, ketone and acetoin (Selli 2007). Together with narirutin, hesperidin and ferulic acid, orange wine was an abundant source of phenolic chemicals, i.e., hydroxycinnamic acids, hydroxybenzoic acids and flavanones. Orange juice was discovered to have a higher anti-oxidant content than orange wine (Kelebek et al. 2009). Kozan wine was found to contain 75 volatile substances, i.e., terpenes, esters, alcohols, volatile phenols, ketones, acids, lactones, aldehydes, and C13-nor isoprenoids 2-phenyl ethanol, isoamyl alcohol, linalool, ethyl-4-hydroxy butanoate and terpinene-4-ol, were among these; ethyl hexanoate, citronellol, ethyl octanoate and eugenol were also significant contributors to the aroma of the wine (Selli et al. 2003). Various yeast (*S. cerevisiae, S. carlsbergensis*) strains employed for the fermentation of orange juice were taken from yam, sugarcane molasses, etc. The maximum concentration of ethanol was found in orange wine made with *S. cerevisiae var. ellipsoideus*, while the lowest was found in wine made with *S. cerevisiae* (from sugarcane molasses) (Oluwanisola & Adediran 2007).

10.4.24 Palm

Tropical trees of the *palmae* family include the oil palm (*Elaeis guineensis*), raffia palm (*Raphia hookeri*) and coconut palm (*Phoenix dactylifera*). They are native to the tropics, primarily in South America, Africa and Asia (Uzochukwu et al. 1994). Palm wine, commonly known as toddy, is a popular alcoholic beverage in many tropical regions of the world. Palm wine is an alcoholic beverage made from the sugary sap of palm trees that have undergone spontaneous yeast-lactic fermentation. It has a one-day shelf life and is a hazy, white liquid with a pleasant alcoholic taste. The sap from the flower buds of a variety of palm species, including the oil palm, coconut palm, date palm and several others, is harvested to make palm wine. A jar is attached to the clipped flower bud to collect the sap. To avoid fermentation, the sap is then quickly transferred to a collection receptacle. A starting culture of wild yeasts is introduced to the sap to start the fermentation process. It usually takes between 2 to 24 hours for fermentation to complete, depending on the temperature and other environmental factors. The palm wine is prepared for consumption or additional processing once fermentation is finished (Adebowale & Adeyemi 2014). A complex blend of organic acids, sugars, alcohols and other volatile substances makes palm wine more delicate and delicious. Depending on the type of palm, the circumstances surrounding the fermentation, and other environmental factors, the chemical composition of palm wine varies. The normal alcohol content of palm wine is 3–6% by volume, and its pH ranges from 3.5 to 4.5. The characteristics of the palm wine range from sweet and unfermented to tart, fermented and vinegary. The same commodity comes in a wide variety and there is no specific way to turn it into vinegar (Joshi et al. 2019). The primary controls involve obtaining a high production of palm sap while minimizing contamination by spoilage microorganisms and storing the sap properly to allow for natural fermentation. The circumstances under which the sap is collected affect the ultimate wine's quality. Also, the item needs to be stored in a cool location out of the direct sun (Hui et al. 2012).

A sensory evaluation of the palm wine was conducted, and high-resolution GC-MS of solvent extracts and headspace samples were used to analyze the main aroma volatiles. The earthy 3-isobutyl-2-methoxypyrazine, the buttery acetoin, the fruity compounds ethyl hexanoate and 3-methyl butyl acetate, and the popcorn-like smelling 2-acetyl-1-pyrroline were observed as important odorants (Lasekan et al. 2007). The presence of acetic and lactic acid bacteria in a variety of yeast species (*S. ludwigii, Hanseniaspora uvarum, Zygosaccharomyces bailii, Candida fermentati, Candida parapsilopsis, Pichia fermentans*) revealed that a multi-starter fermentation occurs in a natural, semi-continuous way during the palm wine processing (Stringini et al. 2009). The biochemical and microbiological characteristics of various week-old fresh palm saps were examined to determine the quality changes resulting from the storage of the fermented saps. The findings demonstrated that all samples of palm wine had an alcohol increase during storage, along with concomitant lactic and acetic acid fermentation. The populations of lactic acid bacteria and yeasts also altered the safety, biochemical makeup and nutritional value of palm wine. Thus, additional research on the presence of these microbes in wine and their impact on food safety will be necessary for wine to be commercialized (Karamoko et al. 2012).

10.4.25 Papaya Wine

Papaya (*Carica papaya*), is one of the most important fruit crops grown in South East Asia. Papayas are frequently eaten either raw or used to process other products, i.e., jellies, jams and juices. Papaya is a fruit that is widely produced in many nations and contains soluble sugars in the forms of fructose, glucose and sucrose. Papaya trees produce fruit all year round in tropical regions like Nigeria (Hasan & Zulkahar 2018). Due to its affordability and accessibility, it has been suggested as a potential renewable energy source for the industrial manufacturing of alcohol (Sharma & Ogbeide 1982). Papain, an enzyme that facilitates digestion, is also present in it. Although papaya has many varieties, however, Red Lady, Tainung No. 2 and Honey Gold are the most frequently utilized for making wine. The sweet flavor and high juice content of these varieties are well-known (Lay et al. 2015).

Researchers extracted different yeasts from rotten papaya fruit, including *S. cerevisiae, S. uvarum, S. bayanus, S. italicus, S. pastorianus, Schizosaccharomyces pombe* and *Zygosaccharomyces*. Tolerance tests for sugar and ethanol were used to determine whether they were suitable for wine production. *S. cerevisiae* was found the most biochemically effective strain for making wine from the papaya variety CO_2. The CO_2 papaya produced the greatest (11.59%) alcohol content after 1 month of fermentation, with residual sugar (1.87%). The research showed that biochemically described yeast strains might be used to produce wine from fruit that is readily available locally utilizing straightforward, affordable and adaptable technologies (Hasan & Zulkahar 2018). Researchers also investigated the effect of adding amino acids (l-leucine, l-valine, l-isoleucine and l-phenylalanine) in the fermentation process on the evolution of aroma compounds in papaya wine fermented using various yeast strains *Williopsis saturnus var. mrakii* NCYC2251. Using *W. saturnus* var. mrakii NCYC2251, regardless of the addition of amino acids, yeast growth, °B, changes in sugars, organic acids and pH were comparable. However, the addition of l-leucine enhanced the concentration of isoamyl alcohol and some esters i.e., isoamyl butyrate, isoamyl acetate and isoamyl propionate. The addition of l-valine somewhat raised the levels of isobutyl alcohol and isobutyl acetate. The addition of l-phenylalanine led to a decrease in the formation of most esters while increasing the formation of 2-phenylethyl acetate, 2-phenylethanol and 2-phenylethyl butyrate. In conclusion, modifying the aroma of papaya wine through the fermentation of papaya juice with *W. saturnus var. mrakii* NCYC2251 and the inclusion of specific amino acids may be possible (Lee et al. 2011b). The impacts of successive inoculation of the yeasts *W. saturnus var. mrakii* NCYC2251 and *S. cerevisiae var. bayanus* R2 on the volatile profiles of papaya wine were examined in a different study. The papaya wine with more fruitiness and acetate esters was achieved by inoculating *S. cerevisiae* with *W. saturnus* after 7 days of fermentation (simultaneous inoculation). However, when *W. saturnus* was inoculated with *S. cerevisiae* after two days of fermentation, the majority of the volatiles were equal in concentration to the control, except for the increased amount of ethyl esters. The fermentation was dominated by the initial yeast inoculation (Lee et al. 2011b; Shaheel et al. 2015).

A variety of volatile chemicals evolved when fusel oil, a by-product of the alcohol distillation industry, was added to papaya wine fermented with yeast *Williopsis saturnus var. mrakii* NCYC2251. Researchers determined °Brix, pH, and yeast growth characteristics except for those that contained 0.5% fusel oil, which impeded yeast development. Fusel oil was added at a concentration of 0.1%, which increased the synthesis of ethanol and acetate esters while decreasing the formation of acetic acid and other unwanted volatiles. This study showed that fermenting papaya juice with the same yeast as before and adding fusel oil in small amounts can be a different way to control the production of compounds that give papaya wine its distinctive aroma (Lee et al. 2011a).

10.4.26 Pineapple

The pineapple (*Ananas comosus*, Family: *Bromeliaceae*) is one of the most widely grown and consumed subtropical fruits in the world. The fruits are mostly utilized in the food industry to make canned fruit, jam, concentrated juice, as well as wine. They are also consumed fresh. Pineapples are the potential source for producing wine since they have good sugar content. TSS in pineapple juice typically ranges from 12 to 15°B; by adding sugar, the TSS is increased to 22 to 25 °B to create a wine with 12 to 13% alcohol. Unfortunately, pineapple's flavor is not consistent as oxidation can happen quickly (Amerine 1980). There are many different cultivars of pineapples, however, the cultivars Smooth Cayenne, Queen and Red Spanish are among the most widely used for making pineapple wine (El-Sayed & Ismail 2015). After making wine from the two cultivars Kew and Queen, it was discovered that phosphate and nitrogen play major roles in the production of high-quality pineapple wine. The highest alcohol content was found in wine made from the Kew variety and inoculated with *Saccharomyces ellipsoideus* 101 (8.40%), followed by wine made from the Queen variety (8.35%). A 20-point scoring system was used by a group of judges to conduct an organoleptic examination. When both N and P sources were added, wine made from the Queen variety scored the highest (16.75 out of 20), followed by wine made from the Kew variety (14.75 out of 20), while wine made from the Queen variety scored the lowest (13 out of 20) (Santoshkumar 2006). Using single as well as mixed non-Saccharomyces yeasts, (Chanprasartsuk et al. 2012) formulated pineapple wine and proposed that the Queen variety would provide an excellent substrate for yeast fermentation. The mixed starting cultures made from the yeast isolates *Saccharomycodes ludwigii* and *Hanseniaspora* could carry out the necessary alcoholic fermentation for the production of Queen pineapple wine. According to Ayogu (1999), pineapple fruit wine can be produced using a *S. cerevisiae* species that was isolated from the fermenting sap of palm wine. This yeast isolate demonstrated increased ethanol tolerance as seen by its high ethanol yield of 10.2% compared to the control wine yeast, which produced only 7.4% alcohol content. The yeast for commercial winemaking from pineapple fruits could also come from Nigerian palm wine, which contains a lot of these yeasts (Ayogu 1999). Youravong et al. (2010) employed a microfiltration technique to clarify pineapple wine. It was found that reducing reversible fouling rather than irreversible fouling was possible by increasing the gas sparging rate. After microfiltration, the turbidity of pineapple wine was reduced, and a clear product with a bright yellow color was produced. It was also noted that the wine's alcohol concentration decreased as a result of gas sparging (Youravong et al. 2010).

During the fermentation process, the freshly crushed pineapple juice from Australia and Thailand reveals the presence of *Pichia guilliermondii* and *H. uvarum* yeast strains (Chanprasartsuk et al. 2010). While *H. uvarum* populations increased from 5 log CFU/ml to 8 log CFU/ml after 6 days of fermentation, *P. guilliermondii* populations were dominating and constantly present during the early stages of the fermentation. In Thai samples, the maximum ethanol concentration of 2% was discovered after two days of fermentation. The ethanol %age in the Australian samples, in contrast, increased steadily throughout the 6-day fermentation period, reaching 3–4%. Whereas Australian samples had constant concentrations of various acids throughout the fermentation process. Thai samples revealed a large increase in lactic acid levels and a considerable fall in citric acid levels. Contrary to many other fruit juices, none of the sample fermentation systems had any Saccharomyces yeasts or other wine yeasts. It was determined as a result that freshly crushed pineapple juice might have an impact on other autochthonous yeasts involved in alcoholic fermentation (Chanprasartsuk et al. 2010). When the HS-SPME and GC/MS analyses were combined, 18 volatiles were found, including 13 esters, 4 alcohols and 1 acid. The main ingredients were ethyl octanoate, 3-methyl-1-butanol, ethyl acetate and ethyl decanoate. Yet, according to OAV, the chemicals ethyl acetate, ethyl octanoate and ethyl 2-methylpropanoate significantly influenced the scent of pineapple wine (Pino & Queris 2010).

10.4.27 Pomegranate

Pomegranate (*Punica granatum*, Family: *Lythraceae*) is a promising source of phytochemicals, i.e., ellagitannins and anthocyanins (Navarro et al. 2014). Due to improper handling and over-ripening, which makes pomegranate fruits susceptible to microbial deterioration, vast numbers of pomegranate fruits are discarded. To reduce production losses and increase profitability, pomegranate fruits are typically processed into jams, juices, etc. The wine was made from overripe pomegranate fruits, specifically from the *Mollar de Elche* and Wonderful cultivars. Pomegranate wines were abundant in anthocyanins, which gave the wine its color and anti-oxidant properties.

Thus, pomegranate wine offers a new processed product with positive health impacts (Mena et al. 2012). They also reported melatonin (N-acetyl-5-methoxytryptamine) in the wine produced from the *Mollar de Elche* and Wonderful pomegranate cultivars. Neurohormone melatonin (N-acetyl-5-methoxytryptamine) is linked to a wide range of physiological processes and has demonstrated therapeutic effects. Certain foods have been thought to act as an exogenous source of melatonin that can raise blood levels when consumed because they contain the hormone as well (Mena et al. 2012).

10.4.28 Cashew Apple

Cashew (*Anacardium occidentale*) is one of the most commercially significant plantation crops in Nigeria, Brazil, Vietnam and India (Mohanty et al. 2006). Contrary to the cashew nut kernel, which has an undeniably unique fine taste and its commercial appeal, the cashew "apple" is essentially unknown to consumers despite having a high nutritional value (vitamin C and minerals, i.e., P, Ca and Fe). The thick container or "fake fruit" that the cashew nut or actual fruit is attached to is known as the edible cashew apple. In Goa, India, a common native brandy known as "fenny" is made from cashew apples and contains 40% alcohol (Batista-Silva et al. 2018). After fermenting the juice with the wine yeast strain NCYC 125 and *S. cerevisiae var. bayanus*, Mohanty et al. 2006 and Osho & Odunfa, 1999 attempted to make wine from cashew apples and compared the sensory qualities with commercial grape wine. The formation of lactic acid, vinegar and ethanol from cashew apples was investigated. The highest documented ethanol production rate had a yield of 55% (Silva et al. 2007). By examining the effects of the juice's initial sugar concentration on the wine's composition and sensory qualities, Attri et al. (2009) found that each *S. cerevisiae* strain is different and that the fermentation process's conditions and initial sugar concentration have a significant impact on the amount of ethanol produced. The panelists gave cashew wine a somewhat lower overall score than commercial grape wine in a preliminary sensory evaluation analysis, but they gave it a "like much" rating for taste, aftertaste, and color/appearance. Nonetheless, the panelists ranked aroma and flavor between 2.5–3.0 ("like moderately"–"like greatly"), perhaps because of the high tannin concentration in cashew wine, which lends a somewhat astringent flavor. Yet, all of the panelists approved of the cashew wine (Attri 2009).

10.4.29 Coconut Wine

The coconut palm (*Cocos nucifera*), a member of the *Arecaceae* family, produces coconut fruit. Coconut is typically found near the sea coasts in tropical areas between 22°N and 22°S of the equator. In an experiment, the sterile tender coconut water with 10% sugar was infected with a purified yeast strain derived from Bangalore blue grapes. Each fermenting bottle received either fresh red rose petals or mint leaves. One control with the same quantity of sugar was kept for comparison. After complete fermentation, filtering, and six months of maturing at 4°C, the soft coconut water gave the wine a nice rose flavor, a crimson color and a tartness similar to that of grape wine. Mint leaves also flavored the wine, although they had no color. When compared to the mint-flavored and control wines, the rose-flavored wine had a more appealing color, better flavor and higher alcohol level, according to the organoleptic and alcohol tests. Mint-flavored wine only had 4% alcohol, compared to up to 8% for rose-flavored and control wines (Emmanuel & Odoyo 2011; Souza et al. 2013).

10.4.30 Sapota Wine

An important tropical fruit from the *Sapotaceae* family is the sapota (*Achras sapota*). Since the fruit ripens after 8 to 10 days of harvest, it has a limited shelf-life and post-harvest loss. An estimated 20–30% of India's total fruit production is lost. The pulp has a high latex concentration and is yellowish, granular, soft, juicy and sweet. Mature sapota possesses good amounts of carbohydrates (21.4%), dietary fiber (10.9%) and tannins (3.45%). Researchers have examined the production, content and taste characteristics of sapota wine in light of its higher production, nutritional relevance and shorter availability period. Sixteen panelists conducted a sensory evaluation on a range of characteristics, including taste, aroma, flavor, color/appearance and aftertaste. The findings revealed that the panelists had a strong preference for the flavor, taste and scent of sapota wine and that there was little difference between the two replicates for the majority of sensory characteristics (Panda et al. 2014).

10.4.31 Carambola

The *Oxalidaceae* family includes the star fruit known as carambola (*Averrhoa carambola*). When ripe, fruit can range in color from light yellow to deep amber. The Indo-Burma region uses it as a commercial fruit (Das et al. 2013). Vitamins A and C are abundant in them. It is an extremely delicate, thin-skinned fruit with a very short shelf-life (Napahde et al. 2010). About 25% of star fruits do not match the required market standards because of their size, shape and appearance. Using inferior fruits that cannot be sold through the processing of star fruit, wine might help decrease waste and boost crop value. Alcohol at its highest yield (10.98%) can be produced (Aye et al. 2014).

10.4.32 Ber

Ziziphus species, often known as ber or Indian jujube, are members of the *Rhamnaceae* family. In semi-arid regions of the world, it is a fruit crop that is underutilized. It provides a plentiful supply of carbohydrates, vital minerals and ascorbic acid. It contains more protein, minerals and vitamin C than an apple or a mango. Ber fruits have a very limited shelf-life. Juice from the ber fruit has an alcohol production of up to 2.24% (Das et al. 2013). Indian jujube wine, often referred to as ber wine, is a widely consumed alcoholic beverage throughout South Asia, particularly in Bangladesh, Pakistan and India. The fermenting of fruit juice, primarily from the ber fruit (*Ziziphus mauritiana*), is a traditional method for making ber wine (Singh et al. 2018). For the fermentation of beer wine, a variety of yeast strains, including *Candida utilis, Saccharomyces cerevisiae* and *Saccharomyces bayanus* can be utilized. Nonetheless, *S. cerevisiae* is the yeast that is most frequently used to make ber wine because of its capacity to create large amounts of alcohol and give the wine flavorful characteristics. Ber fruit comes in a variety of forms, each with a distinctive flavor and perfume. Umran, Bari and Neelam are some of the well-liked cultivars of ber fruit utilized in wine production (Panda et al. 2015). Due to the inclusion of numerous bioactive substances, i.e., polyphenols, flavonoids and carotenoids, ber wine provides several health advantages. These substances' anti-inflammatory, anti-cancer and anti-oxidant capabilities can help prevent or lower the risk of chronic illnesses like cancer, diabetes and cardiovascular disease. The cultivar, the circumstances surrounding the fermentation, and the amount of aging affect the chemical makeup of ber wine. Alcohol, volatile substances (esters, alcohols, and aldehydes) and organic acids (mostly malic and tartaric acids) all contribute to the flavor and aroma of ber wine. The flavor profile of ber wine is distinctive and can be characterized as fruity, flowery and somewhat acidic. The aromatic perfume has hints of berries, tropical fruits and spices like clove and cinnamon (Kader et al. 2017). Due to the alcohol content, the flavor is warming and sweet with a hint of acidity. Due to the high sugar content of the fruit juice, which can cause sluggish or incomplete fermentation, low alcohol concentration and spoiling, the creation of ber wine can be difficult. Moreover, improper fruit handling and sanitation can result in microbial contamination from bacteria and fungi, which can degrade the wine's quality and safety (Rehman et al. 2015).

10.4.33 Karonda

The *Apocynaceae* family includes the evergreen spiky shrub known as karonda (*Carissa carandas*). It may grow in a variety of climatic settings. When mature, karonda fruits are served as a dessert. Mature fruits have a high pectin content and are used to make a variety of products, including jelly, jam, squash, sauce and syrup. The unripe fruits can be used to make pickles and chutney since they are tart and astringent. The alcohol content of wine made from ripe fruits ranges between 14.5 and 15% (Singh et al. 2014). The yeast strains *Saccharomyces cerevisiae, Candida pulcherrima* and *Hanseniaspora guilliermondii* are frequently used for karonda wine production. These varieties are renowned for producing fine wine with a fruity flavor. The karonda fruit comes in a variety of cultivars, including Arka Sahan, Arka Anamika and Arka Suman. Each cultivar has a unique flavor, which can affect the quality and flavor profile of the wine (Gopalakrishnan & Lakshmi 2016).

10.4.34 Aonla

The Indian gooseberry, or aonla (*Emblica officinalis*), is a member of the *Euphorbiaceae* family and is widely utilized in Ayurveda for its therapeutic benefits. Aonla is less favored for direct eating because of its high acidity and astringent flavor. As a result, it is eaten after being properly processed or blended with other fruits. Fruits that have undergone thermal processing lose important phytochemicals, including ascorbic acid, polyphenols, antioxidants, etc. Therefore, fermentation as an alternate processing method is advantageous, its usage is restricted to wines due to its strong acidity and astringency. Thus, it is believed that combining two or more fruit juices and associated beverages is a practical way to use it (Ravani & Joshi 2014). The varieties, which are renowned for producing fine wine with a fruity flavor, include Banarasi, Francis, Chakaiya, NA-7 and Krishna. Each cultivar has its own flavor and taste that might affect the wine's quality and flavor profile (Mohammed 2019).

10.5 Conclusion and Future Perspectives

The present study is an amalgamation of studies on the production of wine from a variety of fruits, as grapes are not only fruit from which wine can be manufactured nowadays. According to studies, wine can be made from a variety of nutritious, perishable and underutilized fruits. This can help to extend the shelf-life of fruits by lowering post-harvest losses, increasing consumption and export, increasing fruit cultivation and commercialization, and providing profits to grape growers and the wine industry as a whole. The majority of studies have been conducted on grape wine rather than on non-grape fruit wines, although during the past few years, remarkable progress has been made in wine biotechnology, particularly in wine yeast improvement, the development of genetically modified yeast and lowering alcohol concentration in wine. However, current and anticipated advancements in improving aroma volatiles through the use of improved strains of yeast, and advanced chemometric analysis, led to broader commercialization of non-grape fruit wines, thus contributing more to the wine industry's economy.

References

Adaramola, F. B., & Benjamin, G. (2018). Antimicrobial and antioxidant activities of crude methanol. *Food Science and Biotechnology*, *5*(October 2019), 0–3.

Adebowale, A. A., & Adeyemi, I. A. (2014). Chemical composition and sensory evaluation of palm wine from Elaeis guineensis. *Journal of Food Processing and Preservation*, *38*(2), 711–717.

Akubor, P. I., Obio, S. O., Nwadomere, K. A., & Obiomah, E. (2003). Production and quality evaluation of banana wine. *Plant Foods for Human Nutrition*, *58*(3), 1–6. https://doi.org/10.1023/B:QUAL.0000041138.29467.b6

Altuntas, E., Cangi, R., & Kaya, C. (2011). Physical and chemical properties of persimmon fruit. *International Agrophysics*, *25*(1), 89–92.

Alves, J. A., de Oliveira Lima, L. C., Nunes, C. A., Dias, D. R., & Schwan, R. F. (2011). Chemical, physical-chemical, and sensory characteristics of Lychee (Litchi chinensisSonn) wines. *Journal of Food Science*, *76*(5), 330–336. https://doi.org/10.1111/j.1750-3841.2011.02188.x

Amerine, M. A., Cruess, W. V., & Berg, H. W. (1972). The technology of wine making: 3d Ed. In Maynard Andrew Amerine (Ed.), *AVI, Westport, Connecticut: Vol. 4th Ed.* AVI Publ. Co.

Amerine, M. A., Kunkee, R., Ough, K. C. S., & AD, W. (1980). The technology of wine making. In Maynard Andrew Amerine (Ed), *AVI, Westport, Connecticut: Vol. 4th Ed.* AVI Publ. Co.

Attri, B. L. (2009). Effect of initial sugar concentration on the physico-chemical characteristics and sensory qualities of cashew apple wine. *Indian Journal of Natural Products and Resources*, *8*(4), 374–379.

Aurore, G., Parfait, B., & Fahrasmane, L. (2009). Bananas, raw materials for making processed food products. *Trends in Food Science and Technology*, *20*(2), 78–91. https://doi.org/10.1016/j.tifs.2008.10.003

Aye, S., Linn Ko, T., & Oo, C. C. (2014). Processing and characterization of wine by using an underutilized fruit. In *2nd Myanmar Korea Conference Research Journal*. Star Fruit.

Ayogu, T. E. (1999). Evaluation of the performance of a yeast isolate from Nigerian palm wine in wine production from pineapple fruits. *Bioresource Technology*, *69*(2), 189–190. https://doi.org/10.1016/S0960-8524(99)00162-5

Ballesteros, L. C., Carmona, M., & González-Álvarez, J. (2021). An overview of cider production: Raw materials, fermentation, and post-fermentation processing. In A.G.H. Lea, J.R. Piggott (Eds.), *Fermented Beverages* (pp. 109–138). Springer.

Barrett, D. M., & Lloyd, B. (2012). Advanced preservation methods and nutrient retention in fruits and vegetables. *Journal of the Science of Food and Agriculture*, *92*(1), 7–22. https://doi.org/10.1002/jsfa.4718

Basu, A., Du, M., Leyva, M. J., Sanchez, K., Betts, N. M., Wu, M., Aston, C. E., & Lyons, T. J. (2010). Blueberries decrease cardiovascular risk factors in obese men and women with metabolic syndrome. *Journal of Nutrition*, *140*(9), 1582–1587. https://doi.org/10.3945/jn.110.124701

Batista-Silva, W., Nascimento, V. L., Medeiros, D. B., Nunes-Nesi, A., Ribeiro, D. M., Zsögön, A., & Araújo, W. L. (2018). Modifications in organic acid profiles during fruit development and ripening: Correlation or causation? *Frontiers in Plant Science, 871*, 1689. https://doi.org/10.3389/fpls.2018.01689

Bchir, B., Bouaziz, M. A., Ettaib, R., Sebii, H., Danthine, S., Blecker, C., Besbes, S., & Attia, H. (2020). Optimization of ultrasound-assisted osmotic dehydration of pomegranate seeds (Punica granatum L.) using response surface methodology. *Journal of Food Processing and Preservation, 44*(9), 14657. https://doi.org/10.1111/jfpp.14657

Bisht, A., & Barman, S. C. (2013). Mango wine production and its quality evaluation. *Journal of Food Science and Technology, 50*(5), 981–987. https://doi.org/10.1007/s13197-011-0574-4.

Byaruagaba-Bazirake, G., & Rensburg, P. (2013). Characterisation of banana wine fermented with recombinant wine yeast strains. *American Journal of Food and Nutrition, 3*(3), 105–116. https://doi.org/10.5251/ajfn.2013.3.3.105.116

Chambers, P. J., & Pretorius, I. S. (2010). Fermenting knowledge: The history of winemaking, science and yeast research. *EMBO Reports, 11*(12), 914–920. https://doi.org/10.1038/embor.2010.179

Chandel, R., Kumar, V., Kaur, R., Kumar, S., Kumar, A., Kumar, D., & Kapoor, S. (2023). Bioactive compounds, health benefits and valorization of Pyrus pyrifolia (sand pear): A review. *Nutrition & Food Science*. https://doi.org/10.1108/nfs-09-2022-0306

Chanprasartsuk, O. O., Prakitchaiwattana, C., Sanguandeekul, R., & Fleet, G. H. (2010). Autochthonous yeasts associated with mature pineapple fruits, freshly crushed juice and their ferments; and the chemical changes during natural fermentation. *Bioresource Technology, 101*(19), 7500–7509. https://doi.org/10.1016/j.biortech.2010.04.047

Chanprasartsuk, O., Pheanudomkitlert, K., & Toonwai, D. (2012). Pineapple wine fermentation with yeasts isolated from fruit as single and mixed starter cultures. *Asian Journal of Food and Agro -Industry, 5*(02), 104–111. http://scholar.google.com/scholar?hl=en&btnG=Search&q=intitle:Pineapple+wine+fermentation+with+yeasts+isolated+from+fruit+as+single+and+mixed+starter+cultures#1

Cheirsilp, B., & Umsakul, K. (2008). Processing of banana-based wine product using pectinase and α-amylase. *Journal of Food Process Engineering, 31*(1), 78–90. https://doi.org/10.1111/j.1745-4530.2007.00152.x

Cho, J. Y., Kim, S. J., Lee, H. J., Kim, J. Y., Lym, I. J., Kang, S. K., Park, K. H., & Moon, J. H. (2011). Isolation and identification of low molecular volatile compounds from ethyl acetate layer of Korean black raspberry (Rubus Coreanus Miq.) Wine. *Korean Journal of Food Science and Technology, 43*(5), 558–563. https://doi.org/10.9721/KJFST.2011.43.5.558

Chowdhury, P., & Ray, R. C. (2007). Fermentation of jamun (Syzgium cumini L.) fruits to form red wine. *International Food Research Journal, 14*(1), 15–23.

Craig, J. T. (1988). A comparison of the headspace volatiles of Kiwifruit Wine with those of Wine of Vitis vinifera Variety Müller-Thurgau. *American Journal of Enology and Viticulture, 39*(4), 321–324. https://doi.org/10.5344/ajev.1988.39.4.321

Das, J. N. (2009). Studies on storage stability of jamun beverages. *Indian Journal of Horticulture, 66*(4), 508–510.

Das, S. C., Prakash, J., Deb, A. K., & Biswas, T. (2013). Medicinal value of underutilized fruits in hilly Tripura. *Acta Horticulturae, 972*, 135–141. https://doi.org/10.17660/ActaHortic.2013.972.17

Ding, P. (2016). Tropical fruits. In Brian Thomas; Brian G Murray; Denis J Murphy *Encyclopedia of Applied Plant Sciences* (Vol. 3). CAB International. https://doi.org/10.1016/B978-0-12-394807-6.00185-4

Duarte, W. F., Dias, D. R., De Melo Pereira, G. V., Gervásio, I. M., & Schwan, R. F. (2009). Indigenous and inoculated yeast fermentation of gabiroba (Campomanesia pubescens) pulp for fruit wine production. *Journal of Industrial Microbiology and Biotechnology*, *36*(4), 557–569. https://doi.org/10.1007/s10295-009-0526-y

Duarte, W. F., Dias, D. R., Oliveira, J. M., Teixeira, J. A., de Almeida e Silva, J. B., & Schwan, R. F. (2010). Characterization of different fruit wines made from cacao, cupuassu, gabiroba, jaboticaba and umbu. *Lwt*, *43*(10), 1564–1572. https://doi.org/10.1016/j.lwt.2010.03.010

Duarte, W. F., Dias, D. R., Oliveira, J. M., Vilanova, M., Teixeira, J. A., e Silva, J. B. A., & Schwan, R. F. (2010). Raspberry (Rubus idaeus L.) wine: Yeast selection, sensory evaluation and instrumental analysis of volatile and other compounds. *Food Research International*, *43*(9), 2303–2314. https://doi.org/10.1016/j.foodres.2010.08.003

El-Sayed, M. A., & Ismail, M. M. (2015). Production of pineapple wine by Saccharomyces cerevisiae from waste core and cannery waste. *Journal of Food Science and Technology*, *52*(2), 1283–1293. https://doi.org/10.1007/s13197-013-1042-2.

Emmanuel, I. O., & Odoyo, O. (2011). Studies on wine production from pawpaw (Carica papaya). *F Brewing and Distilling*, *2*(November), 56–62.

Fan, W., Xu, Y., & Yu, A. (2006). Influence of oak chips geographical origin, toast level, dosage and aging time on volatile compounds of apple cider. *Journal of the Institute of Brewing*, *112*(3), 255–263. https://doi.org/10.1002/j.2050-0416.2006.tb00721.x

Ghosh, A. (2015). Persimmon wine: Production and sensory quality evaluation. *Indian Journal of Traditional Knowledge*, *14*(2), 295–300.

González-SanJosé, M. L., Jiménez-López, C., & Borrás-Linares, I. (2021). Peach wine and its health benefits. *Beverages*, *7*(3), 37.

Gopalakrishnan, B., & Lakshmi, S. (2016). Karonda (Carissa carandas Linn.) wine production and evaluation of its physico-chemical and sensory characteristics. *Journal of Food Science and Technology*, *53*(8), 3195–3201. https://doi.org/10.1007/s13197-016-2283-4

Gresele, P., Cerletti, C., Guglielmini, G., Pignatelli, P., de Gaetano, G., & Violi, F. (2011). Effects of resveratrol and other wine polyphenols on vascular function: An update. *Journal of Nutritional Biochemistry*, *22*(3), 201–211. https://doi.org/10.1016/j.jnutbio.2010.07.004

Hager, T. J., Howard, L. R., & Prior, R. L. (2008). Processing and storage effects on monomeric anthocyanins, percent polymeric color, and antioxidant capacity of processed blackberry products. *Journal of Agricultural and Food Chemistry*, *56*(3), 689–695. https://doi.org/10.1021/jf071994g

Halpern, G. M. (2008). A celebration of wine: Wine IS medicine. *Inflammopharmacology*, *16*(5), 240–244. https://doi.org/10.1007/s10787-008-8024-9

Hao, J., Wang, Y., Yue, T., Chen, J., & Zhang, M. (2019). The health benefits and production technology of cherry wine: A review. *Food Science and Technology Research*, *25*(6), 861–869.

Hasan, N. A., & Zulkahar, I. M. (2018). Isolation and identification of bacteria from spoiled fruits. *AIP Conference Proceedings*, *2020*(2), 849–852. https://doi.org/10.1063/1.5062699

Hernández, J. Á. G., González, D. M., Castillo, M. P., & Falcõn, T. F. (2013). Use of a specific anti-stretch mark cream for preventing or reducing the severity of striae gravidarum. Randomized, double-blind, controlled trial. *International Journal of Cosmetic Science, 35*(3), 233–237. https://doi.org/10.1111/ics.12029

Hill, M., & Ganeshananthan, K. (2008). Expert opinion on drug delivery – increasing frequency. *Expert Opinion on Drug Delivery, 5*(1), 1285–1298.

Hiscott, R. (2015). Neuro now. *Neurology Today, 15*(16), 25–26.

Hui, Y. H., Evranuz, E. Ö., Arroyo-López, F. N., Fan, L., Hansen, Å. S., Jaramillo-Flores, M. E., Rakin, M., Schwan, R. F., & Zhou, W. (2012). Handbook of plant-based fermented food and beverage technology: Second edition. *Handbook of Plant-Based Fermented Food and Beverage Technology, Second Edition, 21*, 1–813. https://doi.org/10.1201/b12055

Hwang, I. W., Jeong, M. C., & Chung, S. K. (2011). The physicochemical properties and the antioxidant activities of persimmon peel powders with different particle sizes. *Journal of Applied Biological Chemistry, 54*(3), 442–446. https://doi.org/10.3839/jksabc.2011.068

Hwang, Y. C., & Shin, K. S. (2011). Isolation and characterization of immunostimulating polysaccharide in Korean black raspberry wine. *Journal of the Korean Society for Applied Biological Chemistry, 54*(4), 591–599. https://doi.org/10.3839/jksabc.2011.089

Jagtap, U. B., & Bapat, V. A. (2015). Wines from fruits other than grapes: Current status and future prospectus. *Food Bioscience, 9*, 80–96. https://doi.org/10.1016/j.fbio.2014.12.002

Jagtap, U. B., Waghmare, S. R., Lokhande, V. H., Suprasanna, P., & Bapat, V. A. (2011). Preparation and evaluation of antioxidant capacity of Jackfruit (Artocarpus heterophyllus Lam.) wine and its protective role against radiation induced DNA damage. *Industrial Crops and Products, 34*(3), 1595–1601. https://doi.org/10.1016/j.indcrop.2011.05.025

Jamir, L., Kumar, V., Gat, Y., Kumar, A., & Kaur, S. (2019). Wine: A potential source of antimicrobial compounds. *Journal of Wine Research, 30*(3), 220–237. https://doi.org/10.1080/09571264.2019.1652151

Jeandet, P., Vasserot, Y., Liger-Belair, G., & Marchal, R. (1997). Sparkling wine production. In V. K. Joshi (Ed.), *Trends in Food Science & Technology* (Vol. 8, Issue 9, p. 315). Asia Tech Publishers, INC. https://doi.org/10.1016/s0924-2244(97)80273-7

Jeong, J. H., Jung, H., Lee, S. R., Lee, H. J., Hwang, K. T., & Kim, T. Y. (2010). Antioxidant, anti-proliferative and anti-inflammatory activities of the extracts from black raspberry fruits and wine. *Food Chemistry*, 123(2), 338–344. https://doi.org/10.1016/j.foodchem.2010.04.040

Johnson, M. H., & Gonzalez de Mejia, E. (2012). Comparison of chemical composition and antioxidant capacity of commercially available blueberry and blackberry wines in Illinois. *Journal of Food Science, 77*(1), 141–148. https://doi.org/10.1111/j.1750-3841.2011.02505.x

Johnson, M. H., Lucius, A., Meyer, T., & Gonzalez De Mejia, E. (2011). Cultivar evaluation and effect of fermentation on antioxidant capacity and in vitro inhibition of α-amylase and α-glucosidase by highbush blueberry (vaccinium corombosum). *Journal of Agricultural and Food Chemistry, 59*(16), 8923–8930). https://doi.org/10.1021/jf201720z

Joshi, V. K. (2009). Production of wines from non-grape fruit. *Natural Product Radiance, 8*(94), 313–469.

Joshi, V. K., Gill, A., Kumar, V., & Chauhan, A. (2014). Preparation of plum wine with reduced alcohol content: Effect of must treatment and blending with sand pear juice on physico-chemical and sensory quality. *Indian Journal of Natural Products and Resources, 5*(1), 67–74.

Joshi, V. K., & Kumar, V. (2011). Importance, nutritive value, role, present status and future strategies in fruit wines in India. In P. S. Panesar et al. (Eds.), *Bio-Processing of Foods* (pp. 39–62). Asia Tech Publishers, New Delhi.

Joshi, V. K., & Kumar, V. (2017). Influence of different sugar sources, nitrogen sources and inocula on the quality characteristics of apple tea wine. *Journal of the Institute of Brewing, 123*(2), 268–276. https://doi.org/10.1002/jib.417

Joshi, V. K., Kumar, V., & Kumar, A. (2012). Physico-chemical and sensory evaluation of wines from different citrus fruits of Himachal Pradesh. *International Journal of Food Fermentation Technology, 2*(2), 145–148.

Joshi, V. K., Sandhu, D. K., & Kumar, V. (2013). Influence of addition of insoluble solids, different yeast strains and pectinesterase enzyme on the quality of apple wine. *Journal of the Institute of Brewing, 119*(3), 191–197. https://doi.org/10.1002/jib.79

Joshi, V. K., Sharma, R., Kumar, V., & Joshi, D. (2019). Optimization of a process for preparation of base wine for cider vinegar production. *Proceedings of the National Academy of Sciences India Section B – Biological Sciences, 89*(3), 1007–1016. https://doi.org/10.1007/s40011-018-1019-4

Kadam, S., Gaikwad, V., & Khedkar, C. (2015). Lychee wine production technology and its sensory evaluation. *Journal of Food Science and Technology, 52*(5), 2705–2712. https://doi.org/10.1007/s13197-014-1317-6

Kader, M., Shilpi, J. A., & Sultana, S. (2017). A review on nutritional and antioxidant potential of ber (Ziziphus mauritiana Lamk.) fruit. *Journal of Food Science and Technology, 54*(12), 3754–3765. https://doi.org/10.1007/s13197-017-2845-y

Kang, H. J., Choi, H. J., & Kim, D. H. (2014). Nutritional and antioxidant properties of persimmon wine fermented with different yeast strains. *Journal of the Korean Society of Food Science and Nutrition, 43*(5), 665–672. https://doi.org/10.3746/jkfn.2014.43.5.665

Karamoko, D., Djeni, N. T., N'guessan, K. F., Bouatenin, K. M. J. P., & Dje, K. M. (2012). The biochemical and microbiological quality of palm wine samples produced at different periods during tapping and changes which occured during their storage. *Food Control, 26*(2), 504–511. https://doi.org/10.1016/j.foodcont.2012.02.018

Kaur, M., Sharma, H., & Bajwa, U. (2020). Apricot wine: A review of production, quality, and health benefits. *Journal of Food Science and Technology, 57*(10), 3463–3473.

Kelebek, H., Selli, S., Canbas, A., & Cabaroglu, T. (2009). HPLC determination of organic acids, sugars, phenolic compositions and antioxidant capacity of orange juice and orange wine made from a Turkish cv. Kozan. *Microchemical Journal, 91*(2), 187–192. https://doi.org/10.1016/j.microc.2008.10.008

Khan, A. R. (2018). Antioxidant and anti-inflammatory properties of Meldar (Morus nigra) fruit wine. *BMC Complementary and Alternative Medicine, 18*(1), 1–10. https://doi.org/10.1186/s12906-018-2119-y

Koh, E. J., Seo, W. D., Choi, S. Y., Cho, S. H., & Park, H. J. (2016). Nutritional and functional properties of Prunus mume Sieb. Et Zucc. *Journal of Medicinal Food, 19*(8), 749–755. https://doi.org/10.1089/jmf.2015.3589

Kondapalli, N., Sadineni, V., Variyar, P. S., Sharma, A., & Obulam, V. S. R. (2014). Impact of γ-irradiation on antioxidant capacity of mango (Mangifera indica L.) wine from eight Indian cultivars and the protection of mango wine against DNA damage caused by irradiation. *Process Biochemistry, 49*(11), 1819–1830. https://doi.org/10.1016/j.procbio.2014.07.015

Kong, Q., Tao, G., & Zhang, Z. (2019). Production technology, quality improvement, and health benefits of strawberry wine: A review. *Food Science and Technology Research, 25*(3), 373–382.

Ku, C. S., & Mun, S. P. (2008). Optimization of the extraction of anthocyanin from Bokbunja (Rubus coreanus Miq.) marc produced during traditional wine processing and characterization of the extracts. *Bioresource Technology, 99*(17), 8325–8330. https://doi.org/10.1016/j.biortech.2008.03.013

Kumar, V., Jnawali, P., Goud, P. V., & Bhasin, J. K. (2016a). Effect of maturation on physico-chemical and sensory quality characteristics of custard apple wine. *Cogent Food and Agriculture, 2*(1). https://doi.org/10.1080/23311932.2016.1180660

Kumar, V., Jnawali, P., Goud, P. V., & Bhasin, J. K. (2016b). Effect of maturation on physico-chemical and sensory quality characteristics of custard apple wine. *Cogent Food and Agriculture, 2*(1). https://doi.org/10.1080/23311932.2016.1180660

Lasekan, O., Buettner, A., & Christlbauer, M. (2007). Investigation of important odorants of palm wine (Elaeis guineensis). *Food Chemistry, 105*(1), 15–23. https://doi.org/10.1016/j.foodchem.2006.12.052

Lay, J. O., Zhou, M., Chen, J., & Huang, M. (2015). Papaya wine production technology and its sensory evaluation. *Journal of Food Science and Technology, 52*(6), 3226–3234. https://doi.org/10.1007/s13197-014-1466-7

Lee, P. R., Yu, B., Curran, P., & Liu, S. Q. (2011a). Effect of fusel oil addition on volatile compounds in papaya wine fermented with Williopsis saturnus var. mrakii NCYC 2251. *Food Research International, 44*(5), 1292–1298. https://doi.org/10.1016/j.foodres.2010.12.026

Lee, P. R., Yu, B., Curran, P., & Liu, S. Q. (2011b). Impact of amino acid addition on aroma compounds in papaya wine fermented with Williopsis mrakii. *South African Journal of Enology and Viticulture, 32*(2), 220–228. https://doi.org/10.21548/32-2-1382

Li, L., Zhao, C., Zhang, Y., Yao, J., Yang, W., Hu, Q., Wang, C., & Cao, C. (2017). Effect of stable antimicrobial nano-silver packaging on inhibiting mildew and in storage of rice. *Food Chemistry, 215*, 477–482. https://doi.org/10.1016/j.foodchem.2016.08.013

Li, X., Lim, S. L., Yu, B., Curran, P., & Liu, S. Q. (2013). Mango wine aroma enhancement by pulp contact and β-glucosidase. *International Journal of Food Science and Technology, 48*(11), 2258–2266. https://doi.org/10.1111/ijfs.12212

Lim, J. W., Jeong, J. T., & Shin, C. S. (2012). Component analysis and sensory evaluation of Korean black raspberry (Rubus coreanus Mique) wines. *International Journal of Food Science and Technology, 47*(5), 918–926. https://doi.org/10.1111/j.1365-2621.2011.02922.x

Lin, X., Wang, Q., Hu, X., Wu, W., Zhang, Y., Liu, S., & Li, C. (2018). Evaluation of different Saccharomyces cerevisiae strains on the profile of volatile compounds in pineapple wine. *Journal of Food Science and Technology, 55*(10), 4119–4130. https://doi.org/10.1007/s13197-018-3338-0

Mahattanatawee, K., Perez-Cacho, P. R., Davenport, T., & Rouseff, R. (2007). Comparison of three lychee cultivar odor profiles using gas chromatography – Olfactometry and gas chromatography – Sulfur detection. *Journal of Agricultural and Food Chemistry, 55*(5), 1939–1944. https://doi.org/10.1021/jf062925p

Manzocco, L., Anese, M., & Nicoli, M. C. (2017). Fruit wines: characteristics, production, and consumption. *Journal of Food Science and Technology, 54*(6), 1650–1667. https://doi.org/10.1007/s13197-017-2561-8.

Marrelli, M. (2021). Medicinal plants. In Dilantha Fernando *Plants* (Vol. 10, Issue 7). Oxford and IBH Publishing Co. https://doi.org/10.3390/plants10071355

McGovern, P. E., Mirzoian, A., & Hall, G. R. (2009). Ancient Egyptian herbal wines. *Proceedings of the National Academy of Sciences of the United States of America, 106*(18), 7361–7366. https://doi.org/10.1073/pnas.0811578106

Mena, P., Gil-Izquierdo, Á., Moreno, D. A., Martí, N., & García-Viguera, C. (2012). Assessment of the melatonin production in pomegranate wines. *Lwt, 47*(1), 13–18. https://doi.org/10.1016/j.lwt.2012.01.009

Mena, P., Gironés-Vilaplana, A., Martí, N., & García-Viguera, C. (2012). Pomegranate varietal wines: Phytochemical composition and quality parameters. *Food Chemistry, 133*(1), 108–115. https://doi.org/10.1016/j.foodchem.2011.12.079

Mogren, L. M., Andersen, C. M., Laursen, B. B., & Nielsen, N. J. (2019). *Chemical and Sensory Analysis of Red Raspberry Wine: Impact of Fermentation Temperature, Malolactic Fermentation, and Aging* (pp. 5–11).

Mohammed, A. (2019). Nutritional and medicinal importance of goat milk. www.academia.edu/38493518/nutritional_and_medicinal_importance_of_goat_milk

Mohanty, S., Ray, P., Swain, M. R., & Ray, R. C. (2006). Fermentation of cashew (anacardium occidentale L.) "apple" into wine. *Journal of Food Processing and Preservation, 30*(3), 314–322. https://doi.org/10.1111/j.1745-4549.2006.00067.x

Moreno-Arribas, M. V., & Polo, M. C. (2005). Winemaking biochemistry and microbiology: Current knowledge and future trends. *Critical Reviews in Food Science and Nutrition, 45*(4), 265–286. https://doi.org/10.1080/10408690490478118

Mukherjee, D., Datta, S., & Banerjee, R. (2014). Studies on preparation of wine from mango pulp using yeast Saccharomyces cerevisiae. *International Journal of Current Microbiology and Applied Sciences, 3*(5), 231–240.

Napahde, S., Durve, A., Bharati, D., & Chandra, N. (2010). Wine production from Carambola (Averrhoa Carambola) juice using Saccharomyces cerevisiae. *Asian Journal of Experimental Biological Sciences, 1*, 20–23. www.ajebs.com/special/SP-4.pdf

Navarro, M., Amigo-Benavent, M., Mesias, M., Baeza, G., Gökmen, V., Bravo, L., & Morales, F. J. (2014). Pomegranate fruit for health promotion: Myths and realities. *Journal of the Science of Food and Agriculture, 94*(8), 2037–2044.

Okigbo, R. N., & Obire, O. (2009). Mycoflora and production of wine from fruits of soursop (Annona muricata L.). *International Journal of Wine Research, 1*(1), 1–9. https://doi.org/10.2147/IJWR.S4667

Oliveira, J. M., Silva, C. D., & Peres, A. M. (2021). The influence of fermentation conditions on cider quality. *Beverages*, *7*(1), 1–20.

Oluwanisola, W., & Adediran, A. (2007). Quantitation of alcohols in orange wine fermented by four strains of yeast. *African Journal of Biochemistry Research*, *1*(6), 95–100. www.academicjournals.org/AJBR

Orwa, C., Mutua, A., Kindt, R., Simons, A., & Jamnadass, R. H. (2020). Agroforestree database: A tree reference and selection guide version. World Agroforestry Centre, Kenya. www.worldagroforestry.org/sites/treedbs/treedatabases.as.

Osho, A., & Odunfa, S. A. (1999). Fermentation on cashew juice using the wine yeast strain NCYC 125 and three other isolated yeast strains. *Advances of Food Science*, *21*(1/2), 23–29.

Panda, S. K., Panda, S. H., Swain, M. R., Ray, R. C., & Kayitesi, E. (2015). Anthocyanin-Rich Sweet Potato (Ipomoea batatasL.) Beer: Technology, Biochemical and Sensory Evaluation. *Journal of Food Processing and Preservation*, *39*(6), 3040–3049. https://doi.org/10.1111/jfpp.12569

Panda, S. K., Sahu, U. C., Behera, S. K., & Ray, R. C. (2014). Fermentation of sapota (Achras sapota Linn.) fruits to functional wine. *Nutrafoods*, *13*(4), 179–186. https://doi.org/10.1007/s13749-014-0034-1

Panesar, P. S., Kumar, N., Marwaha, S. S., & Joshi, V. K. (2009). Vermouth production technology –an overview. *Indian Journal of Natural Products and Resources*, *8*(4), 334–344.

Patel, V. B., Pandey, S. N., Singh, S. K., & Das, B. (2005). Variability in jamun (Syzygium cuminii Skeels) accessions from Uttar Pradesh and Jharkhand. *Indian Journal of Horticulture*, *62*(3), 244–247.

Patil, S. S., Thorat, R. M., & Rajasekaran, P. (2012). Utilization of jamun fruit (Syzygium cumini) for production of red wine. *Journal of Advanced Laboratory Research in Biology*, *2*(2), 200–203. www.sospublication.co.in

Pino, J. A., & Queris, O. (2010). Analysis of volatile compounds of pineapple wine using solid-phase microextraction techniques. *Food Chemistry*, *122*(4), 1241–1246. https://doi.org/10.1016/j.foodchem.2010.03.033

Pino, J. A., & Queris, O. (2011). Characterization of odor-active compounds in guava wine. *Journal of Agricultural and Food Chemistry*, *59*(9), 4885–4890. https://doi.org/10.1021/jf2011112

Pinto, A. C. Q., Cordeiro, M. C. R., Andrade, S. R. M., Ferreira, F. R., Filgueiras, H. A. C., Alves, R. E., & Kinpara, D. I. (2005). Annona species. In J.T. Williams; R.W. Smith; A. Hughes; N. Haq and C. R. Clement (Ed.), *Annona Species* (Vol. 1, Issue 1). International Centre for Underutilized Crops.

Pretorius, I. S., & Høj, P. B. (2005). Grape and wine biotechnology: Challenges, opportunities and potential benefits. *Australian Journal of Grape and Wine Research*, *11*(2), 83–108. https://doi.org/10.1111/j.1755-0238.2005.tb00281.x

Qiu, W., Zhang, X., Zhang, H., Liang, C., Xu, J., Gao, H., Ai, L., Zhao, S., Wang, Y., Yang, Y., & Zhao, X. (2020). Discrimination of meat from fur-producing and food-providing animals using mass spectrometry-based proteomics. *Food Research International*, *137*, 109446. https://doi.org/10.1016/j.foodres.2020.109446

Ramalhosa, E., Rivas-Gonzalo, J. C., & De Freitas, V. (2015). Influence of oak wood chips addition on the chemical and sensory properties of cider wine. *LWT-Food Science and Technology*, *62*(1), 455–462.

Ravani, A., & Joshi, D. C. (2014). Processing for value addition of underutilized fruit crops. *Trends in Post Harvest Technology*, *2*(2), 15–21.

Reddy, L. V. A., & Reddy, O. V. S. (2009). Effect of enzymatic maceration on synthesis of higher alcohols during mango wine fermentation. *Journal of Food Quality*, *32*(1), 34–47. https://doi.org/10.1111/j.1745-4557.2008.00234.x

Reddy, L. V. A., & Reddy, O. V. S. (2011). Effect of fermentation conditions on yeast growth and volatile composition of wine produced from mango (Mangifera indica L.) fruit juice. *Food and Bioproducts Processing*, *89*(4), 487–491. https://doi.org/10.1016/j.fbp.2010.11.007

Rehman, S. U., Ali, M. A., & Hussain, S. (2015). The effect of yeast strains on the quality of jujube (Ziziphus mauritiana Lamk.) wine. *Journal of Food Science and Technology*, *52*(5), 2815–2822. https://doi.org/10.1007/s13197-014-1339-9

Requena, T., Monagas, M., Pozo-Bayón, M. A., Martín-Álvarez, P. J., Bartolomé, B., del Campo, R., Ávila, M., Martínez-Cuesta, M. C., Peláez, C., & Moreno-Arribas, M. V. (2010). Perspectives of the potential implications of wine polyphenols on human oral and gut microbiota. *Trends in Food Science and Technology*, *21*(7), 332–344. https://doi.org/10.1016/j.tifs.2010.04.004

Santoshkumar, P., & Patil, a B. (2006). Wine production from pineapple must supplemented with sources of nitrogen and phosphorus. *Karnataka Journal of Agricultural Sciences*, *19*(3), 562. http://micro189.lib3.hawaii.edu:2048/login?url=http://search.ebscohost.com/login.aspx?direct=true&db=lah&AN=20073187791&site=ehost-live

Schmitzer, V., Veberic, R., Slatnar, A., & Stampar, F. (2010). Elderberry (Sambucus nigra L.) wine: A product rich in health promoting compounds. *Journal of Agricultural and Food Chemistry*, *58*(18), 10143–10146. https://doi.org/10.1021/jf102083s

Sevda, S. B., & Rodrigues, L. (2011). Fermentative behavior of Saccharomyces strains during Guava (Psidium Guajava L) must fermentation and optimization of guava wine production. *Journal of Food Processing & Technology*, *02*(04), 118. https://doi.org/10.4172/2157-7110.1000118

Selli, S. (2007). Volatile constituents of orange wine obtained from moro oranges (citrus sinensis [L.] osbeck). *Journal of Food Quality*, *30*(3), 330–341. https://doi.org/10.1111/j.1745-4557.2007.00124.x

Selli, S., Cabaroglu, T., & Canbas, A. (2003). Flavour components of orange wine made from a Turkish cv. Kozan. *International Journal of Food Science and Technology*, *38*(5), 587–593. https://doi.org/10.1046/j.1365-2621.2003.00691.x

Setyabudi, D. A. (2022). Postharvest Physiology and Technology of Tropical Fruits (Indonesia). *IOP Conference Series: Earth and Environmental Science*, 1024(1), 1741–1751. https://doi.org/10.1088/1755-1315/1024/1/012052

Shaheel, S. K., Swami, D. V., Kumar, B. P., & Krishna, K. U. (2015). Effect of storage period on physical and organoleptic properties of RTS prepared from karonda (Carissa carandas L.) blended juices. *International Journal of Processing and Post Harvest Technology*, 15(1), 241–246.

Shahidi, F. (2009). Nutraceuticals and functional foods: Whole versus processed foods. *Trends in Food Science and Technology*, *20*(9), 376–387. https://doi.org/10.1016/j.tifs.2008.08.004

Sharma, S., Joshi, V. K., & Abrol, G. (2009). An overview on strawberry [fragaria × ananassa (Weston) duchesne ex rozier] wine production technology, composition, maturation and quality evaluation. *Indian Journal of Natural Products and Resources*, *8*(4), 356–365.

Sharma, S., & Mahant, K. (2015). Improved Persimmon wine with enhanced antioxidant activity and standardized method of production. *Indian National Patent Filling, 3884.*

Sharma, V. C., & Ogbeide, O. N. (1982). Pawpaw as a renewable energy resource for the production of alcohol fuels. *Energy, 7*(10), 871–873. https://doi.org/10.1016/0360-5442(82)90036-6

Shepherd, M., & Waterhouse, A. (2020). JSFA welcomes Digital technologies and integrated research to meet global food challenges. *Journal of the Science of Food and Agriculture, 100*(5), 1841. https://doi.org/10.1002/jsfa.10299

Silva, M. E., Torres Neto, A. B., Silva, W. B., Silva, F. L. H., & Swarnakar, R. (2007). Cashew wine vinegar production: Alcoholic and acetic fermentation. *Brazilian Journal of Chemical Engineering, 24*(2), 163–169. https://doi.org/10.1590/S0104-66322007000200001

Singh, D., Kumar, V., Yadav, E., Falls, N., Singh, M., Komal, U., & Verma, A. (2018). One-pot green synthesis and structural characterisation of silver nanoparticles using aqueous leaves extract of Carissa carandas: Antioxidant, anticancer and antibacterial activities. *IET Nanobiotechnology, 12*(6), 748–756. https://doi.org/10.1049/iet-nbt.2017.0261

Singh, S., Singh, A. K., Meghwal, P. R., Singh, A., & Swamy, G. S. (2014). *Karonda. Tropical Sub Tropical Fruit Crops: Crop Improvement and Varietal Wealth.* Jaya Pub House Ind.

Souza, S. D., Kanchanashri, B., & Suvarna, V. C. (2013). Development of bioflavoured tender coconut water wine using native grape yeast. *International Journal of Biotechnology and Bioengineering Research, 4*(6), 553–554.

Srikanta, A. H., Kumar, A., Sukhdeo, S. V., Peddha, M. S., & Govindaswamy, V. (2016). The antioxidant effect of mulberry and jamun fruit wines by ameliorating oxidative stress in streptozotocin-induced diabetic Wistar rats. *Food and Function, 7*(10), 4422–4431. https://doi.org/10.1039/c6fo00372a

Stringini, M., Comitini, F., Taccari, M., & Ciani, M. (2009). Yeast diversity during tapping and fermentation of palm wine from Cameroon. *Food Microbiology, 26*(4), 415–420. https://doi.org/10.1016/j.fm.2009.02.006

Stull, A. J., Cash, K. C., Johnson, W. D., Champagne, C. M., & Cefalu, W. T. (2010). Bioactives in blueberries improve insulin sensitivity in obese, insulin-resistant men and women. *Journal of Nutrition, 140*(10), 1764–1768. https://doi.org/10.3945/jn.110.125336

Swami, S. B., Thakor, N. J., & Divate, A. D. (2014). Fruit wine production: A review. *Journal of Food Research and Technology, 2*(3), 93–100. www.jakraya.com/journa/jfrt

Thole, J. M., Kraft, T. F. B., Sueiro, L. A., Kang, Y. H., Gills, J. J., Cuendet, M., Pezzuto, J. M., Seigler, D. S., & Lila, M. A. (2006). A comparative evaluation of the anticancer properties of European and American elderberry fruits. *Journal of Medicinal Food, 9*(4), 498–504. https://doi.org/10.1089/jmf.2006.9.498

Towantakavanit, K., Park, Y. S., & Gorinstein, S. (2011). Bioactivity of wine prepared from ripened and over-ripened kiwifruit. *Central European Journal of Biology, 6*(2), 205–215. https://doi.org/10.2478/s11535-010-0118-x

U.S.D.A. (2022). *Pear Wine* (p. 10844). https://fdc.nal.usda.gov/fdc-app.html#/food-details/171860/nutrients

Uzochukwu, S. V. A., Balogh, E., Tucknot, O. G., Lewis, M. J., & Ngoddy, P. O. (1994). Volatile constituents of palm wine and palm sap. *Journal of the Science of Food and Agriculture, 64*(4), 405–411. https://doi.org/10.1002/jsfa.2740640403

Varakumar, S., Kumar, Y. S., & Reddy, O. V. S. (2011). Carotenoid composition of mango (mangifera indica l.) wine and its antioxidant activity. *Journal of Food Biochemistry*, *35*(5), 1538–1547. https://doi.org/10.1111/j.1745-4514.2010.00476.x

Varakumar, S., Umesh Hebbar, H., & Raghavarao, K. S. (2011). Characterization of carotenoids in ripe mango from Western Indian Region using liquid chromatography connected with diode array and mass spectrometric detection. *Journal of Food Science and Technology*, *48*(4), 417–426. https://doi.org/10.1007/s13197-011-0226-3

Viana, T., Gil-Sánchez, L., Vilanova, M., Oliveira, J. M., & Cunha, S. C. (2020). Apple Juice and C. *Fermentation*, *6*(4), 126.

Voguel, W. (2003). *¿Que es el vino? In Elaboracio´n casera de vinos. Vinos de uvas, manzanas y bayas* (S. A. Acribia (ed.); Vol. 84). Zaragoza, ISBN.

Vyas, K. K., & Kochhar, A. P. S. (1989). Studies on cider and wine from culled apple fruit available in Himachal Pradesh. *Indian Food Packer*, *47*(4), 15.

Wakita, K., & Imura, T. (2018). High internal phase emulsion gels stabilized by natural casein peptides. *Journal of Oleo Science*, *67*(12), 1579–1584. https://doi.org/10.5650/jos.ess18140

Wang, L., Xu, Y., Zhao, G., & Li, J. (2004). Rapid analysis of flavor volatiles in apple wine using headspace solid-phase microextraction. *Journal of the Institute of Brewing*, *110*(1), 57–65. https://doi.org/10.1002/j.2050-0416.2004.tb00181.x

Weber, C., & Liu, R. H. (2002). Antioxidant capacity and anticancer properties of red raspberry. *Acta Horticulturae*, *585*, 451–457. https://doi.org/10.17660/ActaHortic.2002.585.73

Youdim, K. A., Martin, A., & Joseph, J. A. (2000). Incorporation of the elderberry anthocyanins by endothelial cells increases protection against oxidative stress. *Free Radical Biology and Medicine*, *29*(1), 51–60. https://doi.org/10.1016/S0891-5849(00)00329-4

Youravong, W., Li, Z., & Laorko, A. (2010). Influence of gas sparging on clarification of pineapple wine by microfiltration. *Journal of Food Engineering*, *96*(3), 427–432. https://doi.org/10.1016/j.jfoodeng.2009.08.021

Zhang, J., Li, H., Wu, Y., Wu, X., Wang, L., Zhang, Y., & Wang, Q. (2021). The effects of Chinese plum wine on the gut microbiota composition and short-chain fatty acids in rats with high-fat diet-induced obesity. *Food & Function*, *11*(1), 890–898. https://doi.org/10.1039/C9FO02580

Zhang, M., Li, X., Sun, M., Chen, Y., & Liu, P. (2020). Prebiotic potential of pear wines produced from different pear cultivars. *Journal of Food Biochemistry*, *44*(7), 13271.

Zhou, C., Sheng, Y., Zhao, D., Wang, Z., & Tao, J. (2010). Variation of oleanolic and ursolic acid in the flesh of persimmon fruit among different cultivars. *Molecules*, *15*(9), 6580–6587. https://doi.org/10.3390/molecules15096580

Index

FSC
www.fsc.org
MIX
Papier | Fördert gute Waldnutzung
FSC® C083411

Kolibri 360 GmbH
Ferdinand-Jühlke-Straße 7
99095 Erfurt, Deutschland
produktsicherheit@kolibri360.de